U0169311

古生物化学入门

An Introduction to PaleoChemistry

黄大一　著

Timothy D. Huang

科学出版社

北　京

内 容 简 介

本书是一个勇敢的尝试，介绍古生物学和化学结合而成的一个新兴多领域的交叉学科，以“看进骨头内”为思维指导，从更完整的角度，探索地球生命的演化，充实人们对于浩瀚宇宙中这颗小蓝色星球的认知，让我们因为更充分的了解而更加爱护保护它。透过结合先进的传统古生物学和化学，让我们不止更详尽地看进化石样本内部细节，也看到更完整的外围岩内成分与变化的真实情况。如此的结合，把古生物学提升到一个系统性量子跳跃阶层，同时也打开了一个化学界的新领域；通过加入化学方面的努力，可以揭开并了解更多的地球生命演化；这是我们该做的双赢科学进步——让我们从最里面到外面全盘了解，其威力和影响，无可限量。

封面图说：一亿九千五百万年前禄丰龙胚胎四种化学成分（磷灰石、第一类胶原蛋白、方解石、锐钛矿）三维拉曼光谱分布。这应该是首次能看到化石内不同化学成分三维分布的记录。

图书在版编目（CIP）数据

古生物化学入门 / 黄大一著 . —北京：科学出版社，2021. 7

ISBN 978-7-03-069224-5

Ⅰ . ①古…　Ⅱ . ①黄…　Ⅲ . ①古生物学 – 生物化学　Ⅳ . ① Q911

中国版本图书馆 CIP 数据核字（2021）第 113177 号

责任编辑：杨　震　刘　冉 / 责任校对：杜子昂
责任印制：吴兆东 / 封面设计：北京图阅盛世

科 学 出 版 社 出版
北京东黄城根北街 16 号
邮政编码：100717
http://www.sciencep.com
北京虎彩文化传播有限公司 印刷
科学出版社发行　各地新华书店经销
*
2021 年 7 月第　一　版　开本：787 × 1092　1/16
2023 年 1 月第二次印刷　印张：22
字数：520 000

定价：198.00 元

（如有印装质量问题，我社负责调换）

序　一

读黄大一教授的这本著作，不由得想起“折戟沉沙铁未销，自将磨洗认前朝”这两句诗来。几十亿年来，地球上的物种虽经历灭绝与新生，相信还是生生不息地在进化过程中。原想这些曾经存在的生物遗体埋在地下，除了骨头和牙齿等硬组织以外的有机物全部被分解，化为灰烬，过去的宏大而灿烂的生物世界便无从认识了，但是跟着黄教授的足迹和思路，看他们如何把那些化石“磨洗”之后用各种研究方法，认识并还原了远古时代生物圈的场景，对我启发很大。我对于生物矿化颇有兴趣，但是我们所研究的仅仅是活的生物如何通过细胞构建硬组织。这本书向我们展示了从活的生物体的生物矿化到地质年代中生物遗骸的完全矿化（Permineralization）的总结果，甚至于能够在这些化石中指认出黑色素、I型胶原蛋白等有机物的印迹。这些结果与黄大一教授善于利用各种现代仪器分析方法有关。本书关于研究方法的一部分对其他研究领域的人们很有启发和帮助。

古生物学与化学交叉一定会发现新问题，开辟新领域。因为之序。

中国科学院院士

王　夔*

* 王夔，1928 年生，1949 年毕业于燕京大学化学系；1991 年当选为中国科学院学部委员（院士）；生物无机化学及无机药物化学家，化学教育家；王夔院士是中国生物无机化学研究的倡导者和先行者之一，主要研究与医药学有关的生物无机化学；曾任北京大学医学部教授、北京医科大学药学院院长、国家自然科学基金委员会化学科学部主任等职。

Foreword I

When I read this book of Professor Timothy Huang, it reminds me of the poem *Chi Bi* written by Du Mu: "Broken halberds in the sand the iron preserved. I shall grind and wash it to learn the past." I thought that these extinct buried organisms were decomposed except the hard tissues, such as teeth and bones left. We would not be able to recognize their past glories. However, following the trail and thought of Professor Huang, I saw how they "wash" these fossils by using various research methods to enable us to understand and to return to the ancient biosphere scene. This is a great enlightenment for me. I am interested in bio-mineralization. However, this is limited to the living organism through the cellular building of hard tissues. This book shows to us the total results of bio-mineralization from the living organism to the permineralization of the remains of the once living organisms, even including the trace of organic remains such as melanin and collagen type I identified by Professor Huang by skillfully using various modern analytical instrumentations. Some of the research methods described in this book shall be very revealing and helpful to many researchers in other fields.

New questions will be raised when paleontology interacts with chemistry for a brand new territory. Thus, I gladly wrote this foreword.

Academician of Chinese Academy of Science

Kui Wang*

* Kui Wang, born in 1928, graduated from the chemistry department of Yanjing University in 1949. He was elected as an Academician of the Chinese Academy of Sciences in 1991. He is a bio-inorganic and inorganic medicines chemist as well as a chemistry educator. In China, he is one of the pioneers and leaders of bio-inorganic researchers on the bio-inorganic related to medicines. He worked in the Medical School of Peking University, dean of pharmacy of Beijing Medical University, and chairman of the chemistry division of the National Science Foundation.

序二　出新意于法度之中

我与黄大一先生相识于5年前初冬。他与加拿大皇家科学院院士罗伯特·赖兹教授一同应邀来我校讲学，我时任吉林大学常务副校长，代表学校会见了两位学者。临行，黄大一先生送我一本《台湾白色恐怖死难者纪念册》，令我大吃一惊。

纪念册中有一位名叫黄温恭的台湾青年人。他早年学医，是台湾屏东县春日乡卫生所主任，台湾光复后，追求进步，加入了中共地下党。1951年因叛徒出卖，台湾省工委会燕巢支部被破获而被捕，1953年被杀害于台北马场町，年仅33岁。他在狱中给妻儿亲属留下5份遗书，信中说："我的心窝儿，乱如麻，痛楚得如刺，如割。"他表示，希望自己的尸体寄附医疗机构，"此尸如能被学生解剖而能增进他们的医学知识，贡献这些知识，再也没有比这有意义的了。"直到2008年，黄大一先生才读到父亲的遗书。

黄温恭的遭遇触发我们回顾中国现代历史上悲壮的一幕。1949年国民党军队败退台湾，为巩固内部，防范中国人民解放军跨海解放台湾，在台湾实行白色恐怖。因叛徒出卖，中共台湾地下党组织被破获，成千上万的共产党员和进步青年被当局逮捕甚至杀害。2013年，为纪念20世纪50年代牺牲于台湾的1100多名我党隐蔽战线无名英雄，有关部门修建了北京西山无名英雄纪念广场，黄温恭先生的名字就镌刻在纪念碑上。

黄大一先生就是黄温恭先生的儿子，是我认识的第一位中共台湾地下党牺牲人员的遗属。他父亲被捕时，他还只有六岁。后来黄大一发愤读书，先后毕业于台湾中兴大学无机分析专业、美国波特兰州立大学有机化学专业、美国俄亥俄州立大学生药学博士班。他做过化学师、专栏作家、计算机顾问、大学客座教授。

黄大一先生随后被聘为吉林大学唐敖庆讲座教授，参与本校恐龙演化研究中心的研究工作。他虽为客座，然工作认真主动，为人直率坦诚，积极提出各种建议。每来校工作，必与我联系交流，日久天长，彼此已成为朋友。他的大作《古生物化学入门》杀青，诚心诚意邀请我作序。我是文科生出身，只是在中学学过化学。我对化学完全是门外汉，不知化学，何论古生物化学！虽几经推脱，然盛情难却，特别是他作为中共台湾地下党烈士后代的特殊身份，不得已而从命落笔。

但作为哲学和社会学学者，我对当代社会发展和学术发展，倒也比较熟悉。古生物化学属于跨学科研究。跨学科研究是20世纪中后期发展起来的新兴学术研究方法。跨学科研究即运用多学科的理论、方法和成果，从整体上对某一课题进行综合研究的方法，也称“交叉研究法”。回顾人类学术研究发展的历史，从古代的不分化的神话的宗教的哲学的猜测或思辨的方法，到近代科学分化，不断分门别类的专门化和专业化的研究方法，到今天科学在高度分化中又高度综合的研究方法，形成了一个正反合的辩证发展过程。据统计，现在世界上有2000多种学科，而学科分化的趋势还在继续，各学科间的联系也日益紧密。

黄大一先生的《古生物化学入门》，就是这种跨学科研究的多年尝试和结晶。他以自己的化学专业为基础，把化学分析方法运用于古生物研究，先后取得了关于云南埃迪卡拉纪实体化石、距今1.95亿年前恐龙骨头内保存第一类胶原蛋白、距今22亿年前多细胞真核生物发现十八种氨基酸与甾烷等有机残留物等研究成果。作为学者，我深深为年逾古稀的黄大一先生所感动，所钦佩。苏东坡《书吴道子画后》有云：“出新意于法度之中，寄妙理于豪放之外。”这是我作为文科生出身，读完黄大一学术大作《古生物化学入门》后的体会。

吉林大学校务委员会副主任，教授

郇　正

Foreword II New Ideas from the Ordinary

I met Mr. Timothy D. Huang in the early winter 5 years ago. He and Professor Robert Reisz, a Fellow of the Royal Canadian Academy of Science, were invited to our University to give lectures while I was the executive vice president representing our school to greet them. To my surprise, at the end of the meeting, Mr. Huang gave me a book entitled *Memorial Book for the Victims of the Taiwan White Torror*.

In this memorial book, there was a young person in Taiwan by the name of WunKong Huang. He studied medicine and was the director of the health office of ChunZhi village of PingTung County. After the recovery of Taiwan, he searched for advancement and joint the secret Communist Party. In 1951, due to the betrayal, the YanChao chapter of Taiwan Worker's Committee was broken, and he was arrested. He was only 33 years old and executed in 1953 at the MaTing site in Taipei. Before his execution, he wrote five letters to his wife and children before his death. In one of them, he said: "my heart chaos as hell, pain as stings and cuts." He also wished to donate his corpse to the medical school, "if this corpse can be dissected by students and enhance their learning of medical knowledge, there is nothing more meaningful than this." It was until 2008 that Mr. Huang got a chance to read his father's letter before his death.

WunKong Huang's encounter caused us to look back at the recent sad Chinese history. In 1949, KMT's troops retreated to Taiwan. In order to solidify internally to prevent the People's Liberation Army to cross the strait, Taiwan was under the siege of White Terror. Caused by the traitor, the Chinese secret Communist Party organization was cracked, which caused thousands of Communist Party members and elites were arrested and even killed. In 2013, in order to pay attributes to the more than 1,100 unnamed heroes killed in Taiwan for their sacrifices, our government built the Monument of Unknown Heroes in Beijing West Mountain. WunKong Huang's name is engraved on the Monument.

Mr. Timothy D. Huang is the son of Mr. WunKong Huang. He is the very first surviving dependants of a martyr of the sacrificed Taiwan underground party member I met. He was only five years old when his father was arrested. Since then, he was very eager in his studies, graduated from the Department of Chemistry, BS in inorganic, of the Taiwan Chung Hsing University, Portland State University, MS in organic chemistry, and Pharmacognosy Ph.D. graduate school of the Ohio State University. He worked as a chemist, column writer, visiting professor of some universities.

Then, Mr. Timothy D. Huang was appointed as a Tang Aoqing Chair professor of Jilin University and participated in the works of the Dinosaur Evolution Research Center of our University. Although he is a visiting scholar, he works very hard and actively with sincerity to other members and makes many useful suggestions. Every time he came to our University, he contacted and communicated with me. After such a long time, we became friends. When he finished his great work of *An Introduction to PaleoChemistry*, he very sincerely asked me to write a preface. My background is in literature and learned just a little bit of chemistry in middle school. So, I am a layman in chemistry, not to mention PaleoChemistry. I tried to make excuses many times, but his kind feelings are hard to refuse, especially he is one of the unique descendants of the Taiwan secret Communist Party martyrs. I have to write this.

As a scholar of philosophy and sociology, I am familiar with the topics of contemporary society and academic development. PaleoChemistry belongs to the interdisciplinary studies, which started in the late 20th century as a new academic research method. It uses multiple theories, methods, and results. It conducts the study of a particular issue with a comprehensive approach, also called "cross-over study." Looking back at the historical studies and developments of anthropology, it went through those unseparated, mythical, religious, philosophical guesses or thinkings to the modern continuous classification specializations method. The scientific studies now swing back to the highly cross-over approaches and form a process of GWF Hegel's dialectic method. According to statistics, there are more than 2,000 separate disciplines and continuously dividing more. However, the connections of each are getting closer and closer.

Timothy D. Huang's *An introduction to PaleoChemistry* is the trial and result of such multidiscipline of many years. He uses his chemistry background applying to paleontology studies to obtain the excellent results of Ediacaran body fossils, the preservation of collagen type I from the 195 million years old dinosaur embryo, and the 18 amino acids as well as steranes from 2.2 Ga multiple cellular eukaryote organisms, which is the oldest fossil record. As a scholar, I am deeply moved and hence very respectful to the old Mr. Timothy D. Huang. Dongpo Su of Song Dynasty wrote in the *Afterword of the painting of Daozi Wu*: "New ideas from the ordinary, put the ingenious words beyond the bold and unconstrained." This is my understanding after reading Timothy D. Huang's academic great work of "An Introduction to PaleoChemistry".

Vice director of Administrative Committee of Jilin University

Professor Zheng Bing

自　　序

在给本书自序之前，我要特别提出，感谢晏润寒女士的参与协助和文稿编辑。毕竟我成长于中国台湾，两岸在诸多用语习俗方面稍有不同，因此，通过多次意见讨论和修整，尽其所能让本书比较能适应大陆读者的阅读习惯。

古生物化学是个勇敢的尝试，结合传统的古生物学和化学，成为一个新兴多领域的学科，用以从更完整的角度，探索地球生命的演化，充实人们对于浩瀚宇宙中这颗小蓝色星球的认知，让我们因为更充分的了解而更加爱护保护它。

古生物化学这个观念，并非作者天纵英明，一夜间上天恩宠给予异象突然爆发出来的，而是从多年来捞河捞过界玩化石跌跌撞撞，反复修正，终而产生的雏形。书中有几篇心路历程，过去只是个人的记录，但想到或可作为后进参考，故此收入本书，读者可仔细品味其中心路历程转折，或可避免走些冤枉路，是所寄望。

近年来在计算机界甚至蔓延到其他领域的“开放源码 (Open Source)”观念，是笔者从二十世纪七十年代，目睹也参与微处理器与个人计算机从毫不起眼拼装抖凑，到如今人手不止一机的盛况。这其中一个重要的关键因素，就是信息共享，也只有通过公开共享信息，信息才能发挥其最大的威力，科学才能进步，人类社会也因而得到最大的受益。因此本书内有些资料，取自于维基百科或公开网站，还有好些是网络相关论文与报道，也都引用在内，在此一并致谢。也先说明，既然已有很好的可共享信息，就没有必要重新把轮子再发明一遍；世界的科学进步，都是站在别人肩膀上，自己添加一点，别人又添加一些，人们就能看得更远一处，没有人天生下来什么都懂，什么都能自己创作拥有版权，这也是我们每人学习的过程，科学进步的原动力。

此外，读者们可能会发现本书的写作，与常见教科书颇有不同，一是没有老太婆缠脚布式的“参考资料”，另一是充满了个人的风味，笔尖还往往忍不住情感，甚至语带辣味。以前者来说，由于个人的文献搜藏，受限于每篇论文光是从网上借来看就要花上几十块美元的租金，我负担不起，也不屑为之，这完全是违反我对信息公开的理念；另外一点，我也很讨厌那种板起脸来的“言必孔孟”学究式、读起来艰涩咬牙切口的论文。

至于如朱自清般笔尖带着情感的撰写，读者们也须容忍我个人的限制；谁都无法避免大环境的影响，我这从小就在额头上被烙上“台湾黑五类”的白色恐怖受难者家属，在台湾被排挤在社会底层，即便自己不服输，努力发表了两篇《自然》(*Nature*)论文，其中一篇还是封面论文，可是有如好友戏称的，我只是个没有庙的大和尚，这些年来能到处卡油，做出一点成果，能怪我笔锋带感情吗？

日前看美国国家地理频道纪录片专辑系列《世纪天才》(*The Genius*)，介绍爱因斯坦的一生，得知原来他最伟大的几篇论文，如狭义相对论$E=mc^2$，都是在他最艰难的日子发表的，在瑞士专利局当小科员，每天做着无聊的专利案审核，糊口养家。我不敢自比于他，但是越是艰难的外在环境，越是能彰显松柏常青之坚忍生命力，我个人所受到的命运挫折，何尝不是老天的恩赐呢？

这本书是古生物和化学结合成为古（生）物化学的檄文，就如当年个人计算机开始时，简陋到被大计算机公司，如 DEC 嘲笑称连玩具都算不上。不过，今日个人计算机，不就是从那么简陋拉链袋起家的吗？古（生）物化学刚刚起步，我尽了吃奶的力量，整理出这个有系统的雏形，供给大家参考，祈望后进长江后浪推前浪，发扬光大，写出更好的书。因此，此书的缺失，还等着大家一起来努力。

最后，但非最微，本书名《古生物化学入门》(*An Introduction to PaleoChemistry*)，虽然基本上是建构笔者过去投入的古生物研究，但实际上书内的主要思维和操作“看进骨头内”，可以应用于不限于古生物，只要是“古物学（Paleo）”，同样都可以发挥其特效。因此，把本书名改为“古物化学”，也是非常恰当。

感谢：本书之作，受到以下学者人士的鼓励和肯定，在此表达最深感激之意。还要特别感谢晏润寒帮忙编辑整理。

在笔者过去被戏称为“没有庙的大和尚”的曲折人生路程中，这些年来得到了不少鼓励和协助，才能有这本书。台湾俗语说：“吃水果，不能忘了拜树头。”我们做人总该怀着谢恩的心，感激他们。2010年，本人有幸得到加拿大多伦多大学赖兹（Robert Reisz）院士的指导合作，才能有2013年*Nature*封面论文的恐龙胚胎学。在我们进行该课题过程中，台湾同步辐射研究中心（SRRC）李耀昌博士和江正诚先生的鼎力协助，让我们证实了古生物化学的可行性。接下来多次到澳大利亚悉尼的澳大利亚核科技组织（ANSTO）进行中子扫描，承蒙Joseph Bevitt博士鼎力协助，也不能忘记一笔。这几年来，经由孟宪伟女士的媒介，吉林大学给我无数的帮助以建立古生物化学平台，特别是李元元前校长和前常务副校长邴正教授、未来科学中心 (ICFS) 的于吉红院士、化学院徐如人院士，还有我们的“恐龙演化研究中心 (DERC)”的金利勇馆长、陈军老师等等伙伴都敦促鼓励了这本书的完成。当然，这一两年来，提供最顶尖质谱分析实验的东华理工大学陈焕文副校长，加上其他该实验室的好几位成员，也都是我走在古生物化学路途上的好伙伴。最后，非常感谢中国科学院王夔院士作序，这些都给了我莫大的鼓励；当然还有好多没列出来的恩人，在此表达最深的真诚感谢。

吉林大学客座教授、唐敖庆讲座教授、博士生导师

黄大一

撰写于新店

2017年7月至 2019年12月完稿

* 说明：

本书内有些引用网页、维基百科和相关文献等公开资料，都尽量注明出处，外文资料出处以较小字号斜体列出，如：(*title*); (*authors*), (*Journal*)。

Preface

Before my own Preface, I would like to point out, particularly my appreciation for the participation, helps, and text editing by Ms. Runhan Yan. After-all, I grew up in Taiwan, China. Many expressional differences exist. Thus, thru countless discussions, idea exchanges, and revisions, this book can be more suitable for readers.

PaleoChemistry is a brave attempt, combining traditional paleontology and chemistry into a multi-discipline field of science to use more complete viewing angle to explore the life evolution of this Earth, and to enhance our understanding of this tiny blue planet in the vast universe so that we can love and protect it more since we know more about it.

The concept of PaleoChemistry is not from the author's god-send smartness or a vision given from heaven in a night. It is from many years of over-boarding in fossil studies with numerous obstacles and revisions to reach this prototype. There are several personal experiences in the book. These used to be just for my own personal records. However, in consideration of these may be helpful for the new followers, so they are included in this book. The reader can taste the twists and turns I went thru, and hope that can save some unnecessary waste of time.

In recent year in the computer field, and also spreads to more other fields, the concept of "open source" is what the author of this book subscribed dearly since the 70 of last century, in which I participated and saw the microprocessors and personal computers started from very rough form and evolved to the current situation of more than one computer in each person's hand today. One of the critical factor that is information sharing. Only with public sharing of information, it can exert its ultimate power, and science advanced, so that whole humanity benefits the most. Thus, some of the information in this book was taken from WiKi, published papers, news reports, or public websites. Let me express my appreciation and acknowledgment. Since we already have excellent shared information, there is no need to re-invent the wheel again. The world science advancement is one standing on the other people's shoulders with some additions, and other people add a little bit more so that humanity can see farther. Nobody was born of a know-it-all and created every copyright by himself. We learn from others. Moreover, this is the motive force for science to move forward.

Furthermore, the reader may find that the writing style of this book is quite different from the commonly seen textbooks. For one, it does not have a long list of references, and the other one is that this book has strong personal flavor with emotion, even a bit of spicy on my pen tip. For the first one, I can not afford to rent those published paper costing tens of US dollars per paper. I disdain that. Another point is that I hate the egg head's way of quoting many previous papers, which makes the paper very difficult to digest.

As for the writing style of emotion on my pen tip is caused by my personal limitation. So, please bear with me. Nobody can avoid the influences of the large environment we live in. I am a victim of white terror families, person branded on my forehead with the sign of "black five" since my childhood, and pushed to the bottom of the society. Even with my efforts of published two *Nature* papers, one of which is a cover paper, I am still is a "big monk without a residential temple." I accomplished this little bit by begging favors from others. So, can anyone blame me of emotion on pen tip?

Recently I saw the documentary series of "*The Genius*", Albert Einstein on National Geographic Channel, and learned that several of his most important papers, such as General Relativity, $E = mC^2$, were written during his hardest days when he was a little patent clerk making a living for his family. I can not be in comparison with him. However, the harder the outside environment, the better the pine trees show the endurance of life. So, why is not my personal unfortunate a blessing from heaven?

This book is a war proclamation of combining paleontology and chemistry into PaleoChemistry. Just like the very early time of the personal computer, it was so ugly, simple, and laughed at by the big computer company such as DEC as not even a toy. However, did not the personal computer of today start in the zip bags? PaleoChemistry just started in infancy. I used all my milk-socking power to compile out the prototype of this system as a reference for everybody and hoping the new waves will push the old, and carry forward much further. Thus, whatever missing in this book is pending for enrichment and better books from somebody. Let us mutually encourage each other here.

Last but not least. Although the name of this book, *An Introduction to PaleoChemistry*, is built on my personal experiences on paleontology. However, the central theme and operation of "Look into the Bone" can be used not only for paleontology but also for anything paleology. It can be as powerful in these fields as in paleontology. Therefore, it is very appropriate for the Chinese name of this book to be Paleo-Chemistry.

Acknowledgment: The following persons, particularly Ms Runhan Yan for her help on editing and traditional/simplify Chinese conversion, encouraged and provided affirmative supports; most science appreciation is expressed here.

In the past, the author was tagged as "a big monk without residential temple" in his twisted life. However, within these recent years, numerous encouragements and help were received, thus this book. In a common saying, while eating fruits, do not forget the source trees. We shall keep a most grateful heart for them. In 2010, I was fortunate to have guidance and cooperation from Academician Robert Reisz of the University of Toronto of Canada, so that end up with a Nature cover paper on Dinosaur Embryology. During that project, which is still on-going, Dr. YaoChong Li and ChernCherng Chiang of the Taiwan Synchrotron Radiation Research Center (SRRC) provided countless assistance so that the feasibility of PaleoChemistry was affirmed. Then, in my many Australian Nuclear Science and Technology Organization (ANSTO, Sydney, Australia) trips for neutron tomography experiments, Dr. Joseph Bevitt provided the most valuable help without reservation. This shall be noted. Within these recent years, thru Mrs. XianWei Meng's match-making for me to Jilin University, JLU provided me so much helps, particularly, previous President Yuanyuan Li and current Executive Vice President Zheng Bing, Academician Jihong Yu, Academician Ruren Xu of the Chemistry College of JLU, and our Dinosaur Evolution Research Center's Direct or Liyong Jin, and Jun Chen encouraged me very much to finish this book. Of course, within the past two years, Vice President Huanwen Chen of the East China University of Technology provides his most advanced Mass Spectrometry Lab facility and expert members. They are my good companions. Last but not least, Academician Kui Wang of the Chinese Academy of Science wrote such a wonderful Preface for this book; it gives me unmeasurable encouragement. Of course, there are many other people not mentioned. My deepest and sincere thanks to all.

Timothy D. Huang, Visiting Professor and Tang Aoqing Chair Professor,
PhD Instructor of Jilin University, at XinDian,
July 2017 to December 2019

---

* Directions for Use

In this book, there are many quotations from various web sites/pages, Wikipedia, and published literature. The sources are noted. For the English references, they are expressed in smaller font size and italic, as (*title*); (*authors*), (*Journal*).

目　录

目　录

目　录

CONTENTS

CONTENTS

CONTENTS

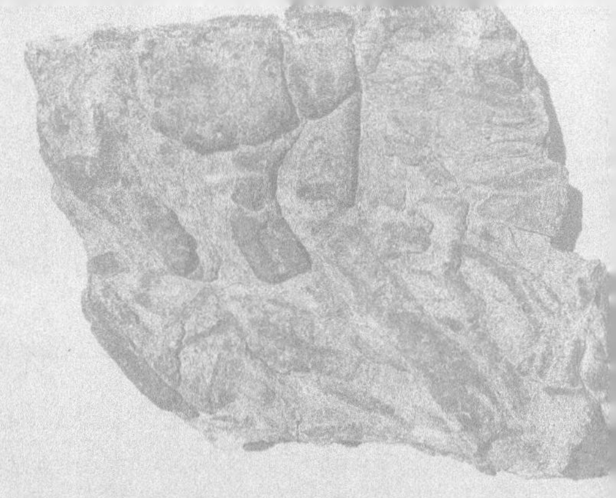

第 1 章　范畴与简介

Chapter 1　Scope and Introduction

本章先叙述古生物化学的范畴与内涵以及从事古生物化学必须注意的化石不均匀本质和微观特性。

This chapter describes the scope and content of PaleoChemistry and the two characteristics of the heterogeneous nature of fossils and micro view, which must be kept in mind.

1.1　古（生）物化学的范畴

古（生）物化学不只是一门新创的古生物学领域，专注于探讨古代生物形成化石过程中和之后、实体与其紧接近临区域围岩的化学与变化，更深入探究古（生）物的化学本质。还基于这些“看进骨头”内外的化学成分与变化探讨，回头了解古生物活着时候和死亡当时与尔后漫长岁月所发生的事，进而了解地球上生命生物演化。它并非取代或排除先进古生物学的显微形态学或组织学独立自主，而是与之相辅相成，从另一个角度深入，进而提供更完整的信息。在此用“古（生）物化学”的用意在于：除了本书主要讨论的化学与古生物学交叉学科之外，其实“Paleo”这个词，也适用于诸多与“Paleo”有关的学科，如

1.1　The Scope of PaleoChemistry

PaleoChemistry is not just a newly created paleontological field. It concentrates on exploring the chemistry and changes during the ancient organism still alive and after the formation of fossils and the surrounding matrix. It is the essence of paleontological exploration. Based on the strategy of "Look into the Bone" on the compositions and their changes, we can go back to understand what happened when the ancient organism still alive, died and the long time after. That will enhance our knowledge of the life evolution of this earth. This is not replacing or excluding the advanced microscopic morphology and histology of paleontology studies. On the contrary, it is complimentary from a different angle to provide complete information sets.

古人类学、考古学等等。

化学，变化之学也，就是探讨宇宙间物质与其变化之学。有趣的是，“Fossil”中文翻译为“化石”两字，不只用到“化”这个字，而且实质上就是探讨古生物石化的结果，透过石化的古生物，探索古生物时代该古生物与其环境可能的情况、地球生命演化等等；然而，目前的古生物科研方法，通常只着重并且限于从外表易见形态系统描述等特征，而少从其内的变“化”切入。因此，古生物化学是将古生物学带回正途的学科，研究古生物学，必须从古生物“化”学的方向着手，深入化石（骨头）里面，探讨该古生物从当年活生生状态变化到如今硬邦邦石块，在此漫长时间内其化学组成细微构造等变化，有什么成分起了什么变“化”？“化”成了什么？如今保留的是什么“化（学）”结果的成分？简言之，古生物化学，即古生物生前死后到如今变化之学也；如此一来，始能从里到外，有个比较完整的认知。

举一个非常实际的例子：从形态学的角度来说，如果想看到骨头化石内的构造、包裹在围岩里面的化石、很小的化石……一定得使用到现代化的几种扫描仪器设备，包括高倍率显微镜、电子显微镜、微型计算机断层扫描、中子断层扫描等等，因为使用不同的照射源，所观察到的结果，可能会有很大的差异。为什么？进一步来说，用X射线为照射源的微型计算机断层扫描所看到的和用中子为照射源的结果不同。为什么？说

Furthermore, PaleoChemistry is not limited to combining paleontology with chemistry as the majority topics of this book, but also for many fields related to "Paleo-", including such as Paleoanthropology, Archeology etc.

Chemistry (化学) is a science of studying changes. It studies the changes of matter in the universe and how they change. Interesting enough, the translation of the word "Fossil" into Chinese "Huashi (化石)" , not only use the same character of "Hua (化)" as in chemistry, the actual works of paleontology are to explore the fossilization of ancient organisms by exploring the paleoenvironment, life evolution on earth, etc. However, the current paleontological studies are usually limited and concentrated on the outside morphological system description with little efforts on how they changed. So, PaleoChemistry is the route to bring paleontology back to the right track. That is for PaleoChemistry studies, the change of composition is the right direction, dive deeply into the (bone) fossils, to explore what minute chemical composition changes from the living condition thru long-buried time. What compositions were changed? What is it changed to? What changed compositions were preserved? In short, PaleoChemistry is the study of the changes in ancient organisms alive or dead until today. By doing so, we can have a richer understanding inside out.

As an actual example: From the morphological point of view, in order to see the internal structure of fossil bones, or fossils inside the matrix, or tiny fossils, … etc., various modern scanning instruments such as high magnification microscope, electron microscope, micro-computer tomography, neutron tomography are used. How-

穿了，原因很简单：因为样本内不同成分的原子对于不同照射源会有不同的反应。搞清楚某块化石内的原子及分子组成，就是古（生）物化学的一部分。

古生物化学是一门结合传统系统描述性 (Systematic Description) 古生物学 (Paleontology)、古代生物学 (Paleobiology)和化学的新兴交叉学科，基本的思维是“看进骨头内 (Look into the bone)”和“异想开天 (Think outside the box)”，从非传统典型的方法（用奇异的想法，解开如登天之难的问题），透过探索化石本体与紧邻围岩内的化学成分与化学变化，试图了解古生物活着时候的情况以及死亡掩埋后发生的种种，并从地球生命演化意义的角度来研究古生物学，从化石本体里面到外面，取得相对完整的内外信息，以至于能更彻底清晰地揭开古生物有趣的奥秘。如此的结合，将古生物学提升到一个新的水平：从文献上所见，古生物与相关学术界，到如今多少有些零星的研究成果，分散于各领域（如地球化学等）期刊，然而却未见有系统性的整合。因此，本新学科是个勇敢的系统性整合尝试，不只限于个别有机残留物、无机物而已，更要探讨两者之间的相互作用。欢迎有兴趣者一起合作，撰写新的历史。

因为这是一门前所未有的全新学科，本书试图做个系统性的整合，在正确的“看进骨头内”思维引导之下，把当下古生物学的成果与课题，借用先进科研仪器设备来探讨，期望能打开一片

ever, due to the difference in radiation source, the result can be very different. Why? The results from X-ray based micro-CT can be quite a difference from that of neutron-based. Why? The reason is straightforward: different compositional atoms inside the exam object have different reactions to different radiation sources. Isn't that understanding the atomic composition inside a fossil PaleoChemistry?

PaleoChemistry is a new multidiscipline course combining paleontology, paleobiology, and chemistry. It uses the thoughts of "Think outside the Box" and "Look into the Bone", and nonconventional methods to explore chemical composition and changes inside the fossils and immediate surrounding matrix for attempting to understand the circumstance of ancient organisms alive and the events after death. It also looks at the meaning of life evolution to study paleontology. It will examine from the inside to outside of the fossils under-studying to obtain relatively more complete information so that it can reveal much more mysteries of ancient life more clearly. With such a combination, it will elevate paleontology to a new level. There are some isolate papers published in various sub-categories of related fields, such as geochemistry journals. However, no comprehensive systematic was proposed. Therefore, this is a brave systematic attempt, not limited to only organic remains or inorganic, but also the interactions between them. Interested people are welcomed to work together to write the new history.

Since this is a brand new course, this book attempts to provide a systematic combination under the correct thought of "Look into the

新天地：这也就是说，本书所讨论到、引用到、使用到的课题、文献、科研方法等等，只是目前的一张张缺失而令人困惑的拼图。从实际的角度来说，我们靠着这些来日会觉得是古董的航海图（设备），勇敢地出发航向一个未知的水域，我们现在还不知前面的航行会碰到什么惊奇风暴暗礁乱流。不过有一件事情可以保证，随着科技的进步，日新月异的科研仪器分析方法出现等等，肯定会有好些新发现，否定过去的（错误）认知。举一个实际的例子来说，一般传统的思维，认为化石就是古生物变化成为石头，从有机体变成无机的矿物质，有机体不可能保存成千万上亿年，然而我们团队却在二亿年前的禄丰龙胚胎骨头内，发现了被保存下来的原生第一类胶原蛋白 (native collagen I)。甚至我们还可以大胆推测，如果用正确的思维、对的研究方法、够灵敏的仪器，发现实体化石内保存着有机残留物，会是一个普遍现象。这就如陶渊明所说的：“觉今是而昨非。”今日之我否定昨日之我，科学的进步，不就是如此一句话吗。

虽然本书是以古生物为主要讨论的课题，但是本书所提到的诸多科研方法与设备，没有任何理由不能“捞河捞过界”，从而应用在其他的领域。比方说，来个“考古化学”，不是也很恰当吗？没有任何的理由，同样的方法与设备，不能再度跨领域交叉学科啊，欢迎也等着你来加入！

Bone", using the current results and topics by using the most advanced instrumentations, and hoping to open up a new territory, ie., whatever discussed and referred topics, literature, and methodologies now, are just a confusing puzzle containing many missing pieces. From the practical point of view, we are using this very incomplete navigation chart which will become antique soon to sail into the uncharted water. We do not know what current, sandbar, or storm will come. However, one thing can be sure. Along with the progress of science and technology, new instrumentation and methodology will become available very shortly. Great discoveries will come to negate the past (mis-) understanding. As an example, in conventional thinking that fossils are ancient organism transformed into a rock, changed from organic to inorganic. Organic matter cannot be preserved for such millions of years. However, our team found native collagen type I inside almost 200 million-year-old *Lufengosaurus* embryonic bones. We can even say that if the right thought with the correct method and sensitive enough instrument are used, preservation of organic remains inside body fossils is a common phenomenon. This is like what the poet Yuanming Tao said: "Knowing the right of today and negating the wrong of yesterday." Is not such a saying the advancement of science?

Although this book is mainly in paleontology studies, there is no reason what so ever that the same methodology and facility mentioned in this book cannot be used overboard and used for other fields. For example, how about "ArcheoChemistry" is not it a right field? There is no reason whatsoever that the same ways

看进骨头内，恐龙胚胎学

本节以一个实例来说明古生物化学的诸多“眉角”。

“**恐龙胚胎学** (Dinosaur Embryology)”是我们团队透过在 2003 年中华玩石家地科协会举办到云南的“百战天龙”活动，在接受中央电视台 (CCTV) 采访过程中无意间捡到，后来在 2010 年 3 月机缘巧合下又被重新发现的世界上最古老的恐龙胚胎化石，从而建立起来的恐龙学研究中的一个新领域。我们能够有幸首先创立恐龙胚胎学，透过各种以下所述“看进骨头内”研究方针，使用各种先进的科研仪器设备和方法，深入了解为什么恐龙胚胎发育过程会如此快速，恐龙为什么会长那么大等等。初期的部分成果，我们已经在 2013 年 4 月 11 日的国际顶尖多领域科学期刊*Nature*封面论文（图 1.1.1）内叙述了一部分，同时目前我们团队还正在进行几个很有趣的课题，进一步充实“恐龙胚胎学”的内涵。

“看进骨头内”是我在国际联合团队研究这个世界最古老恐龙胚胎过程中，考虑台湾过去没有好好培养古生物学专家，而不得不“异想开天”所提出来给台湾科研界的一条新路。这个“看进骨头内”为台湾学者们打开了一个新思维领域：台湾从光复以来没有培养古生物学者，所以我只能放弃传统古生物学那种“看骨头外表（系统形态描述）”的做法，把这些看骨头外表的事项，交

and means cannot be used for multi-discipline studies. You are welcomed and waiting for you to join.

Look into the Bone, Dinosaur Embryology

In this sub-section, I am going to use a real case to show the multitudes of PaleoChemistry.

A new branch of dinosaurology, Dinosaur Embryology was established on an unintentionally picked up of a “strange rock” during a CCTV reporter interview in our dinosaur winter camp of “DinoDragon Project” in 2003 at Dawa, Lufeng, Yunnan, China, and then, in March of 2010, the site was rediscovered. We were lucky enough to establish dinosaur embryology first by using the thought of “Look into the Bone” and various advanced instrumentation and methodology to understand why the embryos developed and grew so fast and why dinosaur became so big. Parts of our initial result was published on internationally well-known multidisciplinary journal Nature as cover paper (Figure 1.1.1). We are conducting several fascinating topics to enhance the content of dinosaur embryology now.

“Loo into the Bone” is a concept forced out from my mind on the fact that no paleontologist was trained in Taiwan for the past half a century. I had to “think outside the box” to resolve the dilemma I was facing. It paved a new way for Taiwan scholars. It opens a new domain for them. Since no real paleontologist was trained and available since Taiwan return to China after World War II, no conventional “Look outside the Bone” systematic description paleontologist can

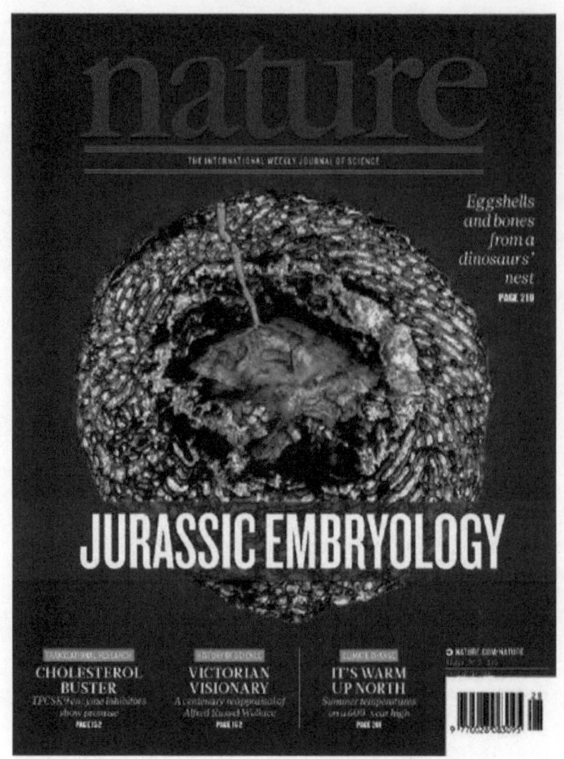

◄ 图 1.1.1　台湾从来没有过的记录。五个台湾学者专家的恐龙胚胎论文被刊登，又荣获为封面论文，一箭双雕

◄ Figure 1.1.1　It is a record for Taiwan. Five scholars published a dinosaur embryology cover paper, the double whammy

给团队里诸多世界顶尖的科班古生物学者，他们才是训练有素的学者专家，在这方面，我完全没有说话的机会。虽然如此，台湾学者们也不能妄自菲薄，透过这个“看进骨头内”的思维，请团队里的国际学者们去看骨头外面，台湾学者专家来探看恐龙骨头里面到底是什么，怎么回事。皇天不负苦心人，我们还是能在全球古生物界做出领先傲人的成果来：我们团队所发表的*Nature*封面论文，就是五位台湾专家学者用“看进骨头内”做出来的傲人成果。这五位没有一位是科班的古生物学者，甚至有好几位刚开始的时候，老是搞不清楚“古生物学 (Paleontology)”和“考古学 (Archeology)”的差异，反而随着无知的

be found in Taiwan. So, for these 'outside' matter had to be done by our international experts. In this regard, I do not have any position to open my mouth. However, I do not think Taiwan scholars should give up on this project. By using my concept of "Look into the Bone" strategy, Taiwan scholars can concentrate on the internal of these embryonic bones, while our international teammates worked on the outside of the bones. This collaboration yielded excellent world-class results. In the *Nature* cover paper we published, five Taiwan scholars are co-authors worked under the guidance of "Look inside the Bone". None of these five is a trained paleontologist. Some of them even could not distinguish the difference between Archeology and Paleontology and said some very silly 'archeological this, archeological that' jokes along with the ignorant reporters.

记者，经常考古来考古去，闹了不少笑话。

从恐龙学的角度来说，大约平均每个礼拜或每个月就会有一种新的恐龙品种被发现，加上科幻电影如《侏罗纪世界》等长期以来对于普罗大众的灌输，恐龙热一直维持在相当的高度。相比之下，从来没有其他的古生物（如三叶虫、菊石等）能有此荣幸。但是，在这么多的科学论文和普罗新闻当中，有关于恐龙胚胎的，相当稀少，而且如果有的话，绝大部分都是中晚白垩纪的恐龙胚胎，才七八千万年的“晚辈”！我们这个是1.95亿年前、最早期恐龙的胚胎；再者，过去所发现的恐龙胚胎，“死相”都太漂亮了，完整或几乎完整的恐龙宝贝卷曲在恐龙蛋里面，漂亮得很，可是从另一个角度来说，如此的样本，同时兼具“福”与“祸”，“福”是这么可爱罕见，

For dinosaurology, on average, it is about one discovery of dinosaur per month or several weeks. The science fiction movies such as *Jurassic World* intoxicate the general public to maintain a relatively high heat level. No other paleontological subject, such as trilobite or ammonite, can have such an honor. However, among so many scientific paper and widespread news reports, it is very rare on dinosaur embryo. Almost all reported dinosaur embryos are mid- or late Cretaceous, only younger generation of 70 or 80 million years old. Our discovery of 195 million years old is the oldest dinosaur embryo ever found. Furthermore, the 'death look' of the past discoveries were too beautiful, complete or almost complete dinosaur baby curling inside the egg. Beautiful! Look at this from a different angle. Such a specimen is a mixture of blessing and curse. Blessing is that such a rarely seen beautiful specimen. What is the Curse? Because

◀ 图 1.1.2　这块在 2003 年无意间捡到的怪石头，就是改变人们对于恐龙认知的“始作俑者”。经历十年寒窗，终于飞龙升天

◀ Figure 1.1.2　This strange rock picked up unintentionally in 2003 is the "creator of a bad precedent." After ten years of cold hard works, this "dragon (dinosaur in Chinese)" flew to the sky

“祸”呢？但也因为太难得完整漂亮，断绝了进行深入骨头内研究的路，谁会让你我拔出某根骨头来切割，看到骨头里面去？只能做到传统古生物学的形态描述；再进一步来想想，完整绞合的恐龙胚胎样本所提供的信息，只是该孵化中恐龙蛋被掩埋的那个刹那，在那刹那之前和之后的恐龙胚胎发育，完全无法提供任何信息，所以这些“死相”完美的恐龙胚胎样本，一方面是福气，另一方面也是诅咒。

相对地，在我所发现的这个恐龙胚胎地点，除了图1.1.2这个相对来说比较完整的样本之外，我们还捡拾到很多非绞合(disarticulate)的零散骨头(图1.1.3)，从完整度来说，一堆零散小死骨头，完全没有令人跌破眼镜的漂亮完整“死相”；可是，从另外的角度来说，这个“祸”才是真正的“福”，因为它们提供了比完整恐龙胚胎更多更多的信息，例如，透过测量股骨的长度和直径，我们分析出，在这个采集点的恐龙胚胎，至少有三个不同孵化的阶段，有可能是来自三个恐龙蛋窝，或该地点曾经有三次淹水事件，相对于其他完整绞合的恐龙胚胎样本只能提供孵化过程中某单一瞬间的状态，我们的材料让我们研究恐龙胚胎的某一段期间，有如片段的视像，而非单独一张拍立得照片；又因为这些数量不少的胚胎骨头都是零散的，我们可以选择某些来“牺牲”，做成切片，让我们使用各种先进的科研设备仪器，仔细观察其内部构造，进行“看进骨头内”

it is so well beautifully preserved, it cut off the way for “Look into the Bone.” Who will let you pick out a bone from that to cut and grind to look inside the bone? Only conventional morphological description can be done. Furthermore, the information of the articulate dinosaur embryo can provide is just the snapshot of the moment it was buried, no information about before and after can be obtained. Thus, these perfect “death look” dinosaur embryo specimens are a mixture of blessing and curse.

On the contrast, at the site, I found the dinosaur embryo, besides the specimen shown in Figure 1.1.2 is relatively more complete, we found a great deal of disarticulated scattered bones as Figure 1.1.3. So, from the point of completeness, all these scatter little bones do not make a beautiful death look. However, from a different angle to see it, this ‘curse’ is the real ‘blessing.’ Because they provide much much more information than the complete dinosaur embryo. For example, by measuring the femur diameter and length, we concluded that there were at least three different stages of incubation. They may come from three different nests or three flooding events. The information provided by complete articulate embryo is just one snapshot. However, our material provides the information of a period, or a video clip, not just a single shot. Furthermore, due to the large number of bones were disarticulated, we can choose and pick some of them to ‘sacrifice’ for thin slices which in turn enabled us to use various advanced facilities to look the internal structure for the “Look into the Bone”, and revealed more mystery of the unborn dinosaur baby.

That is thru these very humble scatter bones,

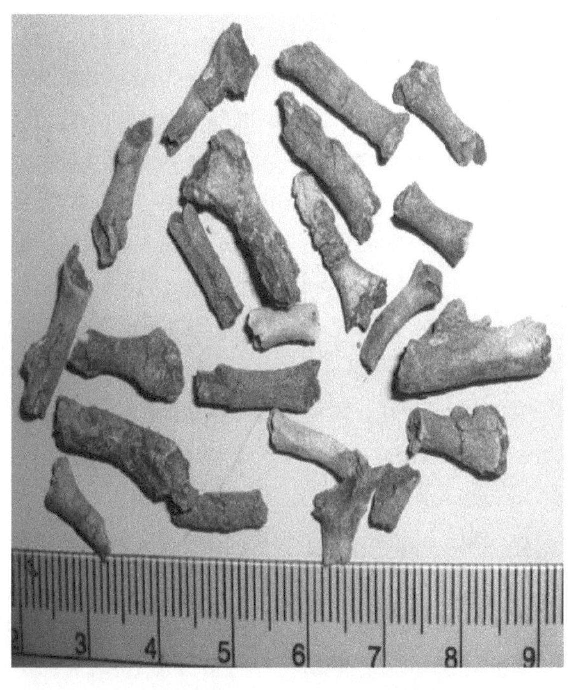

◀ 图 1.1.3 2010 年重回云南禄丰大洼，找到恐龙胚胎点，捡拾这堆非绞合零散的小死骨头样本，提供了我们绝佳的机会，让我们能进行“看进骨头内”。完整死相很好的绞合样本，人家不可能会让你切割研磨探讨。亦即，死相难看，反而是福气

◀ Figure 1.1.3 We returned to Dawa, Lufeng, Yunnan in 2010 and found the dinosaur embryo site. These non-articulated bones were found. They gave us an absolute chance to enable us to "Look into the Bone." Nobody will allow us to cut perfectly articulate embryo for our study. These 'miss fortunate' bones were indeed a blessing for us

的研究，揭开更多恐龙宝贝的奥秘。

亦即，透过这些毫不起眼的零散小死骨头，提供一个绝无仅有的好机会，让我们建立恐龙胚胎学，利用同一生物种的恐龙胚胎化石来研究，探讨（至少是这种）恐龙胚胎的孵化发育成长过程。这种机缘，是那些完美死相恐龙胚胎化石所无法提供的。恐龙胚胎学，至少包括了“古代生物学”和“古生物化学”等主要的学科领域，深入研究恐龙胚胎学，可以提供或解答很多恐龙的奥秘，比方说，禄丰龙在它的时代，是地球上体型最大的动物，体长可达 12 米！这个世界最古老的禄丰龙胚胎，估计孵化出来的时候，体长大约只有 0.2 米，那么，它们如何快速成长到 12 米这么大？早侏罗纪原蜥脚类到中晚侏罗纪的后代蜥脚类恐龙，如梁龙、马门溪龙、泰坦龙等

they provided us an absolutely unique excellent opportunity. They enabled us to establish "Dinosaur Embryology" using the mono taxa dinosaur embryo fossil for the study of the dinosaur, at least this dinosaur embryonic development and growth. This opportunity is not available from the beautiful 'death look' dinosaur embryo fossil. Dinosaur Embryology consists of at least PaleoBiology and PaleoChemistry as the main domains. Deeper into dinosaur embryology will provide information or answers for the mysteries of the dinosaurs. For example, *Lufengosaurus* was the largest animal of their time. Their body could reach a length of 12 meters! The length of hatchling was estimated around 20 cm. Then, how did they grow up so fast to 12 meters? The prosauropods of early Jurassic and the sauropods of the mid-Jurassic period, such as plant-eater dinosaurs Diplodocus, Mamenchisaurus and Titanosaurus could reach 50~60

等吃素的庞然怪兽，体长可达五六十米，体重上百吨，它们是地球从古到今最大的动物，体型如此庞大（gigantism），到底怎么一回事？这是恐龙学中未解之谜，若能求得其解，对于人类社会，会产生重大影响：试想，如果利用上述庞大体型的生长原理来喂养小鸡，让它长得像一头牛那么大，那么对人类的食物供应会有多大的影响？

说来台湾有点悲哀：光复以后，为了求温饱，好多基础科学被忽略了，没有培养古生物学者。从个人的角度来说，这刚好是“福”与“祸”搅和在一起，因为我本身没有古生物的专业训练，在台湾也找不到适当的人选帮忙，在这个重大发现的研究中，台湾成员们根本没有说话的余地，所以只好另外打开一条路，以“看进骨头内”来带着台湾的其他学者参与这项研究，从上述的诸多死骨头，透过必须使用极端破坏性的几种同步辐射设备，观测到好些很有趣的恐龙胚胎化石的内部。如图1.1.4，下方是我们找到的胚胎肋骨切面照片，直径只有0.846 mm，把它切磨成15 μm的薄片后，用同步辐射透射X射线显微（transmission X-ray microscopy, TXM）扫描，把诸多的单格扫描组合成上方的马赛克，可以看到以前从未有人看过的信息，提供很有用的胚胎发育骨头成长信息。

图1.1.5也是在台湾同步辐射研究中心的另一个工作站做出来的，我们利用同步辐射傅里叶快速变换红外光谱显微

meters long with body weight over a hundred tons. There were the biggest record animals in earth history. How and why be such gigantism? This is an unanswered question in dinosaurology. If we can find the answer to this, it will have a very significant impact on humanity. Think, if we can use the gigantism principle to feed little chicken to grow up as big as a cow, what will that impact the human being food supply?

It is sad to say for Taiwan. Since the day she recovered from Japanese occupation, in order to have a full stomach, many fundamental sciences were neglected. No paleontologist trained. From my personal point of view, this is a mixture of blessing and curse. My discovery of this oldest dinosaur embryo, without formal paleontological training and could not find a suitable scholar in Taiwan to help out. There is no room for Taiwan scholars. So, I was forced to find a new way by using the "Look into the Bone" strategy to bring other Taiwan scholars into this research. From such many scatter dead bones, using the very destructive synchrotron radiation facilities, we observed many interesting internal contents. The lower photo of Figure 1.1.4 is the cross cut of a rib of a diameter of 0.846 mm. A thin slice of 15 μm was made and observed under TXM (Transmission X-Ray Microscopy) scans. After the images were stitched together, information has never been seen showed up and provided much useful embryonic growth and development information.

Figure 1.1.5 was also done at a different workstation of Taiwan Synchrotron Radiation Research Center. We used the Synchrotron Radiation FTIR Microscopy to scan these almost 200 million years old embryonic dinosaur femur.

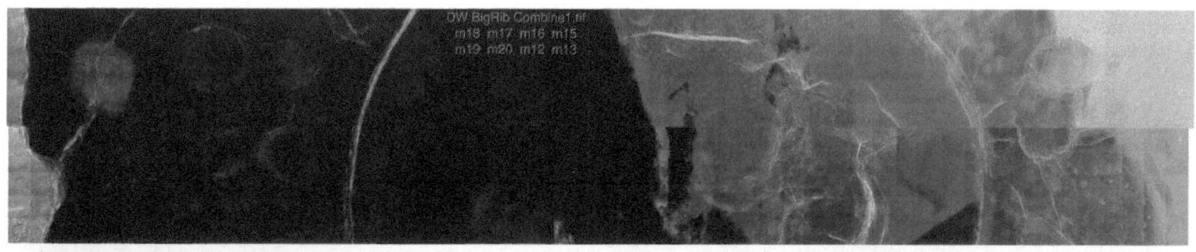

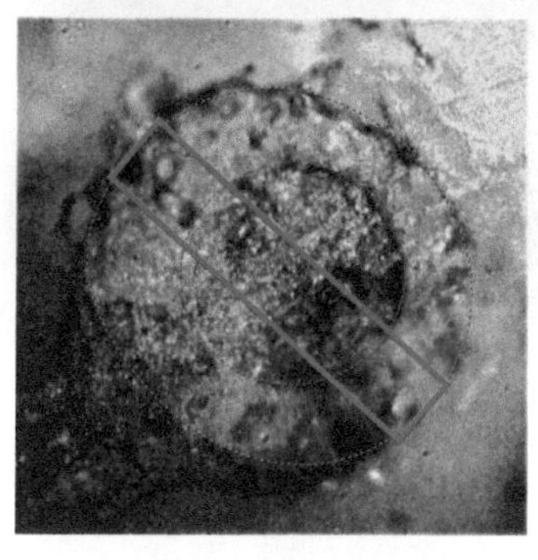

◀ 图 1.1.4 利用同步辐射透射 X 射线显微扫描禄丰龙胚胎肋骨，看到胚胎发育阶段中小肋骨内的结构，真令人咋舌。下方照片内的蓝色框标示扫描区域为 160 μm × 846 μm

◀ Figure 1.1.4 Using the Transmission X-ray Microscope of Taiwan Synchrotron Radiation Research Center to scan the embryonic rib of *Lufengosaurus*, internal development of the tiny embryonic rib can be seen clearly and very shocking. The blue box in the lower photo shows the scan area 160 μm × 846 μm

(synchrotron radiation FTIR microscopy, sr-FTIR)，扫描这些快两亿年前的恐龙胚胎股骨切片，在骨质和诸多原始管状空间 (primary tubular cavity, PTC) 交界处，看到了有机残留物的波峰，图 1.1.5 的光谱是最早期的成果。因此，我们在论文中也只能保守地说找到了保存有机残留物的证据，没有说是第一种胶原蛋白。根据近期的后续分析，终于可说这些有机残留物，就是第一种胶原蛋白与其分解物，同时也在重量级学术期刊发表了论文。

对于提出“看进骨头内”的我来说，这个结果证明了我的思维方向没错，我们团队走在正确的道路上。有趣的是，我们原本投递到《自然》期刊的论文题目，被期刊主编做了修改，新的标题总共有两行，其中第二行：with evidence of preserved organic remains (带着保存有机残留物的证据)，特别强调在这些

Around the intersection of hardy bone and the Primary Tubular Cavity (PTC), we saw the peaks of organic remains. Figure 1.1.5 is the earliest evidence. Thus, in our Nature cover paper, we conservatively said that we found the evidence of organic remains preservation, not collagen type I. In our recent follow-up, we finally can say the organic remains are collagen type I and their degradation product. This was published in a high caliber academic journal.

For me who brought up the concept of "Look into the Bone", the results proved that my strategical thinking was right. Our team works on the right track. Interesting enough, the title we used for *Nature* was edited to two lines. The whole second line "with evidence of preserved organic remains" emphasized the native organic remains inside these ~200 million years old dinosaur embryonic bones. Is not this the most significant confirmation for our Taiwan team? Moreover, a big shot on the arm for my "Look into the Bone," and at the same time, it opened

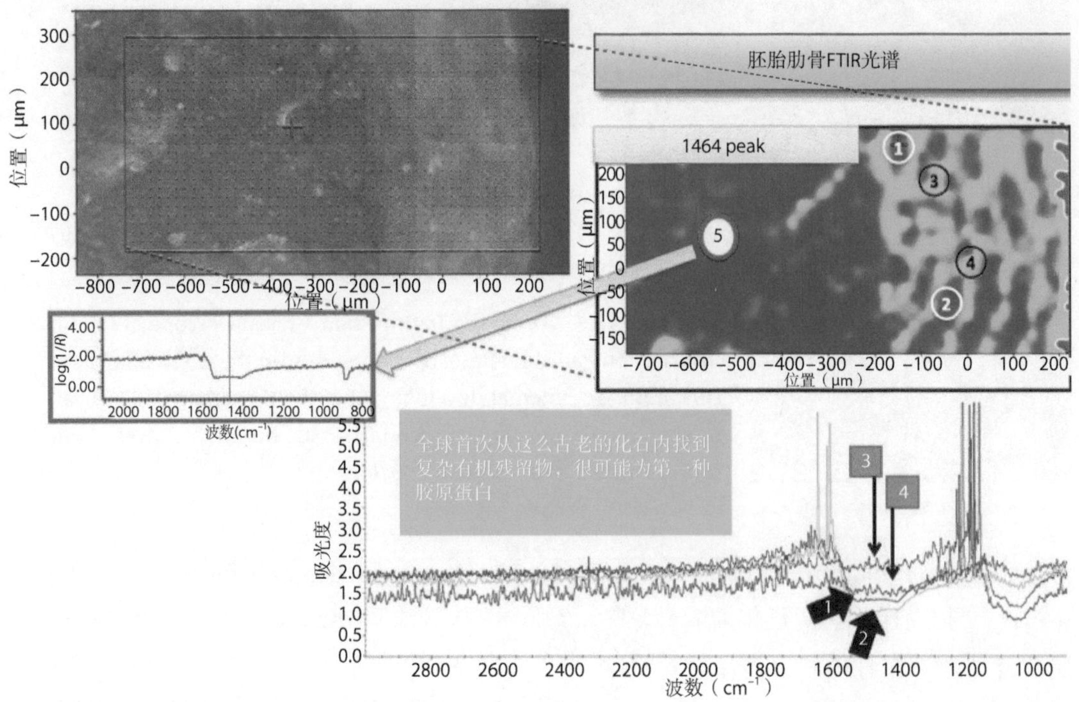

▲ 图 1.1.5 利用同步辐射傅里叶变换红外光谱显微，在这些 1.95 亿年前的恐龙胚胎骨头内，找到了有机残留物的证据，也使我们的《自然》论文题目被修改，其中第二行为：with evidence of preserved organic remains（带着保存有机残留物的证据）

▲ Figure 1.1.5 Using the Fourier Transform InfraRed spectroscopy of Taiwan Synchrotron Radiation Research Center to scan the 195 million years old embryonic bone, evidence of preserved organic remains were found. The second line of the title of our Nature paper was modified to "with evidence of preserved organic remains"

快两亿年前的胚胎化石内，还保存着原本胚胎骨头内的有机物，这不就是给我们台湾团队最大的肯定吗？更是"看进骨头内"的一个有力支撑，同时也给偏重于传统形态描述的古生物学，打开了另外一扇门，通往另外一个更高更宽广的新古生物学范畴。一般来说，通常认为化石是古代的生物已经变"化"成"石"

up a broader door for conventional systematic morphological descriptive paleontology to an even wider and higher new paleontology domain. In general, usually people think that fossils were ancient living things changed into rocks. Rocks are inorganic matter. Thus, it is not or impossible to find organics preserved inside. Our result, besides it is the oldest organic preservation record, it overturned the misconcep-

头，石头是无机的矿物，因此在化石里面，不应该或说不可能还保存着有机物，我们的成果，除了提出这是化石内保存有机物的最古老证据之外，更彻底推翻了这个“化石里面不可能保存着有机物”的错误认知，这是一个抽地毯式的影响。

话说回头到“看进骨头内”和“恐龙胚胎学”两个课题，其实它们都有好多子课题，所需用到的学科包括光学、力学、光谱学、分析化学、统计学、古生态学等等很多不同领域的交叉学科，而研究的样本包括了恐龙胚胎骨头牙齿、蛋壳、琥珀和其他化石等等，它们彼此之间错综复杂，交织成一个广大的网络，以前只说到了我们团队在台湾同步辐射研究中心所做的两种实验扫描，其实我们还需要使用更多其他的科学分析方法。

前面说过，台湾因为没有培养古生物专业的人才，台湾团队只好“异想开天”不按牌理出牌，把看骨头形态外表的部分，拱手恭请团队内国际专家来负责，我们则专注于看到骨头里面的部分，除了已提到的两种同步辐射检测设备之外，基本上还有以下几种秘密武器与方法：超高分辨率的微型计算机断层扫描（ultra high-resolution μ-CT）、多频显微（multiple harmonic generation microscopy, MHGM）、同步辐射快速X射线荧光扫描(synchrotron rapid scan X-ray fluorescence, srs-XRF）、电感耦合等离子体质谱（inductively coupled plasma mass spectrometry, ICP-MS）、扫描电子

tion of “no organic preserved inside the fossil.” It is a rug-pulling influence.

Bring the discussion back to the topics of “Look into the Bone” and “Dinosaur Embryology”. Indeed, each of these has many many sub-topics. Moreover, the disciplines from physic, optics, mechanic, spectroscopic, analytical chemistry, statistic, paleoecology, etc. are needed as a multi-discipline science. The studied specimens include embryonic dinosaur bones, eggshell, and tooth. They were intertwined to form a network. The two papers we published are just two types of scans we did at SRRC. Many brilliant scientific methods will be needed.

As mentioned before, Taiwan does not train paleontologists in the past. Our Taiwan team was forced to think outside the box. The job of looking outside the bone was assigned to our international colleagues, and we concentrated on the internal inside the bones. Besides the two mentioned synchrotron facilities, there are several secret weapons we can use, but not limited to: Ultra High-Resolution μ-CT, Multiple Harmonic Generation Microscopy (MHGM), Inductively Coupled Plasma Mass Spectrometry (ICP-MS), Synchrotron Rapid Scan X-Ray Fluorescence (srs-XRF), Scanning Electronic Microscope (SEM), etc.

Speaking from a grand point of view, we hope to use nondestructive examinations for any topic. In other words, all possible nondestructive methods shall be done first. If we need to repeat any one of them, the specimen is still intact. We can redo that experiment. Those destructive experiments shall be done at last, because once the experiment is done, the specimen also destroyed during the process. For the rare and precious dinosaur embryonic

显微镜（scanning electronic microscope, SEM）等等，但是可用的仪器与方法绝对不止这些。

先从整体来说，我们希望对于某个课题的检测，首先最好用非破坏性的检验，也就说，如果能先做非破坏性的检测，就优先进行，把这些非破坏性的检测先都做完，万一在其中有必要重做某个实验，样本还在，还可重做该检测；那些一定得用破坏性的实验，一定排到最后头，因为一旦做了破坏性检测之后，样本也没了。对于罕见的恐龙胚胎化石来说，万一实验没做好，绝对没有后悔的空间，所以尽量先选用非破坏性的实验，再安排破坏性的实验。其次，针对某同样的样本，尽量避免只用一种检测方法，如果可能的话，尽量使用多几种不同的检测方法（图1.1.6和图1.1.7），所谓原位同点检测 (*in situ in loco* multiple examinations)，如果这些不同的检测结果，都指向相同或类似的结论，那么从我们所研究的课题成果和所做出的推论，就会很铿锵有力，站得住脚，不怕别人鸡蛋里挑骨头。哈！鸡蛋里面的骨头，不就是小鸡胚胎骨头吗。没错，我们的研究，也包含了很多鸡胚胎的骨头，鸡（鸟类）就是恐龙啊！

从理论上来说，超高分辨率的微型计算机断层扫描的原理和医院里用的大台计算机断层扫描 (computerized tomography, CT) 是同样的，透过 X射线照射样本取得影像，再经过计算机运算组合，呈现某一套“切片”的影像，或

bone, there is no room for regret. Therefore, nondestructive ones have higher priority than destructive ones. Then, for the same specimen, try to avoid just only one measurement method. Try to use as many as possible for the same specimen at the same spot (Figure 1.1.6 and Figure 1.1.7), called *in situ in loco* Multiple Examinations. If the results from these various methods on the same spot of the same specimen agree to each other, our result will be very very solid and can stand many criticisms. We do not have to worry that other people can pick a bone in our chicken egg. Ha, is not that bone in the chicken egg embryonic chicken bone? Yes, in our studies, we did many many chicken embryonic bones. Chicken (birds) are dinosaurs!

In theory, the basic principle of higher resolution micro-CT and conventional hospital CT is the same. They use X-ray to image the specimen and use a computer to reconstruct the result as a set of slice images, or even to make a 3D model. The only main difference is the resolution: about 1mm for the medical CT, while μm for the micro-CT. Earlier models could achieve about 10μm. However, as technology advanced, the resolution nowadays can be as high as 0.35μm. The even higher nano-meter resolution also rumored to be available in the near future. This is a nondestructive examination, depends on the different absorption of X-ray by the specimen (also called X-ray density) to see the internal structure of the specimen. Without the destruction of the sample, it provides the purpose of looking into the bone.

For dinosaur embryology, as mentioned before, during the dinosaur embryonic development stage, in comparison to other animals,

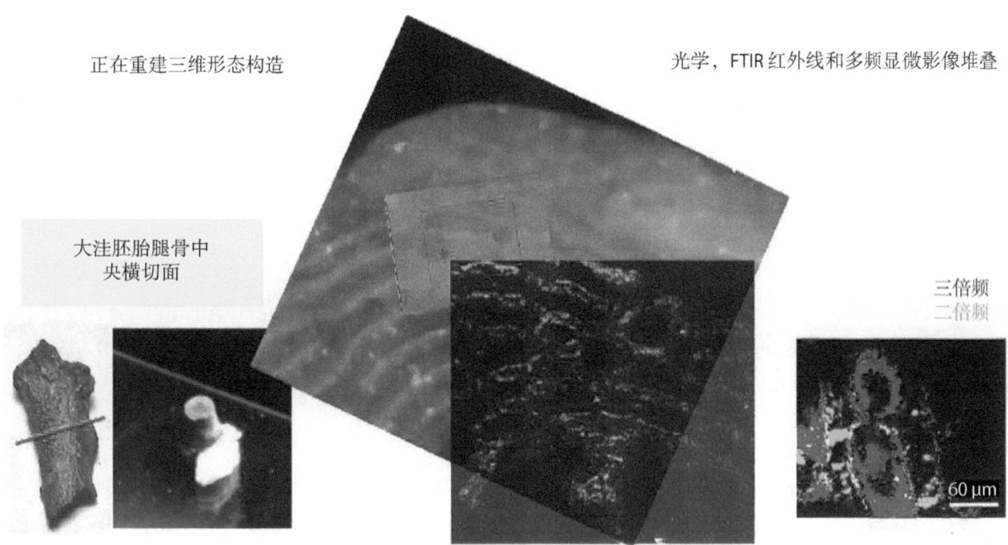

▲ 图 1.1.6 三重光学影像重叠，同样一块大洼禄丰龙胚胎腿骨样本，取得光学显微照片、红外线扫描照片和多频显微照片，这三种光学的影像紧密重叠，同时提供三种不同的信息。其实还可把高分辨率微型计算机断层扫描的影像也加进来，综合四种光学影像的信息，来做多元分析探讨

▲ Figure 1.1.6 Triple optical image superimposed on the same embryonic *Lufengosaurus* femur: Optical, Infrared, and Multiple Harmonic Generation images. These three type images match so well, providing three different information. If the high resolution micro Computerized Tomography can be added to this, it will have four optical information for our multiple analyses

更进一步做成三维影像模型进行研究。不同的是，医疗用CT分辨率大约只有1毫米 (mm) 上下，而超高分辨率的微型计算机断层扫描，通常以微米 (μm) 为度，稍早期的仪器可做到 10 μm左右，但是随着科技的进步发展，如今分辨率已经可达0.35 μm了，最近听说还有更高分辨率纳米级的仪器，已经上市了。这是一种非破坏性的检测，依赖被检测物质对于 X射线的吸收不同，亦即所谓的 X射线密度，分辨出样本里面的结构，达到看进骨头里面结构的目的，而不破坏珍贵的样本。

the development speed is much faster. We saw about 50% of Primary Tubular Cavity (PTC) from the outer rim of the cross cut hardy bone. See the brown bottom image of Figure 1.1.6. In contrast, the vascularity of adult dinosaur bone is only 5%-10%. Therefore, the PTC is a unique phenomenon. It is closely related to the development of the embryo and even the growth after hatch and adulthood. We have to look into this matter more deeper. In order to do a more precise calculation of 3D PTC and making the 3D model as Figure 1.1.8, we did some very high-resolution μ-CT scans. For the spaces in adult and embryonic bones, they are some significant nature differences besides the ratios

对于恐龙胚胎学来说，前面说过，在恐龙胚胎发育阶段，成长速度相对于其他动物来说，发育速度快很多，我们从股骨的横切面看到，在股骨外缘的硬骨圈内（图 1.1.6 压底棕色照片），其实有很多我们称之为“原始管状空间（primary tubular cavity, PTC）”的空洞，大约占了整个硬骨区的 50% 以上。相对来说，成龙骨头的“脉管度”（vascularity，也称为“孔隙度”），大约只有5%~10%而已，所以在胚胎阶段的高原始管状空间，是一个很独特的现象。这和胚胎发

of hardy bones and voids. For the adult bones, the spaces are mainly vascular channels (Haversian System) and Lacunas. However, during the embryonic stage Haversian system was still evolving. Thus, PTC is the spaces for blood vessels and nutrients. Since there is such a big difference between these two kinds of spaces, we coined the term "Primary Tubular Cavity (PTC)" for embryonic bones to distinguish that from vascularity for adult bones.

In March of 2013, I was lucky enough to be invited by the researchers of Manchester University to go to Stanford Synchrotron Radiation

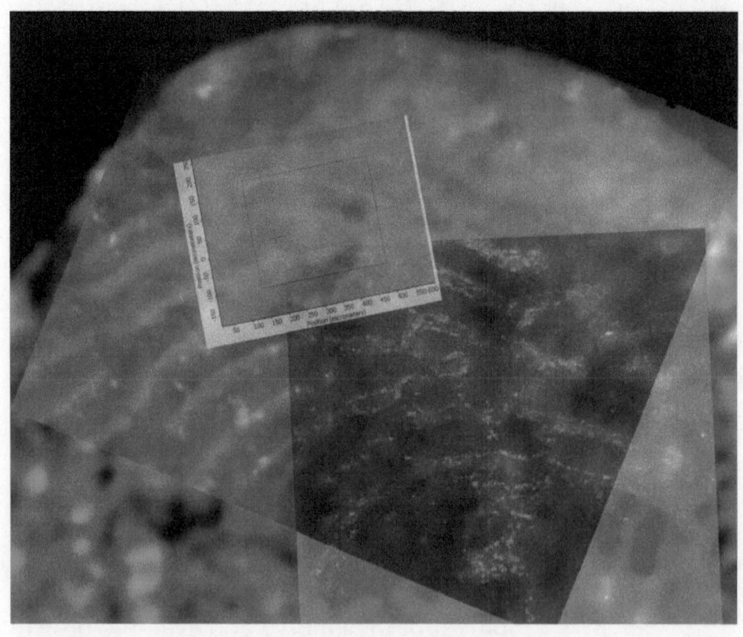

▲ 图 1.1.7　同样禄丰龙胚胎股骨的四重影像重叠，综合上述三种光学影像的信息，再加上 6.7 μm 中子扫描影像，可做更多元分析探讨。因为这些影像数据都非破坏性地取自于同样一个样本同样的位置，达成了真正的“原位同点”分析，免除了类似但不同样本（如不同的显微切片）的差异

▲ Figure 1.1.7　Quadruple image superimposed on the same embryonic *Lufengosaurus* femur from the above three light sources with an addition of a 6.7 μm neutron scan image. This enables us to do more analyses. Since all the images were taken nondestructively from the very same sample at the very same position, it is the real *in situ in loco* analyses, avoiding the variation of the similar sample but a different specimen, such as different microscope slice

育乃至破壳后的骨头成长发育，有很密切的关系，很值得深入探讨。因此，我们团队安排了一些高分辨率的微型计算机断层扫描，计算胚胎骨头里面的三维原始管状空间形态等等相关的项目，做出了三维形态模型，如图 1.1.8所示。对于成龙骨头内的空间和胚胎里面的空间，其实除了所占的面积/体积有很大的差异之外，这些空间的本质也有很大的不同。成龙的空间，主要是哈氏系统 (Haversian System) 和骨腔隙 (Lacunae) 空间，而在胚胎骨头内，哈氏系统还正在形成中，所以这些最早期的空间内应该是血管和液态养分。因为有如此较大的差异，所以对于胚胎，我们特别用“原始管状空间 PTC”来描述，以便和成龙的“脉管度”有所区分。

2013年3月，有幸被英国曼彻斯特 (Manchester University) 的研究员邀请，一起到美国加利福尼亚斯坦福大学的同步辐射实验室(Stanford Synchrotron Radiation Lab/Stanford Linear Accelerator Center, SSRL/SLAC) 去进行一些同步辐射快速 X射线荧光扫描。SSRL/SLAC有两套做同步辐射快速X射线荧光扫描的工

Lab/ Stanford Linear Accelerator Center (SSRL/SLAC) to conduct some X-ray fluorescent (XRF) scans. They have two sets of instruments, one for big specimen scans and one for the micro scans. The basic principle of XRF is by using high energy X-ray to radiate the specimen. The X-ray can kick the electrons from inner orbitals out to the outer orbital. While the electrons return to the deeper orbital, fluorescence will be released with specific wavelength for each kind of atom. Thus, we can obtain the elemental distribution inside the fossil specimens. The paper we published in *Nature* as cover paper relied on

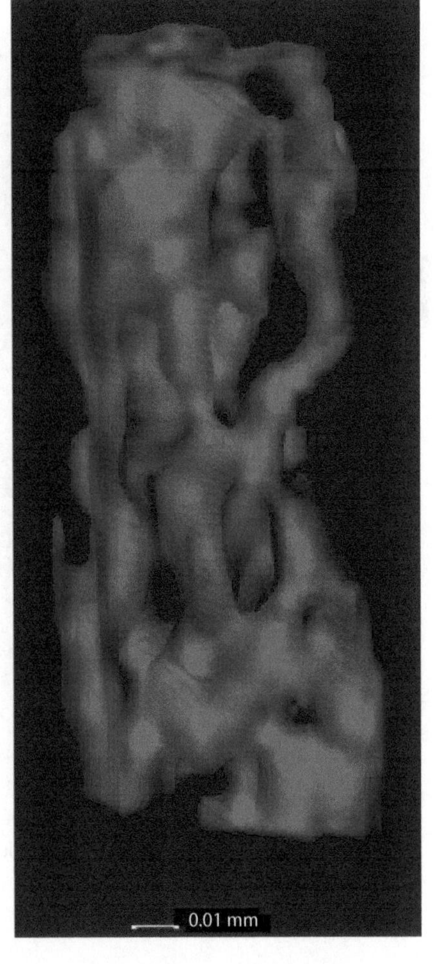

▶ 图 1.1.8 利用很高分辨率微型计算机断层扫描制作出来的恐龙胚胎股骨里面的三维原始管状空间(红色)，黑色看不见的部分就是刚刚形成的硬骨头

▶ Figure 1.1.8 Primary Tubular Cavity (red) in 3D by very high resolution micro Computerized Tomography. The invisible black areas were the hard bone just formed

作站，一套用于大型的样本，一套做显微扫描。SSRL/SLAC 设备的原理是，利用X射线打击样本，元素原子的内层轨道电子被激发，不同的元素会释放出特定波长的荧光，因此，可以从样本所发出荧光的波长来判断样本里面的元素分布。就以我们的样本和在同年4月份发表的《自然》期刊封面论文来说，我们提出这些快两亿年恐龙胚胎骨头里面，还保存着有机残留物，主要是靠台湾同步辐射中心的傅里叶变换红外显微 (sr-FTIR) 获得的数据来支撑。从支撑力度来

SRRC sr-FTIR solely. This may be a little bit weak. However, from the srs-XRF scans, we got yet another strong support for the preservation of organic remains. The major advantages of XRF are not only it can provide the distribution of elements, but also tell us the oxidation state of such element. Inorganic sulfur compounds such as gypsum and pyrite are very commonly seen inside fossil bones. However, organic sulfur can only come from the sulfur-containing amino acids made up of proteins or peptides. The oxidation state of organic and inorganic sulfur is different. Figure 1.1.9 shows the organic sulfur

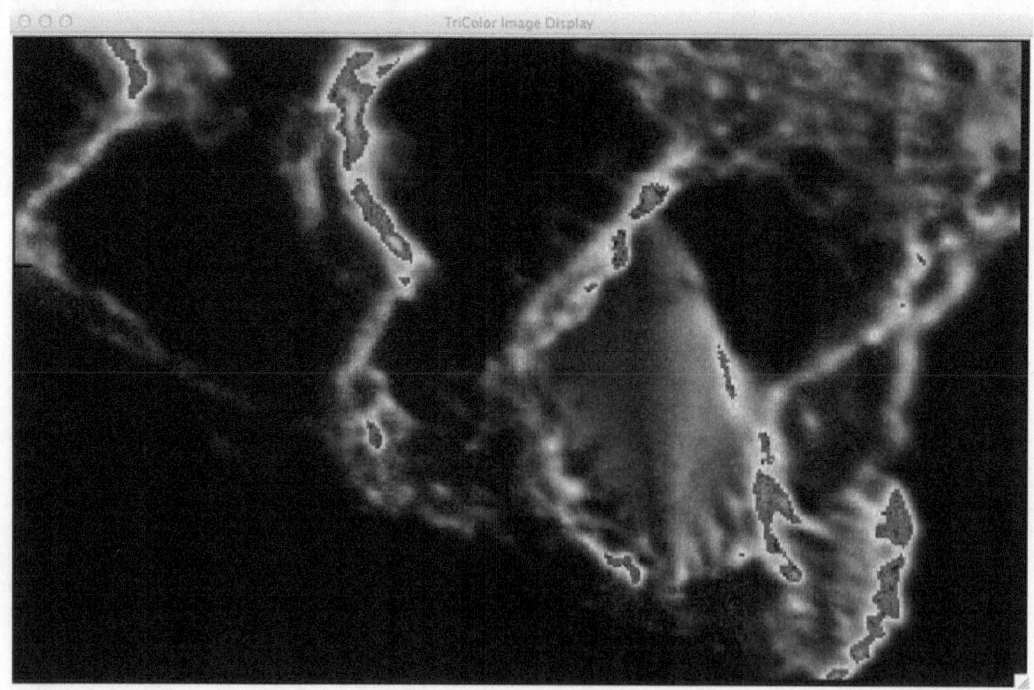

▲ 图 1.1.9 禄丰龙胚胎上颌骨内有机硫的分布，骨头主要成分是磷灰石，所以这两种元素的着色：钙用绿色，磷用蓝色混为底色，有机硫则以红色表达；影像中可见到很多红色的点和区块，表示这些地方含有有机硫

▲ Figure 1.1.9 The distribution of organic sulfur inside the maxilla of embryonic *Lufengosaurus*. The main composition of the bone is apatites: green for calcium and blue for phosphor as background color, and red for organic sulfur. As can be seen, many red dots and areas are meaning these contained organic sulfur

说，有人觉得还是有一点单薄，从斯坦福这次的元素分布扫描，又得到另一个强而有力的有机残留物保存证据。X射线荧光的另一个很重大的优点就是，它除了显示不同元素在样本里面的分布之外，同时也可告知该元素的氧化价数，如化石内的硫元素，可有无机硫和有机硫两类：无机硫化物，如石膏、黄铁矿等矿物，也常见于化石内。而化石内的有机硫，主要来自组成多肽或蛋白质的含硫氨基酸，有机硫和无机硫的氧化价数不同。图1.1.9是一小块禄丰龙胚胎上颌骨含有机硫的分布影像，影像中红硫点处处可见，也就是说，这是另外一个有力的证明，证明在这些恐龙胚胎内还保存着含硫氨基酸甚至是原生的胶原蛋白。可惜在这时候，我们论文已经被接受，进入排版阶段，无法更改把这部分信息加进去。

多频显微是台湾大学孙启光教授的一个领先世界的伟大发明，它的原理是利用脉冲红外线激光照射于样本，产生三种光学反应：二光子(two photon, TPH)、二倍频(second harmonic generation, SHG)和三倍频(third harmonic generation, THG)。这三种光学反应以毫无破坏性的快速方式提供了样本的多种信息，这些信息是其他光学显微镜无法提供的，孙老师过去的样本，主要是活体样本，如皮肤样本等等。当我知道有这么一个“武林至宝”之后，死缠活缠他一起来试试死体样本（化石），所得的结果（图1.1.10），令人大吃一惊，也

distribution in red of a *Lufengosaurus* embryonic maxilla. We saw many many red dots/areas. This means that this is yet another compelling supportive evidence for the preservation of sulfur-containing amino acids or native collagen. Unfortunately, at that time, our paper was accepted and in the process of layout for publication. We could not put this information into our paper.

Multiple Harmonic Generation Microscope (MHGM) is an excellent world-class invention by professor ChiKuan Sun of Taiwan University. It uses pulse laser on the specimen to produce three different optical responses: Two-Photon (TPH), Second Harmonic Generation (SHG), and Third Harmonic Generation (THG). These three optical responses provide nondestructive quick scan information that other optical microscopes cannot. His specimens in the past were mainly on living tissues such as skin lesion. When I knew such a fantastic tool, I used all my charms to try on some dead objects (fossils), and the result shown as in Figure 1.1.10. We were very stunned and shocked. The initial results were published in high caliber international journals. In our initial trials, we scan the cross cut of mid-shaft of a *Lufengosaurus* femur. The left image shows the 3D of SHG in green, corresponding to the apatites. The middle image shows the 3D of THG in magenta, corresponding to PTC. We learned from sr-FTIR that the organic remains could cause these. The far right image is a portion of the combination of SHG and THG. The exciting thing is that in this superimposed image, we saw some white spots and areas. From the optical principle, we know that green and magenta are complementary color. When

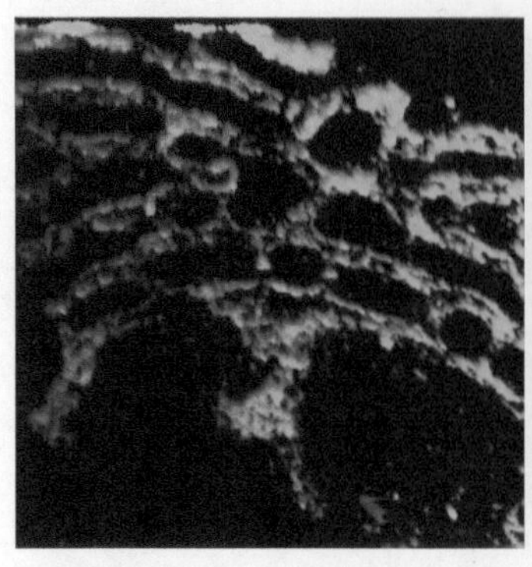
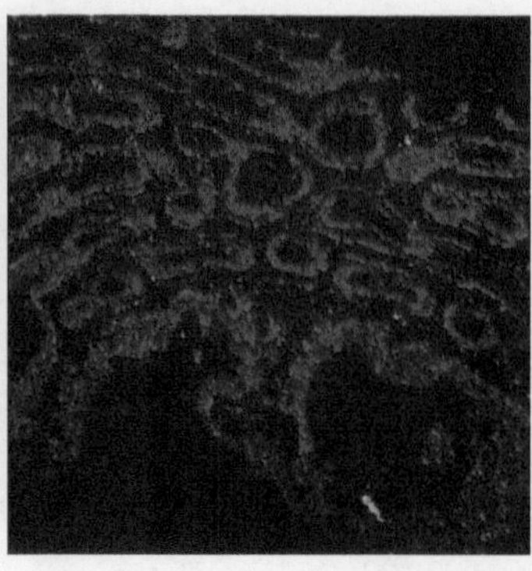
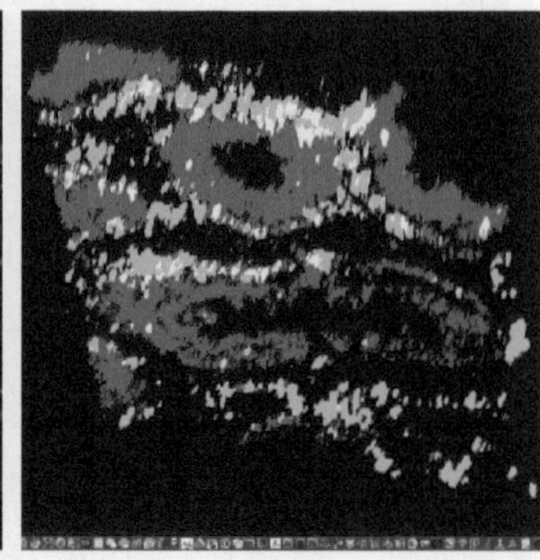

▲ 图 1.1.10　多频显微扫描禄丰龙胚胎股骨，最左边是二倍频影像，中间是三倍频影像，最右边是二倍频和三倍频部分组合（放大）的影像，白色显示另外一个强力的化石中保存有机残留物的证据

▲ Figure 1.1.10　MHGM scans of embryonic *Lufengosaurus* femur. The left is the Second Harmonic Generation (SHG) image, the middle is the Third Harmonic Generation (THG). And the right shows the enlarged combination of SHG and THG. The white areas are strong evidence of organic preservation

大开眼界，最初期的成果，已经二度发表在国际顶尖光学学术期刊上。在初步的尝试，扫描了禄丰龙胚胎的大腿骨切面，得到很有趣的结果，如图 1.1.10 所示：最左边的是禄丰龙胚胎股骨中央横切面的二倍频立体影像，以绿色表示，相对应于磷灰石群矿物的硬骨头部分；中间的是同样的样本同时扫描时的三倍频立体影像，以洋红表示，对应于前面说的原始管状空间，同时从同步辐射傅里叶变换红外显微分析得知，这些很可能就是有机残留物所造成的；最右边的影像是将部分二倍频和三倍频重叠组合的结果，最有趣的是在此影像中，看到了一些白色的斑点和区块，回到光学来说，绿色和洋红是互补色，两者加在一起，就变成白色；亦即，这些白色的区域，同时有二倍频和三倍频的反应！这就有趣啦！什么意思？如果上面的假设

these two are mixed, the result is white. That is to say, the white areas we see on this image means both SHG and THG responses. This is very interesting! What does this mean? If the above assumption stand (SHG shows the hardy bone of apatites and THG for organic remains), then, this is yet another very compelling evidence for the preservation of organic remains (collagen I) in these almost 200 million years old dinosaur embryonic bone endogenously. Bingo, another evidence to support our claim of the preservation of organic remains.

Furthermore, thru ICP-MS analyses for the Rare Earth Elements inside the dinosaur bones, we found that the adult bones contain ten times more apatite minerals. This is a total destructive method, but it provided us valuable information to enrich our Look into the Bone Dinosaur Embryology.

With the establishment of Dinosaur Embryology by Look into the Bone, we brought the paleontology to a new level of multi-disciplines.

成立（二倍频显示磷灰石群矿物的硬骨头，三倍频显示有机残留物），那么，这就是另外一个很强而有力的证据，证明在这些快两亿年前的恐龙胚胎骨头内，还保存着原生的有机残留物（第一类胶原蛋白）！ Bingo，又多了一个证据支持我们的有机物保存论点。

此外，我们通过电感耦合等离子体质谱 (ICP-MS) 分析恐龙骨头化石内的稀土元素，也证明了成龙的骨头的磷灰石群矿物含量，是胚胎骨头的十倍，虽然这是一个绝对破坏性的分析方法，但是也从另外一个角度提供了宝贵的资料，让我们看进骨头内的恐龙胚胎学，更加充实丰富。

以“看进骨头内”思维领导建立恐龙胚胎学，今日古生物学已经被我们推高到全方位的交叉学科，所以必须是大家齐心努力，共同无私地合作！

1.2　形态学

在传统的古生物学中形态描述是主要的系统描述部分，叙述某古生物遗骸的外观形状尺寸等等（图 1.2.1），进而和其他相近属种做比对，判断是否为新属种。比较先进的古生物学，会从某根特定的骨头长度直径等数据，推论该古生物的成长，也有进行切片，从组织学的角度，探讨其骨头结构，甚至比对于现生已知生物，来研究推论该古生物的成长过程（图 1.2.2）。

通常，传统古生物学除了照相之外，

We have to work together without a selfish attitude.

1.2　Morphology

In conventional paleontology morphological description is the main part of the systematic description to describe the skeletal remains of the ancient organism (Figure 1.2.1), and then compare with other related material to see if that is a “new” genus and or species. The more advanced paleontology go further to the growth, i.e., the length and diameter of a particular bone to infer the growth of such animal. Thin slice under microscopic histological examinations and compare with extant animal bones are also standard these days (Figure 1.2.2).

Besides the camera photos would be taken, line drawing with remarks are also conventionally used, as Figure 1.2.3.

There is absolutely nothing wrong with these. They are the most fundamental requirements used by serious scholars. See, when some bones were found, the first thing is to identify them, and see if it is a new taxon, being described? etc. These are the most critical issues that must be clarified at first.

Even by using thin slices for microscopic internal morphological and histological observations, how to interpret the image saw under a microscope? What differences between fossil and extant bones? Can the view of extant biology be directly used for the fossilized bones? What was seen under the microscope? Moreover, the more important question is “What’s NOT seen?” What morphological and chemical changes took placed during the fossilization

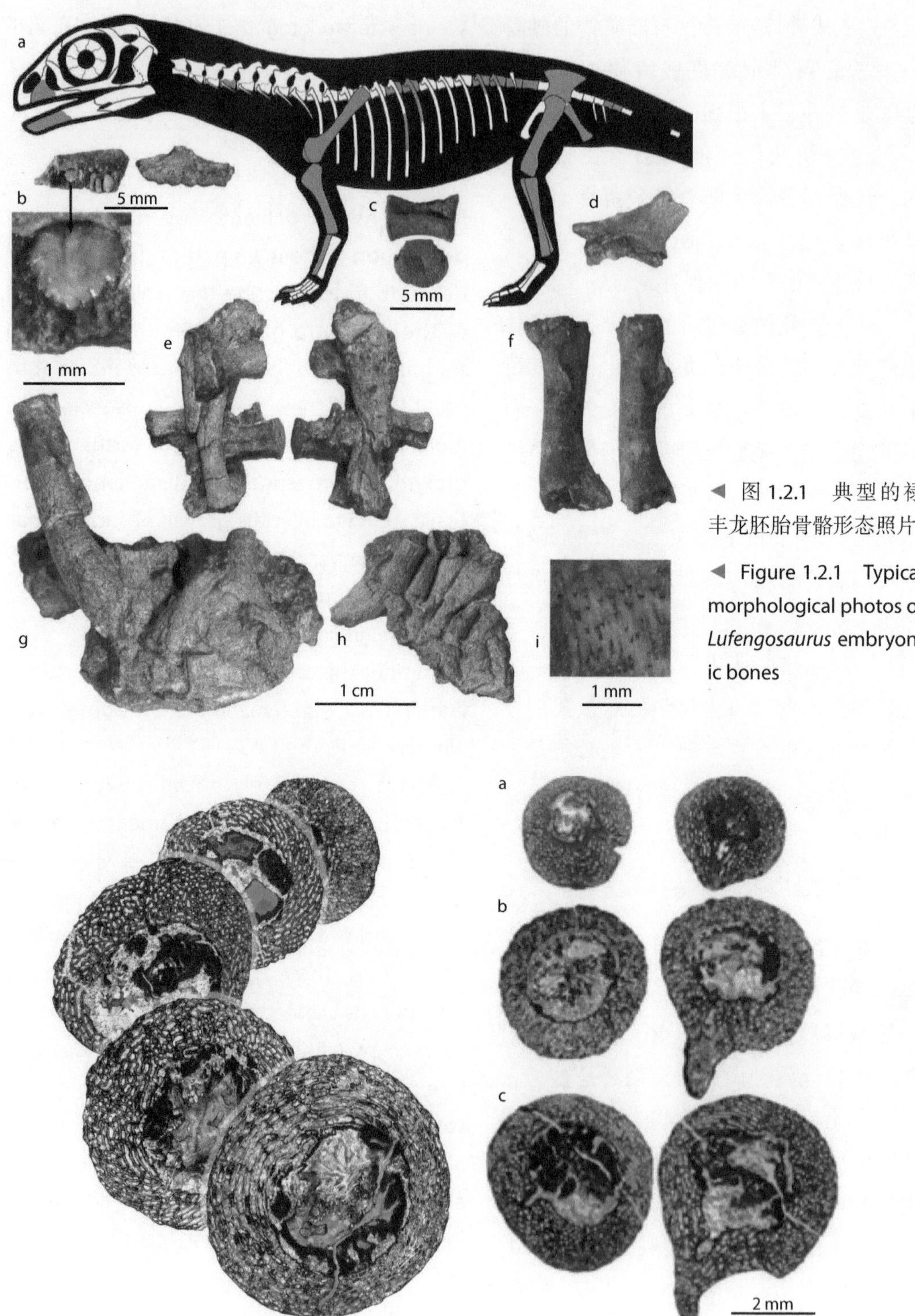

◀ 图 1.2.1　典型的禄丰龙胚胎骨骼形态照片

◀ Figure 1.2.1　Typical morphological photos of *Lufengosaurus* embryonic bones

▲ 图 1.2.2　比较先进的形态学，进行骨头切片，进行组织学探讨，求得成长过程。禄丰龙胚胎股骨

▲ Figure 1.2.2　Advanced morphology by thin slices for the histological study to understand the growth process of *Lufengosaurus* embryonic femur

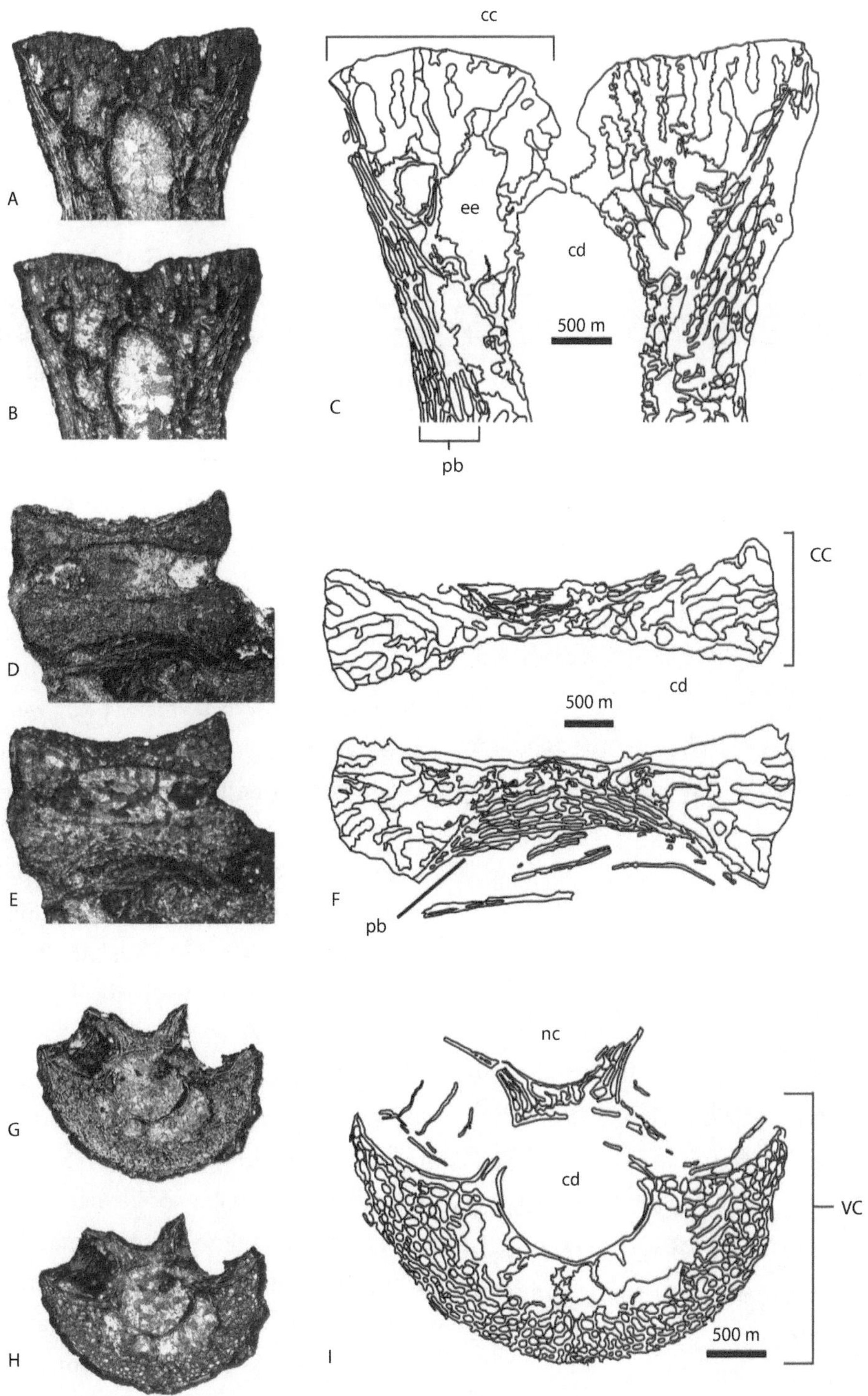

▲ 图 1.2.3　禄丰龙胚胎脊椎骨的形态描述

▲ Figure 1.2.3　Morphological description of *Lufengosaurus* embryonic vertebrae

还会把骨头（含骨头切片）的形态，用线条描绘出来，加上标示，注明某根骨头某个位置是什么构造（图 1.2.3）。

如此的古生物研究，说来没错，而且是绝对必要的基本功夫，大部分治学严谨的学者，在这方面也都很认真努力；试想，当发现某种骨头，首先当然要得知到底它是什么东西，依照生物分类是哪种古生物的化石，到底是不是一个“新”品种，以前没有被发现过，等等基本问题，这些都是必须首先厘清的关键项目。

即便采用切片等先进的手段来进行内部形态组织学观察，在显微镜下面看到的影像，如何解读呢？与现生骨头的切片有什么差异？能否就直接把现生动物生物学的看法，直接搬过来解读骨头化石呢？显微镜下看到了什么？更重要的问题是“没看到什么”？在石化的过程中，到底发生了哪些形态和化学变化？下面举三个例子来讨论：

（1）把恐龙骨头横切，会看到像树木年轮的圈线，每年一轮线，叫做成长停止线 (line of arrested growth, LAG)，学者用来计算恐龙的年龄。粗看之下，哇！太棒了，能从有几条这种线来计算那只恐龙活了几岁，解决了一个困扰恐龙界很久的问题。图 1.2.2 的禄丰龙胚胎，没看到这种线，很合乎此说啊！因为还在胚胎阶段，不到一岁，所以当然没有成长停止线！不过，从另外一个角度来看，骨头是活的组织，由破骨细胞 (osteoclast) 和成骨细胞 (osteoblast) 相互

process? Let us look at the following three examples:

(1) Dinosaur bone cuts transversally will show the ring lines like the annual growth rings of the tree. These annual lines are call Line of Arrested Growth, LAG. It looks wonderful! By counting how many LAG, the age of that dinosaur can be obtained. This resolved a long hanging question. In Figure 1.2.2, the embryonic *Lufengosaurus* bone, no such line can be seen. Perfectly OK, because the baby dinosaur was less than one year old, so there was no LAG! However, looking from another point of view. Bones are living tissues consisting of interactions of osteoclast and osteoblast to maintain the health of bone. In other words, during the growth period, the inner side of hard bone close to bone marrow, how many LAGs were eaten away by the osteoclast, and how many new LAGs were formed around the outside of the bone rim? Do the number of eaten LAG equals to the new form LAG? Even if the ratio of these two is 1:1, how can we be sure how many were eaten? Thus, if we find say ten LAG lines, can we say that dinosaur was ten years old? Questionable and debatable.

(2) The second example (Figure 1.2.4) is the incremental von Ebner lines (A arrow) inside the dentin, one line every day. Some people using this to say that dinosaur incubation would take a very long time, closer to that of reptiles which are not as others observed rapid embryonic development as birds. What is wrong with this saying? Simple mathematics can be used. In that paper, it stated that the distance between to von Ebner lines is between 6 to 3.5 μm. Let us use the average of these two num-

作用，维持骨头的健康。也就是说，在成长过程中，靠近骨髓腔部位（硬骨内侧），无法确认破骨细胞吃掉了多少成长停止线，新的生长停止线产生于硬骨外缘，里面被吃掉和外缘新增的生长停止线，是否是一对一？即便是一对一，如何确定有多少条已经消失的生长停止线，因此，如果算出十条线，能否就确认说这头恐龙活了十岁（而已）？这里面还有些争议。

（2）第二个例子（图 1.2.4），从牙齿切片，在牙本质内可看到 von Ebner 线（A 箭头），每天长一条，甚至有人以此推论出恐龙胚胎的孵化需要很长的时间，接近于爬行类的孵化，而不是其他研究者所观察到认为的恐龙胚胎快速发育，接近于鸟类。此说的问题何在？用简单数学来算一下，两条 von Ebner 线的距离，拿其文内数据 (6+3.5)/2 μm为平均值，灌一点水，算成每天 5 μm 好了，那么一根牙本质厚度 5 mm 的牙齿，就需要1000天的成长时间，即便把中生代每年日数算为现在的每年365天，那么，这根新牙齿需要三年才长成！可是，恐龙一生中不停地换牙齿，有人说平均每根牙齿只用半年，就被换掉了！事实上，恐龙有这种花三年准备，却只用半年的牙齿更换吗？从牙床齿洞内的观察，如果把现用的牙齿算第一代，齿洞内会有第二代，甚至第三代牙齿成长中等着接换第一代老牙齿，第一代半年就掉了，后代需要三年的准备，来得及吗？恐龙难道是“没齿难忘”的动物？事实证明

bers and rounded to 5μm per day. Then, the dentin of 5mm thickness will take 1,000 days to form. If we use the current 365 days as one year in Mesozoic, it will take about three years to form a new tooth! However, as well known, dinosaurs change their teeth all the time during their whole life. Some estimate as half of a year per tooth replacement. So, is that reasonable for the dinosaurs to spend three years to make a new tooth to be used just six months? If we look at the dinosaur tooth socket and assign the current tooth as the first-generation, there are second- and even third-generation tooth waiting to replace the old existing first-generation teeth. Can the first-generation teeth fall off in 6 months and the new one come up three years later? Can dinosaurs be the toothless animal? This is not the fact. We found that even in the embryonic stage, without eating any food, there is undeniable evidence of tooth replacement. The first generation embryonic tooth already has been pushed aside, the second generation took the place of the first, and the third generation tooth was formed underneath awaiting to replace the second generation. Do dinosaur eggs need to have three years incubation time?

(3) The third case is: (Figure 1.2.5) A while ago, someone showed me a bunch of photos and claimed that he had found the dinosaur blood with a clear picture of many many individual red blood cells. He was so sure that he found the “dragon blood” never seen before. After I looked at these photos, I reminded him that by just looking at the outer shape and color, particularly the color inside fossil, these were not enough to be sure. The red blood color hematite has a crystal habit of forming small

▲ 图 1.2.4 牙齿牙本质内的 von Ebner 线（http://www.uky.edu/~brmacp/oralhist/module5/lab/imgshtml/image20.htm）：牙本质内的累增线条；牙小管从左到右，细线 A 几乎垂直于牙小管，从顶端到底部；注意到这些现有规则的累增，这些就是 von Ebner 累增线；这是一片研磨过切片，在去矿物质后的切片，也可看到这些线；von Ebner 线条代表牙本质形成的成齿质细胞循环性活动，这些累增线表示每天的牙本质累积，牙冠部位每天大约 6 μm，牙根部位每天 3.5 μm

▲ Figure 1.2.4 Incremental Lines in Dentin: Dentinal tubules cross the field from left to right. Fine lines A run almost perpendicular to the dentinal tubules from top to bottom on the screen. Note that these lines occur at regular increments. These are the incremental lines of von Ebner. This is a ground section, but these lines can also be seen in demineralized sections. The lines of von Ebner represent the cyclic activity of the odontoblasts during dentin formation. These incremental lines illustrate the daily pattern of dentin deposition that progresses at about 6 μm per day in the crown and about 3.5 μm per day in the root

并非如此，何况我们发现，即便在胚胎阶段的恐龙，还没吃过外界的食物，就已经有换牙的证据了，第一代牙齿掉了被推到一边，第二代牙齿正取代第一代牙齿，第三代牙齿已经等在第二代下面蓄势待发，恐龙胚胎需要三年的孵化时间吗？

（3）第三个例子是，曾经有人拿了一堆照片（图 1.2.5）来给我看，说他找到了恐龙之血，红血球一个个清楚得很，

spheres around the size of several ten μm. Furthermore, can the red blood cells be preserved after almost 200 million years? Very hard to say, and there was no evidence to support such a claim. In the past, Mary Schweitzer published the papers and stated that she found heme in the *T. rex* femur. Unfortunately, the overheated reporters exaggerated it to be red blood cells. It created a roaring storm. She did many analyses and used the words very carefully. She did not say red blood cells, but heme. There is a big

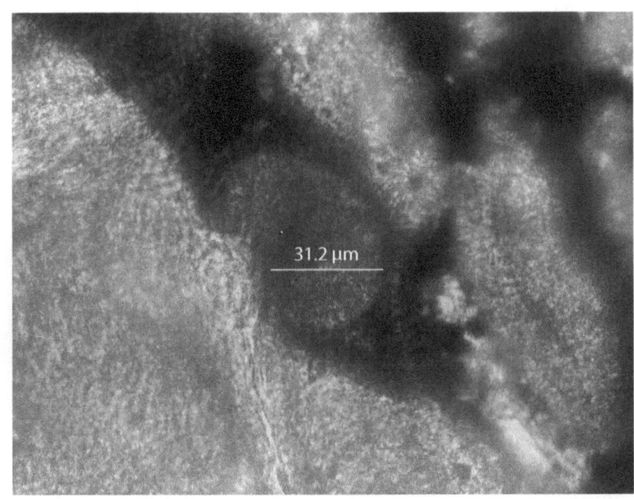

◀ 图 1.2.5 恐龙之血？

◀ Figure 1.2.5 Dragon Blood?

而且大小和动物红血球也相当，他非常肯定说他找到了前人未曾找到过的恐龙之血。我看了照片提醒他，光是看外表形态和颜色或者是化石里面的颜色，很难作为准确鉴定的要素：血红色的赤铁矿有一种晶癖，就是结成小球状，大小也在几十微米(µm)，何况红血球能否保存将近两亿年？很难说啊！也没有任何其他证据支持此说；当年Schweitzer最早提出在暴龙腿骨中找到“血基质(Heme)”，被记者误报道为红血球，差一点被学术界打个半死，人家做了很多分析，而且论文用词很谨慎，没说是红血球，而只是“血基质”，毕竟红血球和血基质，是有很大差别的，就得到如此无情的攻击。现在岂能只靠几张照片里很像红血球的“东东”，就跳到“找到恐龙之血”的结论呢？后来进行的同步辐射透射 X 射线显微和拉曼光谱分析，果真如我所说的，这些是赤铁矿的小球（图 1.2.6），而非红血球，勉强来

difference between heme and red blood cell. She got such a crude response. So, how can we jump to the conclusion by just these photos having something look alike? Then, after we conducted Transmission X-ray Microscopy (TXM) and Raman Spectroscopy, as I predicted, these red spheres are hematite (Figure 1.2.6), not red blood cells. Perhaps it may be exaggerated to say that we had found the remnants of red blood cells (hematite). For sure, these are NOT red blood cells themselves. Red blood cells and the remnants of them are very different. Yes, they are related, but not the same thing. So, do not take FengJing as MaLiang (冯京当马凉).

As far as dinosaurs are concerned even up to today, many named dinosaurs were based only on one bone or one tooth. We saw people used only just more than one hundred words to describe a new dinosaur so he can be the first to name it, such as *Chuanjiesaurus anaensis* (*https://en.wikipedia.org/wiki/Chuanjiesaurus*). One even much much worse than that is the so-called "Chuxiongosaurus lufengensis" (*https://en.wiki-*

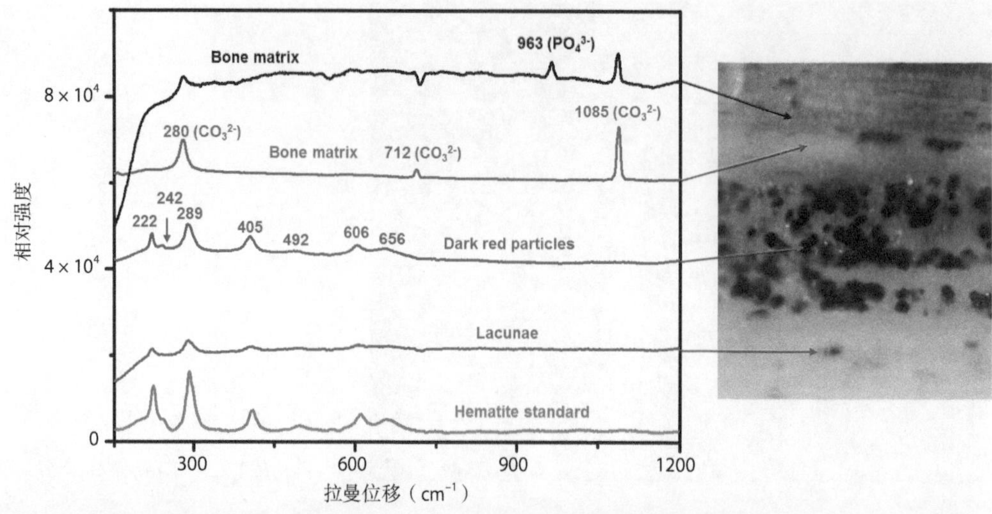

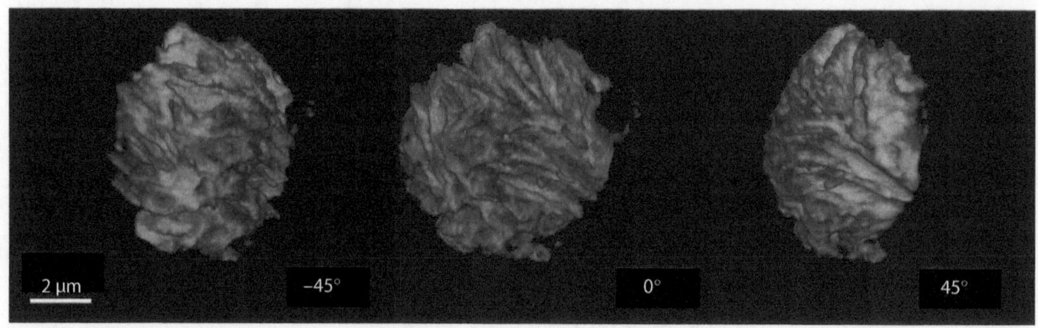

▲ 图 1.2.6 （上）拉曼光谱证明这些小红点，确实是赤铁矿的小球；（下）同步辐射的 TXM 提供赤铁矿小球的形态构造，确认为不是红血球，而是红血球的残留物

▲ Figure 1.2.6 (Top) Raman Spectra showed these tiny red dots are hematite for sure. (Bottom) Synchrotron Radiation Transmission X-ray Microscopy showing the morphology of hematite sphere, and confirm it is not red blood cell, but the remnants of Red Blood Cells

说，可以说是红血球的残留物（赤铁矿），但是绝对不是红血球本身。红血球本身和红血球的残留物，两者有很大的差异，虽然两者有关联，但总不能把冯京当马凉啊！炎黄子孙不是炎黄本人啊！

就以恐龙来说，即便到今日，很多恐龙的鉴定命名，往往仅仅靠着老天爷给的这么一根骨头或一颗牙齿。我们也看过少数人为了抢先发表命名，只用了

pedia.org/wiki/Chuxiongosaurus). The skull and the lower jaw were excavated from two different places at different times. The authors just rushed to name a new dinosaur and made such a “Cow-head-horse-mouth Dragon (牛头马嘴龙).” It is an international joke! We shall use this to learn a lesson. We shall not do such lousy job. It is common in dinosaur field that the later found specimens correct the mis-named previously.

一百多字的描述就想蒙混过关的，如阿纳川街龙（*https://en.wikipedia.org/wiki/Chuanjiesaurus*），还有更荒谬的“禄丰楚雄龙”（*https://en.wikipedia.org/wiki/Chuxiongosaurus*），头颅和下巴是两个分别在不同时间不同地点出土的，作者为抢先命名，搞出这种“牛头马嘴龙”，成为国际笑话，也当借鉴，做学问不能如此鬼混。也因为如此，后来发现更多样本而修订过去错误命名的情况，屡见不鲜。

从我们的角度来说，即便是做得很严谨的形态系统描述，也好像是在擂台上比武，刚把马步跨好，还没使出威力绝招呢！事实上，透过“看进骨头内”的思维，使用现代先进科学仪器与方法，我们可以把化石从里看到外，得到比较完整的信息，针对某古生物做出比较正确的认识。简单来说，要搞古生物化学，对于恐龙或其他古生物的细微形态构造，必须有些基本的认识，才不会白费力气，搞错了方向。

“看进骨头内”有两个层面，一是显微性的形态组织构造观测，可使用先进高分辨率的一些仪器设备与方法取得（如μ-CT、中子扫描等等）；另一层面是探讨其化学成分，甚至推论在漫长的石化过程中，发生了哪些化学变化，有机物和无机物如何互动，形成了当下的这堆化石。

因为如此一来，我们可以把某根骨头（或其他化石）从里到外看光光，因而更容易去探讨该古生物在地球生命演

From our point of view, even with very solid and careful morphological systematic description done, it just like takes the very first firm step on the competition arena. None powerful punch exerted yet. We have to use the “Look into the Bone” strategy with the most advanced instrumentation and methodology to examine the fossil inside out to obtain a piece of complete information set for a more accurate understanding of such fossil. In short, to do PaleoChemistry on the morphological details of dinosaurs or other paleontological material, we have to have some basic understanding of this topic to prevent going to the wrong way.

There are two folds of “Look into the Bone”: one is the observation of microscopic morphological structure. This can be accomplished by using high-resolution instruments, such as μ-CT and neutron tomography. Another one is to explore the chemical composition, even trying to deduct what changed during such a long fossilization period time, what chemical reactions took place, how the organics interact with inorganic to yield the fossil in our hand now.

By doing so, we can see a bone (or other fossils) from inside out thoroughly. That will ease us to explore what role that particular ancient organism played in the life evolution on this earth. After all, we human being are always very interested in finding out the big question of “who am I?”. In order to touch such a serious question, we have to expand our view from that single fossil to the life evolution on this earth. We human beings are just a small branch of countless lives in evolutionary history. We are not the monkey jump out from a rock. So, doing paleontology, we have to pay attention to life evolution. For

化过程中，扮演了什么重要的角色。毕竟,我们人类对于“我是谁”这个大哉问，一直都有很浓的兴趣，想要探讨如此重大的问题，其实更应该扩充到地球生命的演化，毕竟，我们人类也只是地球无数生命演化的一个小分支，肯定不是像孙悟空从石头里蹦出来的。这也就是说，研究古生物，应该着重于生物的演化，比方说，在埃迪卡拉纪 (Ediacaran period) 的生物还没有长出眼睛，也不必在海底泥巴钻洞躲避被猎食，生物的第一只眼睛要到寒武纪三叶虫才出现，从此地球的生物界才出现“猎人与猎物 (hunter and hunted)”的生命猎食法则，改变了整个地球生态。又如，牙齿是怎么演化出来的？四足转变成双足、鳞片到羽毛等等，都是很有趣的生物演化好课题。

1.3 化学成分

西方，至少在美国，有一种说法 ：当你手中有把铁锤，看到的世界到处都是钉子，等着去槌一下。同样的道理，当你是个传统古生物学者，看到化石，就会主动从系统形态描述的角度来看那块化石，从它的外表形态来看它属于哪个古生物物种等等，如果能发现并且描述一个新品种，那就Bingo啦！然而，同样的道理，对于化学工作者来说，我们所看到的宇宙天下任何物质，哪个不是化学品？天天呼吸的空气，主要是氢、氧、氮等化学分子，喝的水是 H_2O，两个氢原子与一个氧原子的化合物，整

example, during the Ediacaran Period, no eye evolved yet. Those creatures did not have to burrow in the mud to avoid predation. The first eye came in the trilobite of the Cambrian Period. Since then, the hunter vs. hunted arm race started and changed the whole earth ecology. Also, how the tooth evolved? Tetrapod to bipod? Scales to feather? All these are fascinating good life evolution topics.

1.3 Chemical Composition

In the Western world, at least in the United States, there is a saying: When you have a hammer in hand, you see nails everywhere awaiting to pond them down. For the same reason, if you are a conventional paleontologist and see a fossil, you will look at it from morphological systematic description angle of view to see what taxon it belongs to. If you can find and describe a new species, beautiful! Also, for the same reason, as a chemist, all matter we see in the universe are chemicals. The air we breath is mainly hydrogen, oxygen and nitrogen molecules. Water is H_2O, two hydrogen and one oxygen atoms combined. All universe, including you and I and all matter, are the product of that Big Bang of 14 billion years ago. All elements reincarnated so many times already. The whole solar system, earth, and people are just Star Dust. Chemistry is the science of studying the changes of this matter, what characteristics they have, and why the reaction takes place. Chemical reactions are everywhere in our daily life. If there are no chemical reactions, we can not survive. For example, to fry an egg is the most common event. Right? How many housewives or chefs will

个宇宙，包括你我和万物，都是将近一百四十亿年前宇宙大爆炸 (Big Bang) 的产物，各种元素在浩瀚宇宙中，都轮回了不知多少次，有人甚至说，太阳系、地球和我们,其实也只不过是“星尘（star dust）”而已。化学就是研究探讨这些物质是怎么回事、有什么特性、产生什么样的化学反应和为什么会有这些反应等等，其实在我们日常生活中，处处都是化学反应，没有化学物质化学反应，你我根本无法存在存活。就拿煎个荷包蛋来说，再也平凡不过了，不是吗？有几个家庭主妇或大厨，会想到其中有趣奥秘的“厨房化学（kitchen chemistry）”呢？原本黏滴滴液态的蛋清蛋黄，一加热以后，就成为好吃的荷包蛋，怎么一回事？熬骨头的高汤，为何那么鲜美？人类的熟食（加热促使化学作用），让人类吃这些东西更容易消化，也避免了很多生食细菌的感染，用火加热食物给人类多大的益处，处处都是不停的化学反应啊！

话说回来，古生物实体化石与其紧邻的围岩，很有可能还保存着有机残留物，但是这方面，过去在传统古生物学界，没有太多注意研究，一般的化石清理员，也很可能把不是骨头化石部分都清理得光溜溜，只剩下很漂亮的化石“死相”。先撇开在特殊环境，如冰冻区域的冷藏猛玛象，或者非常干燥的南美沙漠地区动物木乃伊，一般的情况是，动（植物）死亡后，快速掩埋，软组织腐化被细菌吃掉，只有硬组织被矿物质取代保

think about the mystery and interesting Kitchen Chemistry? The slimy, gooey egg-white and yolk turn to delicious sunny side up. What happens? Why the flavoring soup taste so delicious? Fire cooking (chemical reaction by heat) enables easier digestion and to kill bacteria. What great benefit for the human being by fire cooking? Chemical reactions are non-stop everywhere!

Let us go back to our main topic. The body fossil and the immediate surrounding matrix very likely preserved some organic remains. Unfortunately, no that much attention was paid. The preparaters would clean out the non-bone areas very likely. Only the beautiful fossil left. Let us ignore the freeze Mammoths or dry out South American mummies in the desert areas, in general, the creature died and buried quickly, the soft tissues were eaten by bacteria, and minerals replaced the hard parts and formed a fossil. The paleontologist studies the outside of these dead bones gets some conclusions or deductions and published the paper. However, is the whole matter just this simple?

It is common during the excavation of fossil, and we see the bone, clam, or crab fossils were encased in a shell of the mud of various thickness (Figure 1.3.1). During the preparation, this layer usually has to be removed to see the real face of that fossil. In the field, we usually would see a difference of color around the body fossil and the matrix a bit nearby. Finding the color difference usually is the key to find the fossil. Sometimes, such as finding a dinosaur rib bone, if look across, bone in the center, we would see a white rim layer. Just by looking at the different color of the mud surrounding the bone, we can say that the chemical compositions are differ-

存成为化石，古生物学家靠着研究这些死骨头外表，取得某些结论或推论，发表论文；不过，事情只有这么简单吗？

在挖化石的经验中，经常会发现，化石好像被包裹在一层泥巴壳里面，在骨头或贝壳或螃蟹本身的紧邻外面，会有那么一层东西，厚薄不一（图 1.3.1），清理的时候，需要把这层清除掉，才能看到原本化石的真面貌。在一片同样颜色土地做野外采集过程中，相对于距离化石稍远处和化石紧邻边缘，往往会看到围岩性质显然不同，找到这种不同颜色岩性，通常是找到化石的关键指标。有时候，当挖到如恐龙肋骨，从横剖面看来，骨头在中间，四周有一圈白白的层，单单只从泥巴颜色的不同，就可推论两者之间的化学组成成分不同，要不然就不会有不同的颜色。那么，这一层东西，到底是什么？为什么会形成这一层，它和化石被保存这么久（可达好几亿年），有什么关系？对于传统古生物学者来说，他们所着重的是在这土皮层里面的骨头，我们搞古生物化学者，却认为这其中或有大学问可做。毕竟能从化学的角度，或说古生物化学的角度，探讨化石的石化过程以及化石与相连的围岩内保存着有机残留物的过程，如能揭开残留有机物保存的机制，对于整个人类社会，会产生很深远的影响。

为什么在快两亿年前的恐龙骨头内，还保存着第一类胶原蛋白呢？当年这些死骨头没有用任何的防腐剂，却能保存有机物，而且保存这么久，保存的

ent. Otherwise, we will not see a different color. Then, what is this layer? Why it formed? Why they preserved with fossil together for such a long time, could be several hundred million years? What is the relationship between them? For the conventional paleontologist, they care the bone inside this layer. For us, paleochemists, we think there is something we can do a lot. After all, from chemistry or PaleoChemistry point of view, exploring the fossilization process and the organic remains inside fossil and the immediate surrounding matrix is very interesting. If we can reveal the mechanism of organic remain preservation, it will have a profound implication to the whole human society.

Why collagen type I was preserved inside the almost 200 million years old dinosaur bones? There was no preservative then but was able to preserve the organic matter for such a long time. What is the preservation mechanism? Think as today, how many tons of food were dumped out in the prosperous societies while at the very same time how many children hunger to death in Africa? If the food can be preserved and send to these kids, how nice will it be?

Various chemical analyses can observe different chemical composition. However, they also can be seen by some simple setups. For example, a high power LED laser is very cheap nowadays. No big amount of money is needed to have a Laser Stimulated Fluorescence (LSF) to look at the fossil. The image from the natural light and the LED laser fluorescence could have some surprises. Figure 1.3.2 shows dramatic differences. The top photo was taken under the natural light. The bottom one was the fluorescence produced by 450nm laser with a 480nm

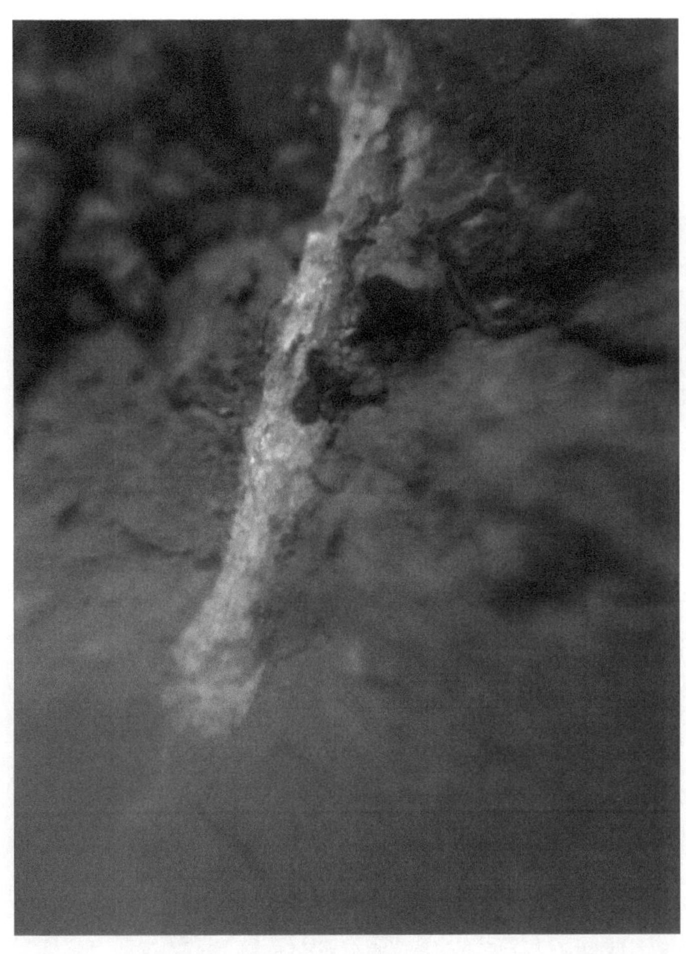

◀ 图 1.3.1 这根在野地现场骨头外表白色的和周围的围岩很不一样，当然是因为化学成分的不同啊！

◀ Figure 1.3.1 This bone in the field is white color coated, different from the surrounding matrix. Of course, this is due to different chemical composition!

机制到底如何？想想当今，富裕的社会地区，每天倒掉多少食物。而在同一时刻，非洲有多少小朋友饿死？如果没吃完的食物能保存下来送给他们吃，免于饥饿，不是更好吗？

不同的化学成分，除了进行各种化学分析以外，其实也可用简单的方法与设备来观察到，比方说，随着科技进步，如今高功率的激光发光二极管(laser LED)已经很便宜。自己在实验室不必花大钱就可以搞一套来用，把化石放在激光照射下，照出荧光反应的相片，对比普通灯

filter. The matrix in the top photo is very mundane, nothing special. However, the fluorescent photo tells a very different story. Furthermore, the yellow fluorescent color showed up around the hind limb palm. Wow! These unseeable things indicate there are different chemical components, very likely organic remnants.

Of course, just such a simple laser stimulated fluorescent response tell us different chemical composition invisible to the naked eyes. Just a tiny step of qualitative information. As for what are the actual composition, further chemical analyses shall be used.

光下所看到的，往往有让人惊喜的发现（图1.3.2），就以此图例来说，上面的照片是在普通灯光下拍摄的，一点也不稀

1.4 Morphology+Chemistry

The new paleontology shall not be bounded

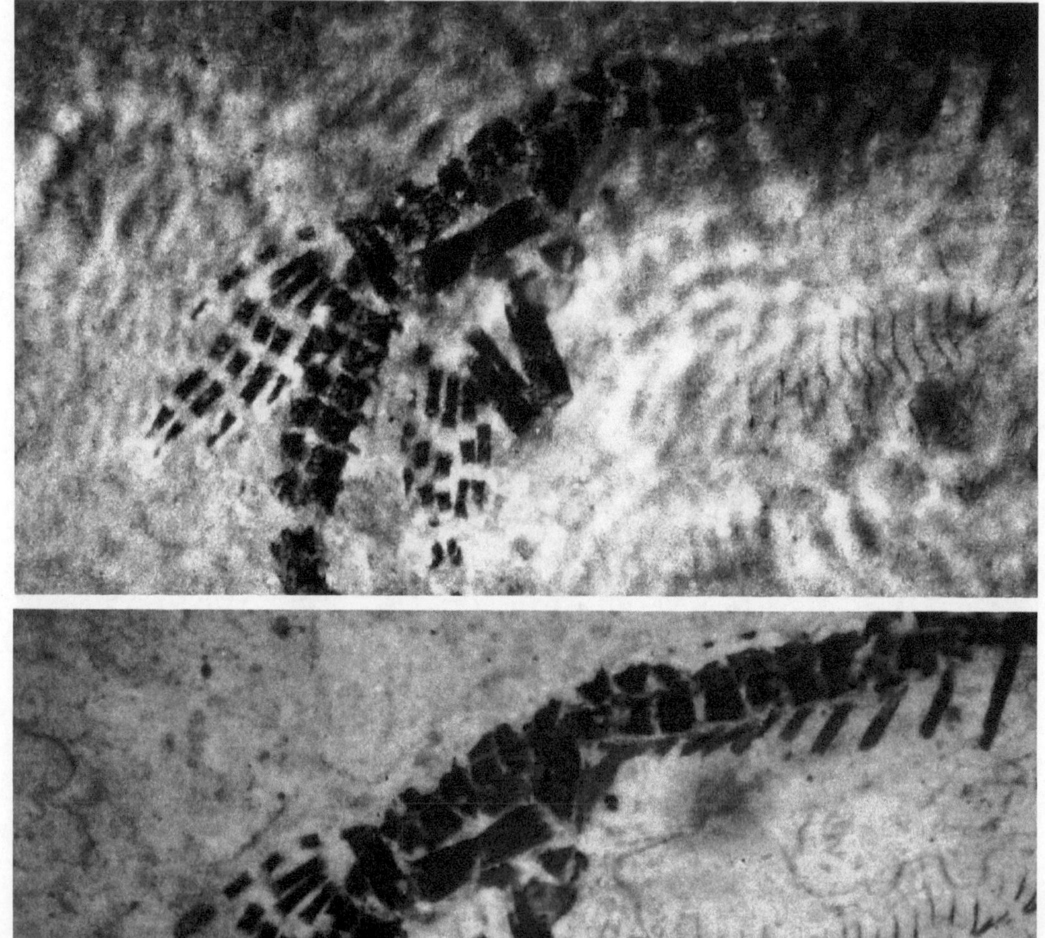

▲ 图 1.3.2 透过激光荧光，可以看到在普通灯（上）光源下所看不到的荧光反应（下），表示有不同的化学成分

▲ Figure 1.3.2 By using laser stimulated fluorescence, we can see something that natural lighting (top photo) cannot see, as the bottom photo. This means different chemical compositions

罕，下面这张，是在450 nm激光/480 nm滤光片组合拍摄的。在上面普通照片中的围岩，没有太复杂的影像，可是激光荧光围岩中，却有点让人眼花缭乱。再者，在两个后肢掌部，出现了黄色的荧光。哈！这些普通灯光下看不到的东西，不就是指出有不同化学成分的物质吗？这很可能也保存了有机残留物。

当然，如此粗浅简单的激光荧光反应，只是告诉我们有肉眼看不出来的不同化学成分，先做到了定性的一小步，至于是什么成分，那就得进一步用各种化学分析的方法了。

1.4 形态学＋化学

新时期的古生物学，不该拘泥于传统的系统描述和形态学，更应该结合古生物化学，从里到外“看透透”，更深入地了解。用比较通俗的话来说，整根骨头从里到外都可以好好看透透，也就英文俗语说的：Turn the inside out，结合里外的信息，更完整地了解该古生物。

前面提过，古生物化学成为“看进骨头内”的关键角色之一，另一重要的支柱是取得显微形态组织学的信息，把这两者结合起来，“形态学＋化学”交叉学科，就是古生物学的新领域与该有的境界。

因此，当我们在野外工作碰到一块化石的时候，不管是否是头新品种的恐龙或未曾被描述过的“新”品种，那种激动的心情当然不用多说。肯定希望能快快把它取出来，回到实验室把围岩去

to the conventional systematic description and morphology. Instead, it shall combine with Paleo-Chemistry to look thoroughly from inside to outside to understand more deeply. In layman term, we shall look the whole bone as "Turn the inside out", and combining the information of inside and outside to have a complete understanding of that ancient creature.

It was mentioned that PaleoChemistry plays a crucial role in "Look into the Bone." Another important one is to obtain microscopic morphological information, i.e., the inter-discipline "Morphology + Chemistry" is the new ground and right way for paleontology.

Therefore, when we encounter a fossil in the field, it does not matter if it is a "new" dinosaur or other fossilized creature, we shall be happy, and hope to extract that as soon as possible and bring that back to the lab for removing the matrix, prepare that bone (creature). Of course, preparation shall have already well trained and conducted with great care. Some of the minute details shall be done very slowly under a dissecting microscope. It is very common to spend many times more than the excavation time - Figure 1.4.1.

However, from the point of view of "Morphology + Chemistry," there are several conflicting vital issues:

(1) The issue of glue: In the conventional excavation, if the fossil were not well solidified, such as those dinosaur bones in Montana. The winter weather affects the bone very significantly. During the night, the temperature drops down to cause water freeze to ice, and water volume expanded and tore the bone apart from inside. Thus, the outside shape may look well;

除掉，清理那根骨头（那古生物）出来。清理的过程，要有严谨的训练，并要非常小心谨慎，有好些细微的地方，还得在立体显微镜下慢慢地清理，花比发掘化石好几倍的时间来清理一件样本，算是很正常（图 1.4.1）。

不过，如果要从“形态学＋化学”的角度来看，传统的发掘和清理方法，有很多相互冲突的致命关键要点：

（1）胶水问题：传统的发掘方法，如果发现化石本身的石化程度不是很好，就如美国蒙大拿州的很多恐龙骨头，经过地表天候作用，夜晚温度下降骨头内所含的水分结冰，体积膨胀，把骨头从里面撕裂开，白天太阳一晒温度提高，冰化成水等等的循环作用影响，往往骨头外表还完整，可是里面已经粉碎掉了，所以，为了取得这样的骨头，保持它的形状，往往就需要在取出之前，灌入很多渗透胶水 (penetrating glue)，常用的如泡在丙酮内的聚乙烯醋酸酯 (polyvinyl acetate, PVAc)；如果碰到断裂的骨头，通常会用三秒胶（氰基丙烯酸酯，

however, the inside is very powder-like. Therefore, in order to extract such bones, a massive amount of penetrating glue (Polyvinyl acetate, PVAc in acetone) was used. When breakage encountered, super glue (Cyanoacrylate) would be used to put them back together. In the lab, various type of glues is used to preserve the shape of the bone. These are very common in traditional paleontology excavation and preparation. Some even more delicate fossils, such as the insect fossils from a locality of Montana are on paper thin shale. Once opened, the glue has to be used to preserve the specimens. These are the correct common practice.

Although using glues in conventional paleontology is an absolute right. However, from PaleoChemistry, let us ask what glues are? They are not only organic matter but also artificial “contaminates” for the fossil and immediate surrounding matrix. They “pollute” the original pure condition. They can cause many headaches during the following analysis. As a very example: Neutron scan is very sensitive and effective to the organic remains (hydrocarbons). The area contains glue will show up as a very bright

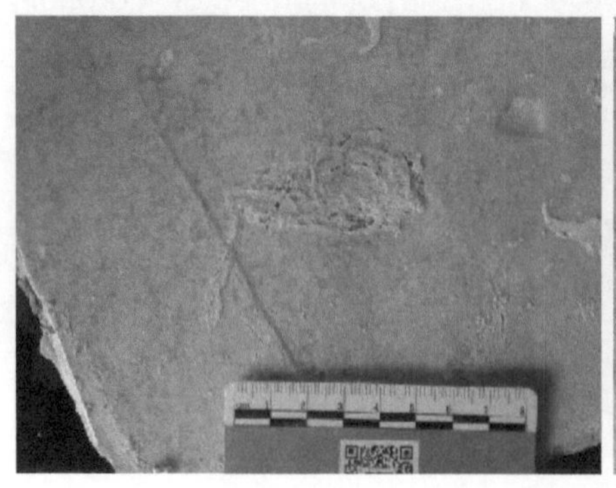

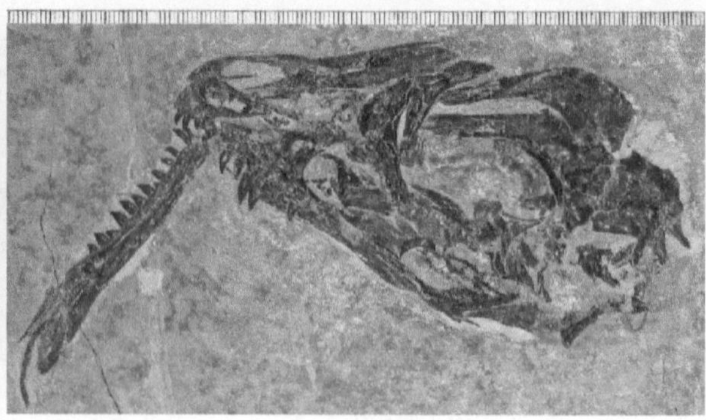

▲ 图 1.4.1　化石清理的功夫，至少要能做到这对相片的水平，否则就不及格

▲ Figure 1.4.1　The passing grade for specimen preparation shall be at least this caliber cyanoacrylate）接回去，回到实验室清理过程中，也常常会用到胶水来保存化石的形态，或将断裂的碎块黏合起来，这些都是传统古生物发掘习惯性的规范处理方法。有些更为脆弱的化石，如美国蒙大拿州一处昆虫化石，产于像纸张那么薄的页岩中，打开之后若没有马上灌入黏着固定的胶水，就很难取下来保存起来，这些都是很正确的惯用手段。

虽然使用胶水于古生物样本，在传统的古生物界天经地义，可是，从古生物化学的角度来看，胶水是什么？它们不仅仅是有机物，而且是化石与邻近围岩的人为“污染物”，污染了原有的化学成分，在样本的后续分析过程中，它们很有可能造成莫大的干扰，举个最简单的例子：中子扫描对于有机残留物（碳氢化合物）特别有效，中子扫描的结果，在化石胶水黏合处会有很强的一道白色区域，妖魔鬼怪无所遁形（图 1.4.2），干扰了我们实际想要看到骨头或围岩内的有机残留物！

这也就是说，在采集和清理化石的过程中，使用任何外来的添加物之前，必须好好地考虑要不要做，To do or not to do！可是，从实际的角度来说，能拿来进行古生物化学分析的样本，当然最好是自己采集的，可以确定没有也不会有人为的污染。但是，很多情况下，分析样本是由别人采集或 / 和清理的，没有办法回到过去重来。那只有一个方法，就是尽量取得发掘和清理过程的资料，多一点这方面的信息，可以省

white patch. Glue devil cannot hide as Figure 1.4.2. That interferes with what we want to see the organic remains inside the bone or in the surrounding matrix.

This is to say that during the fossil excavation and preparation before any additive is used, think carefully. To do or not to do! However, from the practical point of view, unless the fossils were collected by ourselves, which we have full control not to have any artificial contamination. Most of the specimens are very likely collected/prepared by someone else. We can not go back to history. The only way is to have as much information about the excavation/preparation. The more information we can get, the less headache for us later on, and that can produce better meaningful results. Looking from this angle, we have to pay extraordinary attention to those record and mind changing LiaoXi fossils. The skillful farmer excavators would use any trick they can to have better death looking products for more money. Even internationally well-known paleontologists were fooled and even with a documentary film made. The "Archeoraptor" is the classic for such an event.

(Wiki) "Archaeoraptor" is the informal generic name for a fossil from China in an article published in National Geographic magazine in 1999. The magazine claimed that the fossil was a "missing link" between birds and terrestrial theropod dinosaurs. Even before this publication, there had been severe doubts about the fossil's authenticity. A further scientific study showed it to be a forgery constructed from rearranged pieces of real fossils from different species. Zhou et al. found that the head and upper body belong to a specimen of the prim-

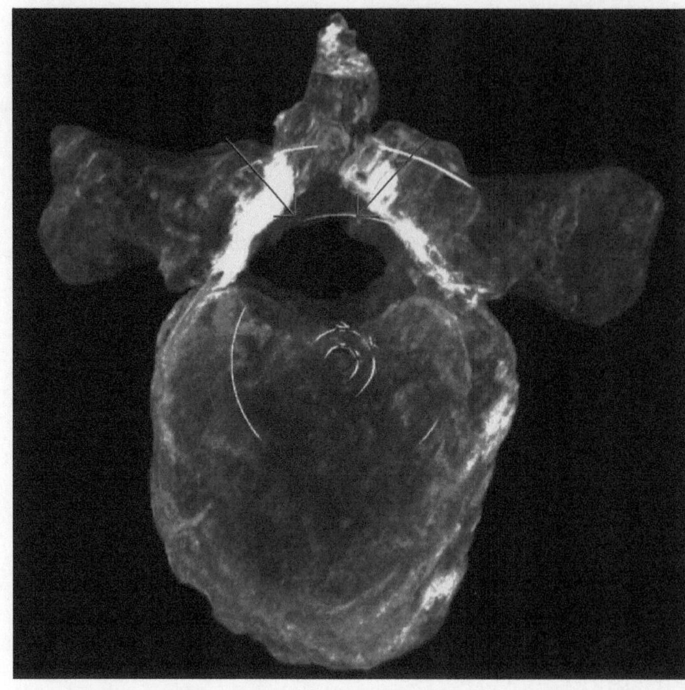

◀ 图 1.4.2 化石断裂，使用快干胶黏回去，即便技术很好，肉眼不容易看出来，可是在中子扫描照妖镜下，无所遁形（红色箭头），圆形亮线是扫描影像处理干扰

◀ Figure 1.4.2 Fossil broke and glued back. Even with very good skill so that the seam couldn't be seen by naked eyes, it shows up very clearly under neutron scan (red arrows). The concentric circles are interferences from image processing software

下很多后面的头痛，也才能得到正确有意义的结果。从这个角度来说，改写诸多古生物学观点的辽西化石，必须特别谨慎小心，那边的老乡为了“死相好看一点”可以多卖一些钱，各种手段都会用上，连国际顶尖的大学者都被蒙骗过去，还做了纪录片的专辑！“辽宁古盗鸟 (Archeoraptor)”就是最典型的实例！

（维基）“古盗鸟”事件发生于二十世纪末期，美国国家地理杂志于 1999 年刊登了一篇来自中国化石的文章，宣称这是陆相兽脚类恐龙和鸟类之间的“迷失环节”。在这篇文章发表之前，已经很被怀疑其真实性，后续的科学研究显示，这是由不同物种排列组合而成的赝货，Zhou等发现头部和身体前端属于原始鸟类燕鸟，2002 年研究又发现尾部是

itive fossil bird Yanornis. A 2002 study found that the tail belongs to a small, winged dromeosaur, Microraptor, named in 2000. The legs and feet belong to an as yet unknown animal. The scandal brought attention to illegal fossil deals conducted in China. Although "Archaeoraptor" was a forgery, many real examples of feathered dinosaurs have been found and demonstrate the evolutionary connection between birds and other theropods.

(2) The issue of the immediate surrounding matrix: Another blind point of traditional paleontology practice is overdone preparation. In order to have a perfect and beautiful skeleton, usually, all the immediate surrounding matrix would be removed completely, or at least significantly. For the platy specimens, most of the matrix would be removed to expose the beautiful skeleton before morphological studies.

属于一种在 2000 年命名小翅似驰龙类的小盗龙，腿和脚又属于另一个未知的动物。这件丑事引发了大家对于中国化石非法行为的注意，虽然“古盗鸟”是个赝品，但是实际上发现了很多带毛恐龙，证明了兽脚类恐龙和鸟类的演化关联。

（2）紧邻的围岩，从古生物化学的角度来说——另外一个传统古生物的盲点，就是过度清理——为了取得完美的骨架，通常围岩会完全被清理干净，完全除去，若是板状的样本，也会把大部分的围岩清除掉，露出很清晰漂亮的骨头骨架，然后进行形态研究描述。

然而，从普通常识的角度来考虑，当动物死亡被掩埋之后，会发生什么情况？细菌吃这些尸体，软组织开始腐化，产生了一些液体的“尸水”，从尸体里面流出很多黏黏的胶质液体，就像餐桌上的肉冻（鸡鸭牛猪肉），这些胶水把紧邻的围岩泥沙胶结起来，形成相对紧密坚硬的结核。这种情况经常在野外采集的时候碰上，比方说，图 1.4.3 就是一坨在云南禄丰地区老乡挖到，被包裹在邻近围岩内的恐龙头颅，如果不把这层包裹物（围岩）清除，则无法见到龙头的庐山真面目。以传统的角度来说，接下来该做的是，要仔细小心地把最后这层包裹围岩清理掉，让整个恐龙头颅骨头完整地显示出来，如果幸运的话，有可能会是一个新品种的恐龙，可以好好发表命名论文等等。

不过，从古生物化学的角度来说，且慢！且慢！在把这层最好的围岩清掉

Let us use common sense for consideration. After the animal died, what would happen? Bacteria will come to chew the corpse. Soft tissue started to decompose, and gooey body fluid will come from the body, similar to aspic (various animals). These fluids will cement the surrounding mud grains together and form a hard concretion. If these concretions are not removed, the bone packed inside cannot be seen. Figure 1.4.3 shows a dinosaur skull excavated by a farmer from Lufeng, Yunnan still packed inside. Unless this concretion matrix is removed, we can not see the real face of this dinosaur. So, from the common point of view and practice is to remove the surrounding matrix to expose the whole skull carefully. If lucky, it could be a new dinosaur and good paper can be written for naming and publication.

However, from PaleoChemistry point of view, hold it, hold it. Before removing this best surrounding matrix, think about what we just said. This matrix layer was formed from the decay body fluid with the surrounding sand and mud. It may contain some organic remains. By analyzing these organic remains, we can get some information about the soft tissues and taphonomy of this dinosaur. In theory, the decomposition could be significant. The importance of the information obtained from Paleochemistry may not be less than that of traditional morphology.

In other words, how to obtain proper conventional preparation without adding PaleoChemistry contaminates to obtain valuable PaleoChemistry information shall be carefully considered. A simple way is to preserve as much immediate surround matrix next to fossil bone or to reserve the whole matrix for the future

◀ 图 1.4.3 典型的化石结核，整个云南禄丰地区出土的恐龙头，被包裹在一层较为坚硬密实的围岩里面

◀ Figure 1.4.3 Typical fossil concretion. A complete dinosaur skull from Lufeng, Yunnan encased with a harder and solid matrix

之前，考虑一下上面所说的，这一层围岩就是当时恐龙软组织尸水与紧邻泥沙的胶结物，里面很可能保存着某些有机残留物，而透过分析探讨这些有机残留物，我们可以取得一部分这头恐龙软组织（理论上可能有较大程度的腐化降解）以及埋藏学的信息，这些信息的重要性，并不亚于传统的形态。

换言之，要如何一方面可以取得好的传统清理结果，又可以不加入会干扰到古生物化学分析的“污染物”，同时可以保存重要的古生物化学信息呢？一个最简单的方法就是，尽量保留骨头旁边紧邻的围岩，或是把围岩保留下来，以便日后进行古生物化学的分析探讨。另外一个更好的方法是，透过计算机断层扫描（图 1.4.4），或是微型计算机断层扫描，透视到含骨头的围岩，进行所谓的“数码清理（digital preparation)”，完全不必动一刀，就可把该化石本体形态等重建出来，甚至利用3D打印，打造出该化石的模型。如果万不得已需要清

PaleoChemistry analyses. Another even better way is to conduct Computerized Tomography (CT) scan (Figure 1.4.4) or μ-CT, looking thru the matrix to do Digital Preparation, without even one cut. The fossil bundled inside the matrix can be reconstructed digitally and even for 3D printing of such specimen. If it has to remove some extra matrix, we can use the digital model as a guide to know where to put the scalpel in order to prevent over preparation. If possible, a neutron scan shall be conducted to obtain the three-dimensional organic remain distribution. Then pinpoint to a spot for further analyses and studies. By doing so, all the information for the whole topic can be obtained without any change or damage to the specimens collected from the field. This is the ideal case, of course - well-plan ahead to win the war afar.

1.5 Heterogeneous Fossil Composition

For the chemical analyses, the specimen is usually in liquid form to obtain meaningful average data. In contrast to the standard chemical

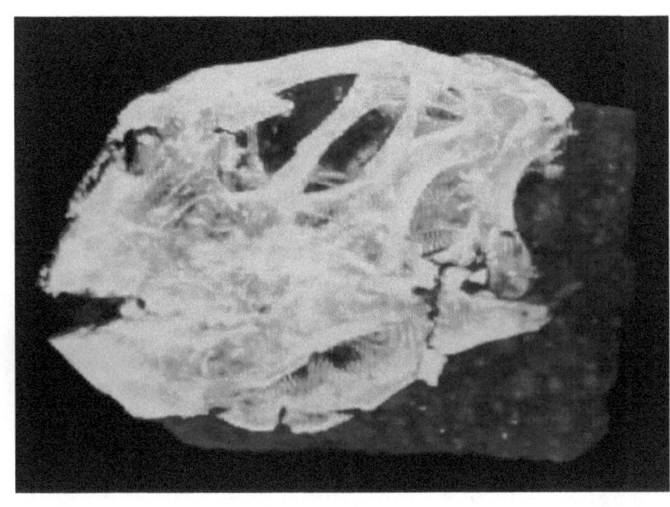

◀ 图 1.4.4　透过（微型）计算机断层扫描，进行“数码清理”，可将骨头和围岩从里到外看透。影像中黄色部分为鹦鹉嘴龙头颅骨头，红色部分则为围岩，分得很清楚

◀ Figure 1.4.4　μ-CT scan for "digital preparation" can see thru the matrix into the inside. The yellow portions are the bones of Psittacosaurus skull, and red area is matrix clearly separated

除某些过多的围岩，也可“看图识字”，事先知道该如何往哪里下手，这样才不会破坏化石或过度清理。若可能的话，更该进行中子扫描，取得有机残留物的三维分布，然后针对某个点，进行后续的研究分析。整个课题做完之后，从野地取回来的样本，外观完全没有破坏，但是所要的信息，也都非常精密完整，这才是最理想的情况！所谓运筹帷幄，决胜千里是也！

1.5　化石的非均匀性成分

一般的化学样本，通常溶解成为液体，来取得比较均匀有意义的数据。然而相对于一般的化学分析，以非破坏性方法检测固体的化石，必须注意到在化石里面的非均匀性，检测位置在这一点和邻近的点，距离可能只有几微米，甚至不到一微米，所检测出来的结果可能就会有很大的不同。这是因为一般来说，

analyses and using non-destructive methods to examine solid fossils, the heterogeneous nature of fossils shall be noticed. The result, such as spectroscopy from this particular spot could be quite different from that neighboring spot just several μm away. This is because in general fossils are very heterogeneous material. Two adjacent spots could have a significantly different chemical composition. This is to say, the resolution of the measuring instrument, particularly spectroscopy or optical equipment, is very crucial. If resolution (radiated/image size) is not good (high) enough, the big radiated area will yield the sum of that whole area, not just a particular point, like to use a cannon to shot a sparrow. Many detail morphology just could not be resolved under such low resolution. When the high resolution is used, then detail structure can be seen, as Figure 1.5.1. Remember, there is no free lunch. In order to have a high resolution, the price has to be paid. High-resolution scans or analyses will yield a colossal data set(s). Thus a bigger and more powerful graphics processing computer will be needed with much bigger

化石是非常不均匀的样本，两个相邻点的化学成分，就可能有很大的差异。也就是说，检测方法，特别是光谱或用到光学仪器设备的分辨率，这个参数就非常重要。分辨率（如照射光点、成像点大小）不够精密（高），照射范围一大块，因此所得到的结果，是那一大块的总和，而非某特定点的真正数据，好多比较细微的形态构造，在较低分辨率之下，根本无法分辨，好比用牛刀杀鸡。但是使用较高分辨率来扫描，则可看到低分辨率看不到的构造（图 1.5.1）。不过，天下没有白吃的午餐，追求高分辨率，当

storage spaces. It may also take more time to process the data.

As an example: Dissolve the specimen (total destructive) for ICP-MS can yield a more meaningful average concentration of each element isotopes. The LA-ICP-MS uses a laser to burn tiny holes on the specimen and then put the vapor into the ICP and MS units. This provides "almost non-destructive" elemental isotope analyses. Judging from destructive vs. non-destructive and the amount of specimen needed, it seems this is a better choice. However, due to the heterogeneous nature of fossil and the laser spots are so small, the result is the data for that particular small spot, not the representative of

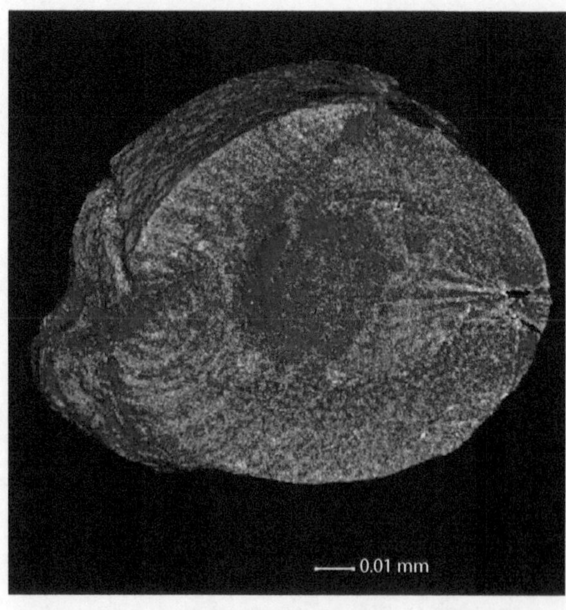

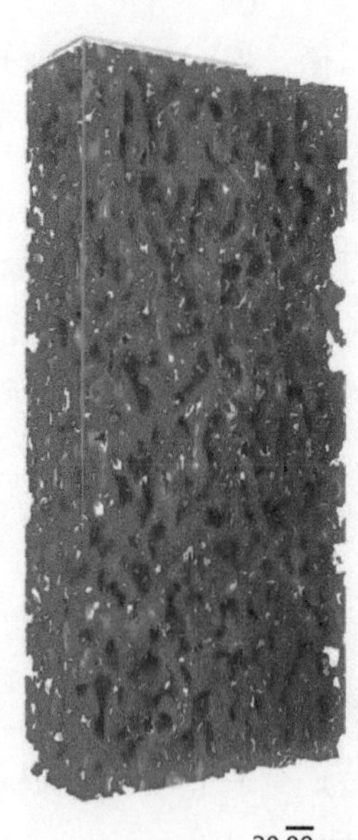

▲ 图 1.5.1　2 µm 分辨率的 µ-CT 扫描，才能解析看清楚云南禄丰龙胚胎股骨内的原始管状空间三维构造。左边照片显示胚胎股骨扫描范围，右边照片撷取硬骨头内一小范围，进一步进行影像处理。在较低分辨率 (9 µm) 之下，无法取得如此详细精确结果

▲ Figure 1.5.1　By using 2 µm resolution µ-CT to see the details of the primary tubular cavity in 3D of the embryonic femur of *Lufengosaurus* from Lufeng of Yunnan province. The left photo shows the scan area. Then a small section (red rectangle) was taken out for further image processing (right). This kind of clarity could not be obtained by lower resolution of the 9 µm scan

然要付出其必要的代价，高分辨率扫描或分析所产生的庞大数据，需要有配套庞大计算机和图像处理的能力，还需要庞大数据储存空间，花更多时间处理数据，那就更不用说了。

以另外一个实际的例子来说，相对于把化石样本（完全破坏性）溶解进行电感耦合等离子体质谱分析可以得到某块化石骨头里面各元素同位素比较有代表性的平均数据。先进的激光剥蚀等离子体质谱 (LA-ICP-MS)，利用激光在样本上烧洞（剥蚀），将样本烧成气体送入质谱仪，提供肉眼几乎看不见“非破坏性”的元素同位素分析，从破坏性与否和样本量需求的角度来说，这是比较好的选择。然而因为化石成分非均匀性的本性，激光点又那么小，所检测出来的结果，只是那某个特定点的元素／同位素数据，而非整体的结果。也就是俗语说的，因为检测方法和相关参数等的不同，很容易导致“以偏概全”。换言之，使用激光剥蚀等离子体质谱来检测骨头化石，激光剥蚀的点要打得够多，才能求取统计学上有意义的真正平均数据。

1.6 着重于化石内部微观与基质

古生物化学所要探讨的，基本上来说，都是显微尺寸，有如郭台铭常说的：“魔鬼藏在细节里面。”从建构化石的基本微观角度来探讨，可以了解更多的精细奥妙。从如此的角度出发，我们所需

the whole thing. This is to say that seeing a part of the whole caused by different measurement method and parameters. Therefore to use laser abrasion, enough spots have to be done to obtain meaningful average data result.

1.6 Concentrate on the Micro View of Fossil Inside and Immediate Surroundings

All the explorations of PaleoChemistry, in general, are micro scale. Just like the boss of Foxcom, Terry Kuo said: “devils hide in details.” We can learn much more intriguing details from the micro view. Thus, the specimen we need does not have to be so big. So we shall start with all possible nondestructive examinations to obtain data for further evaluation of the meaning of these ancient lives.

For general fossil collectors, caused by the limited personal understanding of fossil or social fashion, they usually want big, spectacular, rare, expensive, nice skin deep looking fossil. Therefore the “death look” is essential for them. For a nice looking and a bad looking death look fossil, of course, the nice looking one will have a higher price and more “wows” However, for paleontologist research, it is quite different from the commercial market. Big is not necessarily beautiful. The researcher may not have enough money to buy an expensive specimen for cutting, grinding, etc. What we want is just the small “waste.” That will keep us busy for a very long time.

Even if we have to do the destructive cutting and grinding, the amount we need can be reduced to unimaginable small. Thanks for

要的样本，不必很大，也该从所有可能的非破坏性检测开始，取得所要的数据，研判这些信息所展现的古生物意义。

常人，特别是化石收藏家，限于个人或社会对于化石的有限认知，所注重的往往限在样本要大、要壮观、要罕见、价格昂贵等等“表面功夫”，所以“死相”很关键。同样的化石物种，一个死相较美，另一个丑丑的，当然是死相较美那个会卖得较高价格,得到更多的“啊呜”赞赏声。可是古生物学者的研究工作所需要的条件，通常与市场行情有重大的差别，在研究过程中，硕大不见得为美，科研界也没有钱去买一个昂贵的样本，又拿来切啊割啊的，我们所需要的样本,往往就是那么一小块“下脚料”，就够我们忙个好一阵子，甚至好几年。

即便得把化石做破坏性的切割研磨，随着科技设备的进展，如今所需要的量，已经缩小到难以想象的程度，往往非常小的一小块，就已经很够用了。举个实际的例子来说，以前做化学分析检测某元素含量，取样必须很大，有时候还会超过千克级，可是，随着日新月异发展的仪器分析，好些元素分析所要的样本，只要0.1克或更少。我个人念生药学博士班的时候，以前的学长们从好多有药效的天然产物，花了很多时间，萃取分离出好多不同的成分，教授就把这些分给我们去把化合物的化学式分析出来，每个小玻璃瓶内，大约只有5 mg的量，我们每个人都要把几种物理性质（熔点等等）、红外线与核磁共振光谱测

the advancement of the modern facility. A real example: In my university days, when we want to do elemental analysis, we need big chunks, sometimes on the kilogram scale. However now, along with the advancement of new instrumentation, the amount needed could be 0.1 g or less. When I was in my pharmacognosy Ph.D. graduate school, previous students already spent many times to separate many different components. Our professor gave each us a small vial containing about 5 mg of sample. Each of us has to obtain various physical properties (such as melting point), IR and NMR spectra, and degradation chemical reactions (with spectra). When we hand result back to our professor, we have to provide a pile of spectra, analyses report, and also at least 3 mg of the remaining sample. Otherwise, we can not pass. Our previous seniors spent many efforts to get this tiny 5 mg. How can we fail to identify the chemical structure?

The instruments today are very advanced. Many of them are nondestructive. That enable us to do much better PaleoChemistry studies. We can use specimens smaller than a booger to get many many information inside that fossil. We are now in the Micro scale, μ-, or even Nano scale, n-.

For the resolution of μ-CT, the unit is μm. A VOX (Volume pixel, Voxel) could be several deca μm. The newer instrument can reach even on the nm scale. In other words, the different morphological differences between fossil background and matrix can be easily resolved. That is to say that we can further study and understand the minute difference of the morphological and compositional heterogenous of the fossil back-

出来，同时还需要做出不同阶段的降解反应式（含光谱证明），最后交卷的时候，纸上要写出那个瓶内物质的化学式，附上一大堆各种光谱证明，还需要交回至少3 mg的分析样本，否则就不过关，还要挨教授大骂一顿，前人花了多少代价才分离出这么5 mg，我们学生没有把化学式搞出来，还浪费了那么多的样本！该被骂。

如今的科研仪器设备已经非常先进，还有很多种非破坏性的检测，让古生物化学的科研如虎添翼，让我们可以只要用非常小的样本，就可得到很多化石内的信息。事实上来说，整个科技的进步，不只已经让我们达到微量级 (Micro scale, μ-)，甚至到了纳米级 (Nano scale, n-) 了。

就以微型计算机断层扫描和中子扫描的分辨率来说，如今都以微米（μm）为单位，一个像素体积 (Volume pixel, Voxel, Vox)，十几微米，已经太平常了，甚至更新的设备，已经达到纳米级了。换言之，就在这么小的体积内，化石基质或围岩有什么形态上的差异，都可解析出来，也就是说，化石基质与围岩特色形态与成分的非均匀性，我们可以更细微地研究了解。

一个像素体积是一个像素的三维等量，也就是能区分三维物件的最小元素，在三维空间内，它代表某个特定网格数值。然而，就如像素，像素体积并不含它们在三维空间的位置，而是相对于被指定周围像素体积的位置，因此可把像素体积看成砖块，堆积成更大的结构，在此每块砖块紧邻排列，但没有定义砖块。

ground and matrix.

A volumetric pixel (Volume pixel or Voxel) is the three-dimensional (3D) equivalent of a pixel and the tiniest distinguishable element of a 3D object. It is a volume element that represents a specific grid value in 3D space. However, like pixels, voxels do not contain information about their position in 3D space. Instead, coordinates are inferred based on their designated positions relative to other surrounding voxels. One may compare volume pixels to bricks, which are stacked and used to build bigger structures. In this scenario, each brick is placed next to each other, but the bricks are not defined.

附表A.1　常用中文数量级

个=1x10^{0}	十=1x10^{1}	百=1x10^{2}	千=1x10^{3}
万=1x10^{4}	十万=1x10^{5}	百万=1x10^{6}	千万=1x10^{7}
亿=1x10^{8}	十亿=1x10^{9}	百亿=1x10^{10}	千亿=1x10^{11}
万亿=1x10^{12}	十万亿=1x10^{13}	百万亿=1x10^{14}	千万亿=1x10^{15}
京=1x10^{16}	垓=1x10^{20}	秭=1x10^{24}	穰=1x10^{28}
沟=1x10^{32}	涧=1x10^{36}	正=1x10^{40}	载=1x10^{44}
极=1x10^{48}	恒河沙=1x10^{52}	阿僧祇=1x10^{56}	那由他=1x10^{60}
不可思议=1x10^{64}	无量=1x10^{68}	大数=1x10^{72}	*大一＝+∞，Big_One*

附表A.2　常用中英数量级对比

词冠（Title）	符号（Symbol）	10的因数（10th power）	中文（Chinese）
exa	E	10^{18}	艾（百万万亿）
peta	P	10^{15}	拍（千万亿）
tera	T	10^{12}	万亿
giga	G	10^{9}	十亿
mega	M	10^{6}	百万
kilo	K	10^{3}	千
hecto	h	10^{2}	百
deka(deca)	da	10^{1}	十
deci	d	10^{-1}	分
centi	c	10^{-2}	厘
milli	m	10^{-3}	毫
micro	μ	10^{-6}	微
nano	n	10^{-9}	纳（毫微）
pico	p	10^{-12}	皮（微微）
femto	f	10^{-15}	飞（毫微微）
atto	a	10^{-18}	阿（微微微）
Small_One*		$-\infty$	*小一＝-∞

*: 大一、小一：《庄子·杂篇·天下》：历物之意，曰：“至大无外，谓之大一；至小无内，谓之小一。”

*: Big_One, Small_One: *Zhuāngzi, Misc., The World*: For the meaning of objects, it is said: “Big_one is the one without even bigger, Small_one is the one without even smaller.”

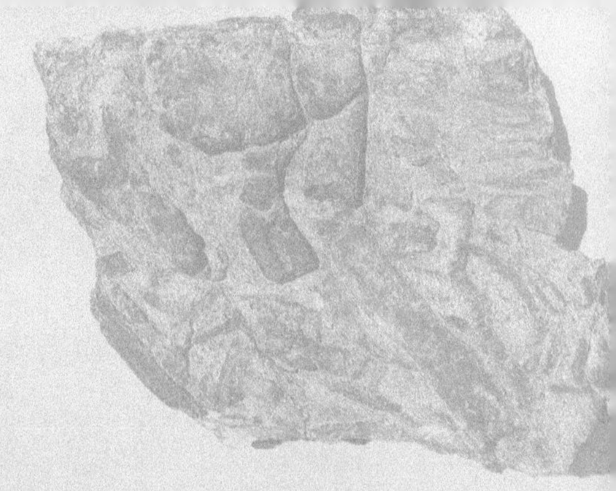

第 2 章　古代生物

Chapter 2　Ancient Organism

本章简单描述地球早期生命的演化，并包含作者以一己之力所进行的埃迪卡拉纪计划的状况，期望能对地球生命的演化，有个粗浅的基础认知。

This chapter describes the life evolution of early Earth including a status of the author's personal effort on the study of my Ediacaran project. Hope to provide the readers with a rough and basic understanding of Earth life evolution.

2.1　远古地球生命演化

2.1　Primeval Life Evolution on Earth

既然我们搞古生物，探讨地球曾有的生物，对于整个地球的生命演化，总该有些基础概略性的了解，对于其他的“古物”，当然也得有些基础的了解，免得闹出外行人说内行话的笑剧。这一章，希望能提供一个很简略的地球生命演化常识。

如果拿一卷卫生纸，假设制造厂商没偷工减料，一卷里面会有五百张小方块，从外面撕掉四十张，剩下四百六十张小方块，则地球起源于最里面那张（四十六亿年前），最外面这一小张的边缘，就是当下现在，每一小张代表一千万年。我们常说，中华悠久的历史六千五百年（不是五千年），试着在这

Since we are working on ancient organisms, exploring existed life, we shall have a general idea and basic understanding of the life evolution on this Earth, as well as other "Paleo" matter, so that we will not make fools of ourselves. This chapter provides a very brief Earth life evolution of common knowledge.

If we take a roll of toilet paper and assume the manufacturer did not cut the corner, then, each roll shall have 500 small pieces. Remove the outer most 40 such squares. The remaining 460 pieces represent the history of Earth. The innermost next to the core is the birth of our Earth (4.6 billion years ago), and the outer most piece is here and now. Each small square rep-

张上面标示出来！想想我们人生百岁多么短暂啊！

resents 10 million years. We Chinese are proud of our 6,500 (not 5,000) years glorious history. Try to mark it on this last piece! How short is our human life even at 100!

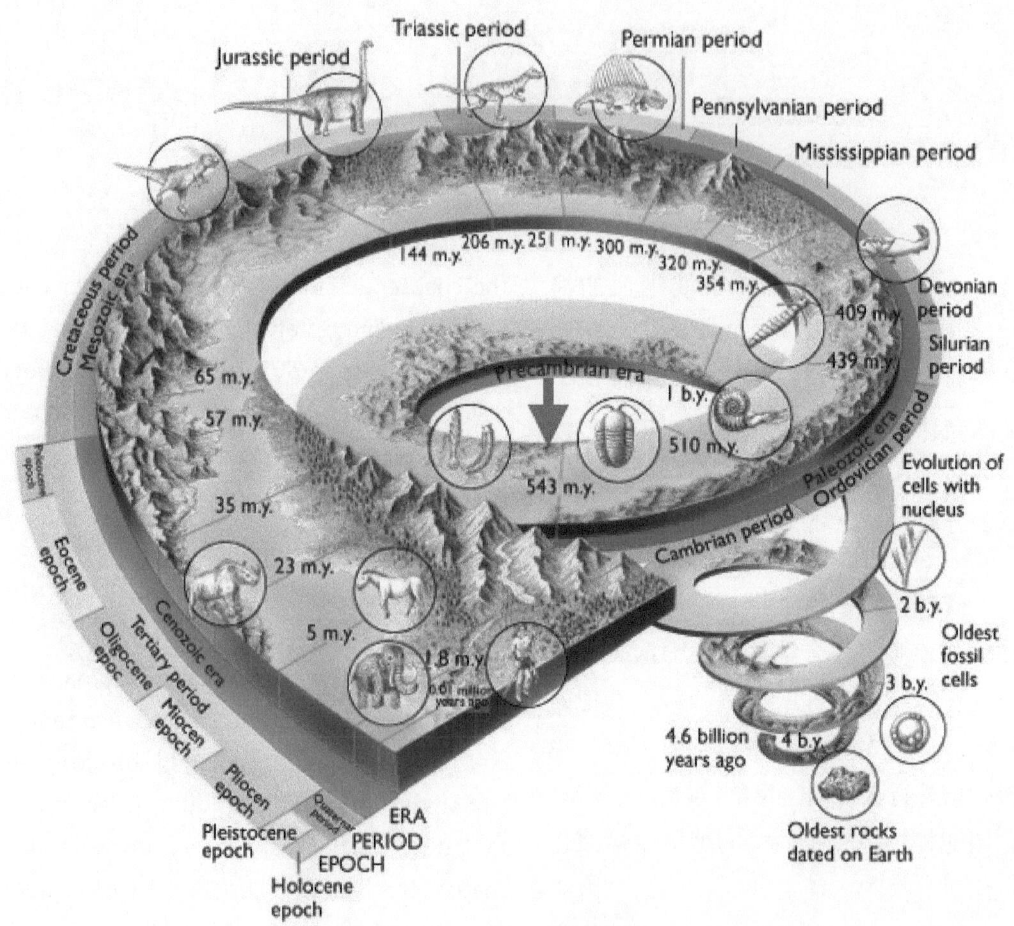

▲ 图 2.1.1 地球生命演化螺旋图；请注意不同生物种出现的时间。取自维基百科

▲ Figure 2.1.1 Spiro diagram of Earth life evolution. Note the different time for different life evolved. From WiKi

表2.1.1 表格式地球生命演化史
Table 2.1.1 Life Evolution on Earth as table

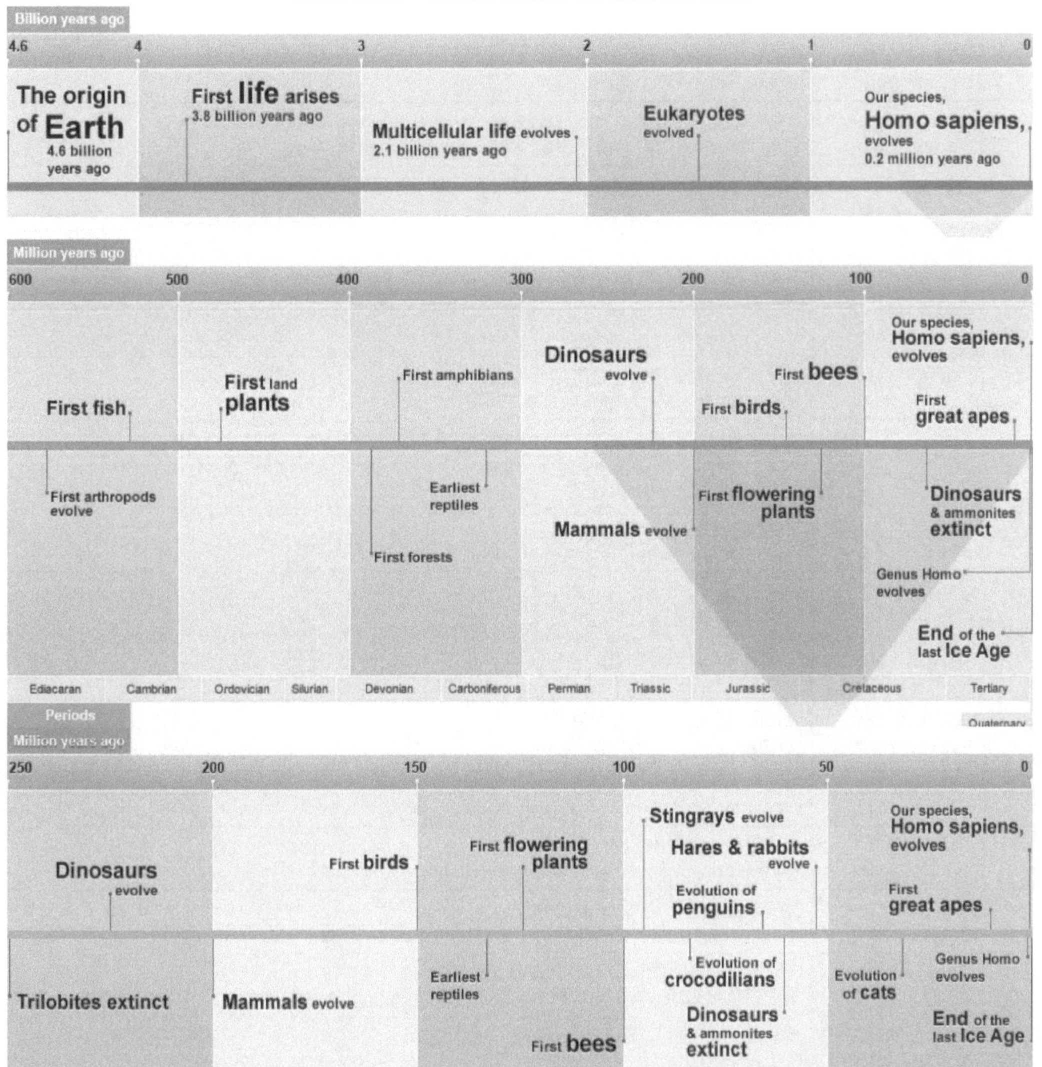

表2.1.2 另一个比较详细的地球生命史
Table 2.1.2 Another more detailed Earth Life history table

时间（Time）	事件（Event）
46 亿年 4.6 billion years	太阳系形成，地球诞生 Solar system formed and Earth born
38 亿年 3.8 billion years	地表熔岩冷却，水开始凝结，每天只有 15 小时 Surface of Earth changed from molten to solid rock. Water started condensing in liquid form. Earth day was 15 hours long
36 亿年 3.6 billion years	最早简单细胞，产氧细菌 First simple cells, oxygen producing bacteria
34 亿年 3.4 billion years	光合作用 photosynthesis

续表

时间（Time）	事件（Event）
22 亿年 2.2 billion years	线粒体生物？ Organisms with mitochondria?
16 亿年 1.6 billion years	复杂单细胞生命出现？原核生物出现 Complex single-cell life appeared? Eukyarote appeared
15 亿年 1.5 billion years	复杂有核细胞出现？ Complex cells containing nucleus appeared?
12 亿年 1.2 billion years	有性生殖出现 Sexual reproduction appeared
10 亿年 1.0 billion years	多细胞生物出现？ Multicellular life appeared?
6.5 亿年 650 million years	雪球地球，70% 海洋植物灭绝 Snowball Earth, Mass extinction of 70% of dominant sea plants
5.8 亿年 580 million years	简单软体生物出现，大气层氧气累积，促成臭氧层 Simple, soft-bodied organisms developed. The accumulation of atmospheric oxygen allowed the formation of the ozone layer
5.7 亿 570 million years	节肢动物出现，昆虫祖先 Arthropods appeared, ancestors of insects
5.6 亿年 560 million years	最早真菌？ Earliest fungi?
5.42 亿年 542 million years	埃迪卡拉 / 寒武纪灭绝 Ediacara / Cambrian Extinction
5.3 亿年 530 million years	鱼类出现，最早的脊椎动物 Fish appeared, the first vertebrate
4.43 亿年 443 million years	奥陶纪大灭绝，49% 生命消失 Ordovician Mass extinction, 49% of life disappeared
4.34 亿年 434 million years	最原始植物和真菌上陆 The first primitive plants and fungi moved onto land
4.1 亿年 410 million years	鱼长牙齿和颌，蜘蛛、蜈蚣出现 Fish developed teeth and jaws. Spiders, Centipedes appeared
3.74 亿年 374 million years	泥盆纪大灭绝，70% 海洋生物消失 Devonian Mass extinction, 70% of marine species disappeared
3.7 亿年 370 million years	最早两栖类 First amphibians
3.63 亿年 363 million years	昆虫、鲨鱼、种子植物 Insects, sharks, seed-bearing plants
3.2 亿年 320 million years	爬行类出现 Reptiles appeared
2.51 亿年 251 million years	二叠纪大灭绝，95% 海洋生物，70% 生物消失 Permian Mass extinction event, up to 95% of ocean species and 70% land species lost
2.25 亿年 225 million years	小恐龙出现 First small dinosaurs appeared
2.01 亿年 201 million years	三叠纪大灭绝，20% 海洋生物死亡 Mass extinction, 20% of all marine species killed
2 亿年 200 million years	哺乳类出现 Mammals appeared

续表

时间（Time）	事件（Event）
1.5 亿年 150 million years	鸟出现 Birds appeared
1.3 亿年 130 million years	开花植物出现 Flowering plants
1.1 亿年 110 million years	鳄鱼出现 Crocodiles appeared
6800 万年 68 million years	霸王龙 *Tyrannosaurus rex*
6500 万年 65 million years	陨石撞击，大灭绝，80%~90%海洋生物种和85%陆相生物种灭绝 Meteor impact, Mass extinction of 80%-90% of marine species and of land species
1400 万年 14 million years	大猩猩出现 The first great apes appeared
440万年 4.4 million years	人种始祖出现 Appearance of *Ardipithecus*, an early Hominin Genus

注：主要灭绝以黄色标示

Note: Major extinction marked Yellow

2.2 埃迪卡拉纪生命爆发

埃迪卡拉（Ediacaran）生物群，以前称为凡登（Vendian）生物群，是由一群形态很令人费解的生物组成——它们大部分是无柄生物的管状体和叶片状形态的生物。它们存活于埃迪卡拉纪（大约6.35亿至5.42亿年前），这些生物的遗迹在世界很多地方被发现，为古老复杂多细胞生物。埃迪卡拉生物群从5.75亿年前阿佛龙爆发 (Avalon explosion) 事件开始辐射开来，那时整个地球刚刚结束了大幅冰期。这群生物在寒武纪生命大爆发生物多样化之际，快速消失了。目前存在的动物体型构造最早出现于寒武纪，而非埃迪卡拉纪。以巨观生物来说，看来寒武纪生物群完全取代了埃迪卡拉纪的化石纪录，然而两者的关系，还有

2.2 Ediacaran Life Explosion

The Ediacaran (|iːdi'ækərən|; formerly Vendian) biota consisted of enigmatic tubular and frond-shaped, mostly sessile organisms that lived during the Ediacaran Period (ca. 635–542 Mya). Trace/cast fossils of these organisms have been found worldwide, and represent the early known complex multicellular organisms. The Ediacaran biota radiated in an event called the Avalon explosion, 575 million years ago, after the Earth had thawed from the Cryogenian period's extensive glaciation. The biota largely disappeared with the rapid increase in biodiversity known as the Cambrian explosion. Most of the currently existing body plans of animals first appeared in the fossil record of the Cambrian rather than the Ediacaran. For macroorganisms, the Cambrian biota appears to have completely replaced the organisms that dominated the Edi-

争议。

埃迪卡拉纪的生物最早出现于6亿年前，繁荣到 5.42 亿年前的寒武纪，然后这些很有特性群聚的化石就全部消失了，最早可能多样化的埃迪卡拉纪群聚，在1995发现于墨西哥索诺纳(Sonora, Mexico)，大约是6亿年前，比5.8亿年前的加斯基斯冰期(Gaskiers Glaciation)还早一些。虽然发现了一些最晚到寒武纪中期的很罕见的化石（5.1亿至5.0亿年前），但是在埃迪卡拉纪结束时，这些最早曾经很繁盛的族群就消失了，只留下令人好奇的片段。有好多不同的假说，试图解释这生物群的消失，包括保存偏好、环境改变、猎食者出现和与其他生物竞争等等。

若要将埃迪卡拉生物放入生命树(the tree of life)，会有相当大的挑战性，它们无法说是动物，也有说是地衣（真菌藻类共生）、藻类、原生生物有孔虫、真菌或微生物群聚，或者介于植物和动物之间的假设中间体。有些物种分类(如朵西绳虫 *Funisia dorothea*）的形态和习惯，看来和多孔动物或刺胞动物有关联，金盘虫（*Kimberella*）可能和软体动物有相似性。其他的生物具有两侧对称，但是还有争议，大部分巨观化石形态和以后的生物形状有明显差异，它们像盘子、管子、泥巴填充的袋子或绗缝床垫。因为难以推论这些生物的演化关系，有些古生物学者认为，这些是完全灭绝的家系，和任何生存的生物都不像，还有位古生物学者甚至认为埃迪卡拉生物群，

acaran fossil record, although relationships are still a matter of debate.

The organisms of the Ediacaran Period first appeared around 600 million years ago and flourished until the cusp of the Cambrian 542 million years ago, when the characteristic communities of fossils vanished. The earliest reasonably diverse Ediacaran community was discovered in 1995 in Sonora, Mexico, and is approximately 600 million years ago, predating the Gaskiers glaciation of about 580 million years ago. While rare fossils that may represent survivors have been found as late as the Middle Cambrian (510 to 500 million years ago), the earlier fossil communities disappear from the record at the end of the Ediacaran leaving only curious fragments of once-thriving ecosystems. Multiple hypotheses exist to explain the disappearance of this biota, including preservation bias, a changing environment, the advent of predators, and competition from other life-forms.

Determining where Ediacaran organisms fit in the tree of life has proven challenging; it is not even established that they were animals, with suggestions that they were lichens (fungus-alga symbionts), algae, protists known as foraminifera, fungi or microbial colonies, or hypothetical intermediates between plants and animals. The morphology and habit of some taxa (e.g., Funisia dorothea) suggest relationships to Porifera or Cnidaria. Kimberella may show a similarity to molluscs, and other organisms have been thought to possess bilateral symmetry, although this is controversial. Most macroscopic fossils are morphologically distinct from later life-forms: they resemble discs, tubes,

若以林奈分类法来说，他们是另外一个生物界 (Kingdom)，以前称为“凡登界 (Vendozoa)”，今重新命名为“凡登生物界 (Vendo-bionta)”。如果这些迷人的生物没有后代，他们奇异的体形，可以看成多细胞生物的“失败实验”，以后的多细胞生物，则是独自从无关联的单细胞生物演化而来。

埃迪卡拉纪“生物群”的观念，当然有点人为，因为无法从地理、地层、埋藏或生物学角度来定义（图2.2.1和图2.2.2）。

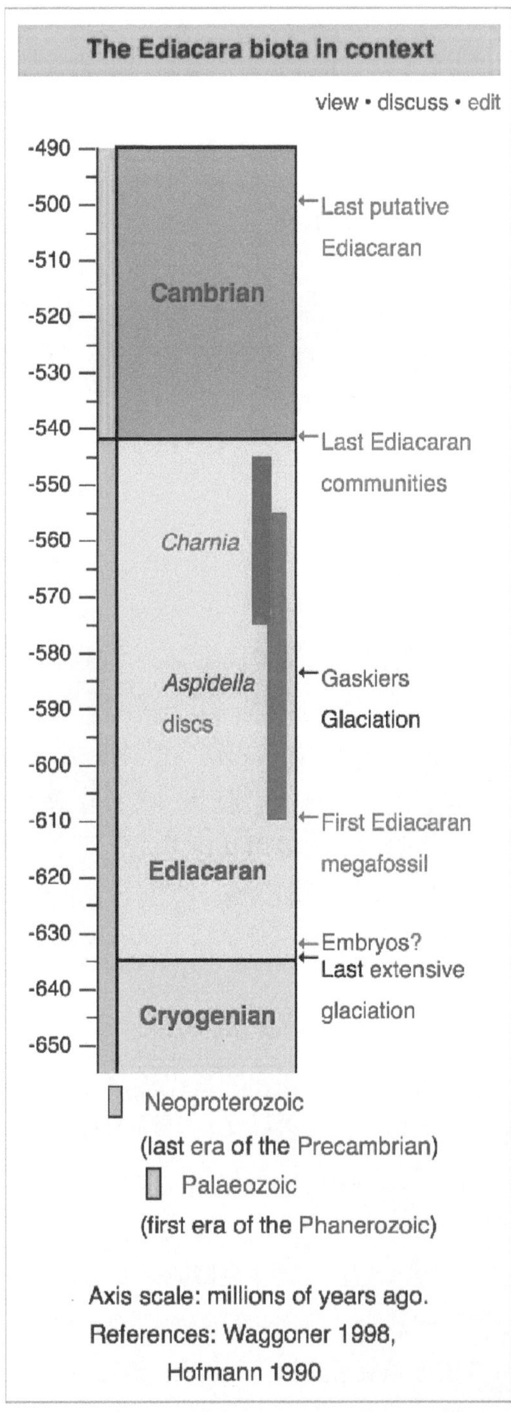

▲ 图 2.2.1　埃迪卡拉纪前后

▲ Figure 2.2.1　Before and after Ediacaran Period

2.2.1　埃迪卡拉的怨叹

这是我个人写的一篇文章，描述自从2003发现这些重要化石以来的一些心路历程和想法，当为个人记录，也希望能找到接棒人，深入探讨研究，揭开地球上很古老的多细胞巨观生物之谜。

2003 年初，玩石家到云南举办恐龙营，在楚雄市行政区内饱满街（小村地名）有个恐龙脚印化石点，我们去的时候顺道前往马路另一边的鱼化石点，找到了一些只有一两厘米的小鱼，当场就有书僮（家长）抱怨说，这么小条，实在没有看头。哈！同行的禄丰恐龙馆小阳（传伟）听到了，私下对我说，他办公室桌子底下，有一块老乡拿来的化石，应该是一条很大的鱼化石，可以给我。过了几天，他就把这块“鱼”化石给了我（图2.2.3），指着上面这里说，你看看这不是鱼的脊椎骨吗，这不是鱼的肋骨

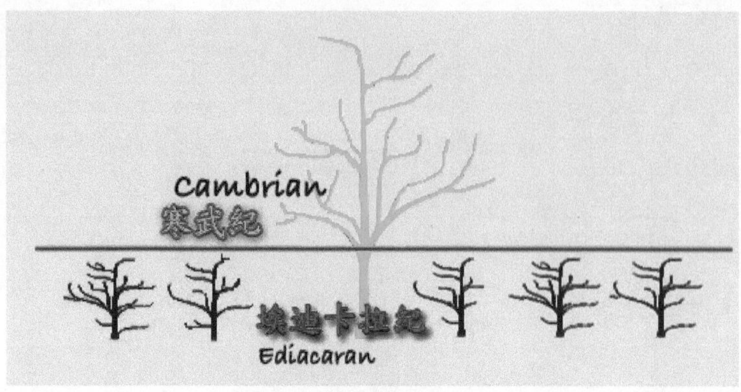

◀ 图 2.2.2 只有一棵生命树？或整片森林？

◀ Figure 2.2.2 Only one tree of life, or the whole forest?

吗？我看这条鱼，应该很不小，至少半米长！

那年春天，我带着美国的古生物学者Jerry Jessen到天龙堡(DinoDragon Castle)，设立了禄丰境内的第一个现代化石清理站，使用我从台湾带来的气动清理工具，教他们如何用这些工具清理、复制化石等等。对他们来说，从不知道世界上竟然有这么好用又快速的气动雕刻笔、喷砂等等化石清理工具，比起他们一向使用小铁锤小钉子清理恐龙骨头化石，把骨头表面敲得坑坑洞洞，这简直是天方夜谭，开了眼界。

期间，我们借住在美国华侨张云茜老太太出钱，在附近购买二百二十亩土地盖房子，并交由中国恐龙王代为看管居住的房子，我们白天走路到天龙堡干活，傍晚就上去聊天休息睡觉。我拿出小阳给的这块“鱼”化石开始清理，把那些还覆盖在化石上面的围岩清理掉，急着想看这条“鱼”的真面目，花了一段好长时间仔细清理，终于清理出来了。可是，一面清理一面看，越看越不

mud-filled bags or quilted mattresses. Due to the difficulty of deducing evolutionary relationships among these organisms, some paleontologists have suggested that these represent completely extinct lineages that do not resemble any living organism. One paleontologist proposed a separate kingdom level category Vendozoa (now renamed Vendobionta) in the Linnaean hierarchy for the Ediacaran biota. If these enigmatic organisms left no descendants, their strange forms might be seen as a “failed experiment” in multicellular life, with later multicellular life evolving independently from unrelated single-celled organisms.

The concept of “Ediacaran Biota” is, of course, somewhat artificial as it can not be defined geographically, stratigraphically, taphonomically or biologically (Figure 2.2.1 and Figure 2.2.2).

2.2.1 The Ediacaran Complaining

This is a personal record describing my venture and thoughts on these important fossils found in 2003. It’s a personal paper hoping to be able to find the successor(s) to study them deeply and reveal the myth of the very early multi-cellular macro-organisms.

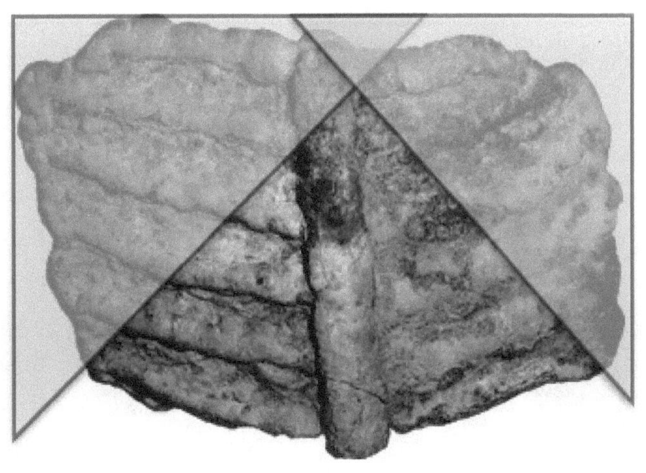

▶ 图 2.2.3 还没清理前的这块怪化石，被图示两个三角形的围岩包裹着，看起来的确像是鱼化石，不过，去除围岩清理后，竟然是“生眼球长眉毛之后都没看过”的某古生物化石，揭开了我研究埃迪卡拉纪实体化石的序幕

▶ Figure 2.2.3 Before my preparation, this weird fossil was covered by the two triangular matrices. Looked like parts of a fish. However, after the top matrices removed, it's a fossil never saw by my eyes. This opened up my study on Ediacaran body fossils

对劲，鱼化石的脊椎骨，应该是一节一节的，怎么会都联成一条管子呢？而且“鱼肋骨”，应该细细直直，怎么会一球又一球串起来呢？最不可思议的是，怎么四周边缘会有管子连串起来呢？特别是“肋骨”终端，都被一条管子连起来，真是奇怪的结构；这肯定是某种生物的化石，但绝对不会是鱼化石。好啦！问题来了，那么，这到底是什么化石呢？这辈子我还没碰到，看见某化石，却无法判断它是什么大类古生物的糗状，进一步鉴定的生物属种可能无法判断，但是大类别应该不会把三叶虫看成螃蟹，Jerry反复看了好几次摇头说，这应该是某种化石，而非某种地质现象所造成的假象，应该没有错，可是真的看不出来是什么。中国恐龙王斟酌了半天，也投降了，如果说是生痕化石——某种古生物在海底泥巴中打的洞的话——这些管子怎么会这么有规则对称性？这应该是某种生物的化石，但是真的从来没看过如此奇怪的古生物。

In early 2003 our Dinosaur Camp to Yunnan, we visited the dinosaur track site at Baomanjie. On the other side of the main road, there is a fish fossil site. The fossil fishes here were so small, one or two cm in length. Parents complained they were too small to see. CW Yang heard that and told me in private that he had one big fish fossil from a farmer under the desk in his office. He can give that to me. Several days later he did, (Figure 2.2.3) and pointed to that and said: "Are not these fish vertebrae, fish ribs? I think this is a big fish of at least a half of meter!"

In the Spring of that year, I brought American paleontologist Jerry Jessen to DinoDragon Castle and set up the first modern fossil preparation lab using the air power tools and teach the locals how to prepare, replicate, etc. To them, they never knew such fantastic air chisel and airbrush tools for fossil preparation. They were used to small hammer with small nails to prepare dinosaur bones, causing too many pits. These opened their eyes.

During this period, we stayed at the 13-hectare house built from the money donated by an American Chinese lady, Ms. YunChien Chang. ZM Dong

这块无法鉴定的化石，躺在我的展示柜内五年多，一直到2008年，实在忍受不了无法鉴定这块化石的窘境和精神压力，就抱着死马当活马医的心态上网去问问看，反正最坏的情况就是没人理我，最后还是无法鉴定。好在网络上有人看了照片后，说有可能是埃迪卡拉纪的化石，这指引了一个搜寻的方向，有了大致的方向之后，当然透过网络到处找这方面的资料，也写Email请各路好汉专家协助，我把相片寄给哈佛大学专门研究埃迪卡拉纪化石的 Andrew Knoll 教授，他的回信，令我非常兴奋：

亲爱的 Timothy，

谢谢你寄来你在云南发现的化石照片。我很确定那个有分枝构造的是埃迪卡拉纪化石，另外一张照片也有可能是埃迪卡拉纪化石，但我难以从形态上评估。

祝你研究幸运！

诚心的 Andrew Knoll，哈佛大学

2008 年 8 月 13 日

这封信是另一个重要的里程碑，终于有国际上这方面的专家确认这是埃迪卡拉纪的化石没错，而不是我自己一厢情愿。

在此期间，小阳和我都很多次尝试着找出这个化石出土地点，每次都因交通或天气障碍，铩羽而归，可是小阳不死心，前前后后跑了十趟，前面九次都没找着，第十次革命终于成功，终于找到了出土地点，捡回了几块给我看的证据样本。最早老乡拿到博物馆给他的

was taking care of this house. We worked during the day at DinoDragon Castle, and return to the house for the night. I took out the "fish" fossil and started to clean out the unwanted matrix, eagerly want to see this "fish". After a long while, the "fish" showed up. However, the more cleaning I did, the more confused I was. The fish vertebra should be a section followed by another section. How could they fuse to become a tube? Also, the "fish ribs" should be thin and straight, why so many knobs strung together? The most unthinkable was that there is a tube surrounding the whole rim, all the "ribs" strung together by a tube. Bizzare! It is some fossil for sure, but not a "fish". Then, the question turned to what is this? Never saw such a challenge. When I saw a fossil, at least I know what kind of animal, the big category. I will not mix up trilobites with crabs. However, this is not the case here. Jerry did not know what this is either. He was sure it is a fossil, not a geological phenomenon, but do not know what it is. Chinese DinoKing Dong could not identify it either after carefully look at it for a while. If this was a trace fossil from an ancient creature making the dwelling inside the mud, it should not be so symmetric. We never saw such a weird fossil before.

This unidentifiable fossil laid in my exhibition cabin for more than 5 years until 2008. I really could not stand the mental pressure of not be able to know what it is. So, with the last hope to surf the internet and expecting no reply as the worse case. Luckily enough, someone saw the photos and told me it could be an Ediacaran fossil. This pointed a direction for me to search. I got some information and sent out emails to experts in this field asking for help. One of them was the Ediacaran expert Professor Andrew

样本，原本以为是恐龙化石，后来也就没联络了，事隔多年，人都不好找。幸好，皇天不负有心人，他终于找到了出土地点，带我第一次到该出土地点，让我过了个最独特的六十大寿（晚两天）（图2.2.4），也采集增添了好些样本。

这些样本，从形态来说，基本上可分为两大类：直径不等的管状体和叶状体 (Fronds)；我到现场所看到的，有很多“飘浮”在地表面（围岩已经被风化，完全只是躺在地上）的管状体和数量稍微少一些的叶状体碎块，胶结在一起的大块样本，需要从大约十来厘米厚的掩埋层中去挖出来，为了保护现场，我们所捡的绝大部分是“飘浮”在地表的，当然也从边缘地区，挖出来几块胶结的团块带回来研究（图 2.2.5）。

这些样本这些年来，我一直面对着几个很严肃又困扰的问题，其中之一就是，到底它们是什么？从网上和各方面收集到的资料显示，这些在寒武纪生命大爆发之前几千万年的巨型多细胞生物形态很奇怪，与以后的生物形态都不同，体形大的可达一二米长，一般也比寒武纪生命大爆发的生物体形来得大，但是它们似乎都没有躲过埃迪卡拉纪/寒武纪大灭绝，至今没有埃迪卡拉纪的生物存活到寒武纪的记录。有人开玩笑说，在埃迪卡拉纪时段内，上帝做的第一次生命实验失败了，只好到了寒武纪再重新来。再者，到目前为止，世界二三十处所发现的埃迪卡拉纪化石，都只是铸模 (Cast)，没有实体化石，即便有些宣称为

Knoll of Harvard University. His kindly reply encourages me profoundly:

Dear Timothy,

Thank you for sending images of the fossils you've discovered in Yunnan. I'm quite convinced that the branching structure is an Ediacaran fossil. The other may be, as well, but I had a hard time evaluating its morphology.

Good luck with your research.

Sincerely, Andrew Knoll,
Harvard University, 8/13/2008

This letter is an important milestone. Finally, I got an internationally well known Ediacaran expert to identification for sure, not just my personal wishful thinking.

During this period of times, Yang and I tried many times to find the place this fossil came from. Each time, our trip was interrupted by road condition or weather. However, Yang did not give up. He made ten attempts with the first nine failed. Finally, during his 10th try, he found the site and brought back several for me as evidence. The old farmer who gave the first specimen to Yang thought it was a piece of a dinosaur fossil. Since then, no further contact. It is difficult to find the guy after these many years. Luckily, the efforts were rewarded. The site was found. He took me to the site for my two-day belated 60th birthday, Figure 2.2.4, and collected more specimens.

From the shapes (morphology) to say these specimens can be separated into two main types: Tubes and Fronds. There were so many tubular "floater" (matrix weathered out with fossils on the surface) and few frond fragments when I

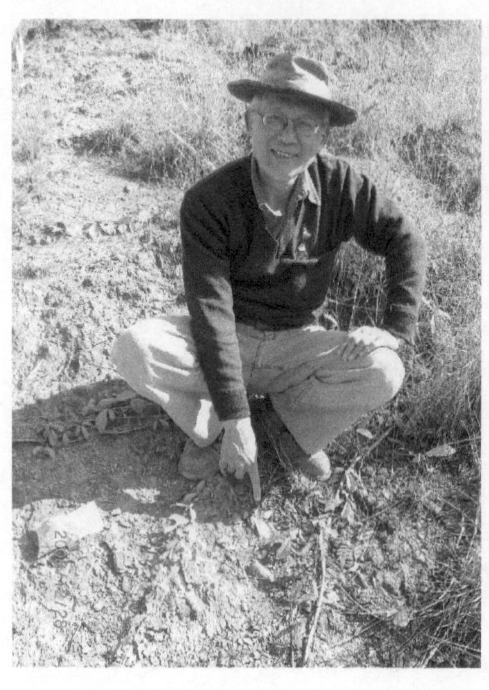

◀ 图 2.2.4　我“六十大寿”是如此特别，晚了两天才第一次拜访到埃迪卡拉实体化石出土点

◀ Figure 2.2.4　Visited the Ediacaran body fossil site two days after my 60th birthday

实体化石者，也都很小很小，仅在毫米 (mm) 范围，没有厘米、分米或米的化石出土过。从研究的角度来说，无法从铸模深入“看进骨头内”来了解这些生物，就如医学院的学生若要学习人体构造，不能拿服装店用灌模的人样模特儿来做大体解剖，一定得用真人尸体，才能学习到真正正确的知识。也因如此，这些生物的分类，也有各种说法，从体形大小来说，可以确定它们不会是生物分类学中微生物的几个界，但是它们到底是动物（界）还是植物（界）？曾经有德国学者 Seilacher 还提出一个已经完全灭绝的“凡登生物界 (Vendobionta Kingdom)”的假说，说这些生物是一个灭绝的生物界，具有缝合气垫构造，属于形态类 (Rangeomorphs)，如查恩盘虫

arrived. Large cemented chunks had to be plied out from the ground. In order to protect the site, most of what we collected were floaters with only a few cemented specimens. See Figure 2.2.5.

For these specimens and these many years, one pressing question bothers me a great deal. What were these? The information I collected thru the internet, I learned that the creatures several deca-million years before the Cambrian Explosion were very weird. Their shapes were very different from the later life. The size could be as long as 1 or 2 meters in length. Moreover, in general, the body sizes are much bigger than that of the Cambrian Explosion. It seems that they did not survive thru the great extinction of Ediacaran/Cambrian boundary. As now, there is no fossil evidence to show any Ediacaran creature survived in the Cambrian Period. So the joke saying that even God failed His first life experiment during the Ediacaran Period, and thus repeated in the Cambrian Period. Furthermore, up to now, all the specimens from more than twenty or thirty Ediacaran sites worldwide were just imprints or casts, no body fossil found. Even if some claimed body fossils were tiny, they are in the mm range. No fossil of cm, deca-meter, or even meter size ever found. For better study and hence understanding of these mysterious creatures, we cannot “Look into the Bone” on these imprints/casts. Just like the medical student

◀ 图 2.2.5　从云南中部埃迪卡拉纪化石点出土的化石块，上有好多种形态不同的生物，主要有管状的和叶状的

◀ Figure 2.2.5　This chunk from central YunNan contains many different shape creatures, mainly tubulars and fronds

(*Charniodiscus*)。因此，埃迪卡拉纪生物到底属于哪个生物界或者哪几个生物界呢？

就像没有庙的大和尚，若需要动用到精密高贵仪器的话，只能到处卡油，靠人家施舍给恩惠。可是我还是尽自己微薄之力，利用自己现有的拼凑工具，做了一点点形态和组织构造方面的小研究。

首先，从图2.2.3来说，这个化石是此地出土点数量比较少的生物，虽然相对诸多几厘米长小片段之管状体（图2.2.6）要来得完整；但也只是这个原本被我命名为“啥米碗糕体”，后来为了纪念最伟大的老妈，从而改名为“黄杨清莲体(*Huangyang chinglian*)”生物的一部分而已。这生物的下半部和另一面，到底长得什么样子？从这块难得完整美丽的样本来说，就不得而知了，除非能有人帮我做计算机断层扫描。幸好，几次到现场，有捡到一些“飘浮”在地表的碎块（图2.2.7和图2.2.8），提供了拼图的重要实证。终于，从这些比较完整和只是零星碎块切片（图 2.2.9和图 2.2.10）

cannot learn human anatomy by dissecting mannequin, real human corps have to be used to obtain the correct knowledge. Therefore, there are many taxa classifications. From the size of these creatures, we can rule out the Microbial Kingdom. What Kingdom do they belong to? The Animal Kingdom? The Plant Kingdom? Once the German scholar Seilacher proposed a complete extinct Kingdom called the Vendobionta Kingdom and said these extinct creatures had quilted mattresses structure and belonged to Rangeomorphs, such as *Charniodiscus*. So, what Kingdom(s) these Ediacaran creatures were?

Jokingly say under a condition of a big monk without a temple, I do not have access to many advanced, expensive instruments. The only way I can is to ask favors from other scholars. So, I did some initial works with my limited tools on the morphological and structural aspects.

Let us start with Figure 2.2.3 This is one of the less from that site. It is much complete in comparison to those several centimeter long tubes, Figure 2.2.6. It was named as “SaMeWanGe (What is this)”, then changed to *Huangyang*

▶ 图 2.2.6 片段管状体，比较粗者的直径可达 6 厘米，一般的一两厘米

▶ Figure 2.2.6 Section of tubulars. The bigger one has a diameter of more 6 cm. Most of them were between 1 to 2 cm in diameter

等的综合拼图，我提出了想象中黄杨清莲体的形态构造，如图 2.2.12。

有人可能会说，透过图2.2.3、图2.2.7到图2.2.10，只有叶片部分，怎么知道这个黄杨清莲生物体的完整形态呢？哈！这就是一般玩石家可能的疏忽，通常在采集化石的时候，只会注意到比较完整的样本，对于诸多破碎的“下脚料”，往往不屑一顾，可是搞古生物的话，会把诸多碎片都一起捡拾回去，因为这些碎片，除了可做各种需要的切片以外，也可作为拼图用，提供重要的形态资料，如图2.2.11，就是埃迪卡拉纪黄杨清莲体的另一端，让我能够把整体形态组合出来！真正的科学研究，必须有更加充足的资料，因此，做切片、溶解等破坏性的实验，当然尽可能不要把“死相”完美的样本拿来切割，而是利用那些没有看头的下脚料当牺牲品，魔鬼通常就藏在这些细节里面。

这个像芭蕉扇的埃迪卡拉生物黄杨清莲体，和已知的埃迪卡拉纪叶状生物，如查恩盘虫，有几个重大的不同：①已知埃迪卡拉纪叶状生物的顶端是尖

chinglian in 2009. This is just the one terminal part. What is the other half look like? Unless I can do some CT scans. From this rare and relatively complete specimen and the specimens collected on the following trips, more floaters were collected from the ground, as Figure 2.2.7 and Figure 2.2.8, provided more evidence for this puzzle. Finally, from the combinations of the cross cuts, Figure 2.2.9 and Figure 2.2.10, I can propose the shape of *Huangyang chinglian* as Figure 2.2.12.

Someone may say that thru Figure 2.2.3, Figure 2.2.7–2.2.10, only the frond portions are known, how can you know the shape of the whole *Huangyang chinglian* ? Ha, this is commonly neglect by amateur collectors. Usually, in their fossil collection, they would look for more complete fossils and ignore the fragments. However, for paleontology, lots of fragments would be bagged back. These fragments can be cut and provide the information on the missing piece of the whole puzzle. Figure 2.2.11 showed the other termination of *Huangyang chinglian*, which enabled me to come up with the complete shape. For real scientific study, one must have enough information. Thus, destructive ways of making thin slices or dissolv-

▲ 图 2.2.7 黄杨清莲体叶片的碎块，“正”、“反”面

▲ Figure 2.2.7 Both sides of *Chinglian huangyang* 2009 Huang fragment

▲ 图 2.2.8 黄杨清莲体叶片的碎块，正反面合并照；右边照片显示中间那根中央管（稍微往右上斜），从这又分岔出二级管；左边的照片，显示另一面如豆干形态的构造；红色线为图 2.2.9 和图 2.2.10 的切割位置

▲ Figure 2.2.8 Both sides of *Chinglian huangyang* 2009 Huang fragment. The right photo shows the central tube (pointing to upper right). Secondary tubes branched out from this. The left photo shows the other side with bean curd morphology. The red line is where the cut for Figure 2.2.9 and Figure 2.2.10

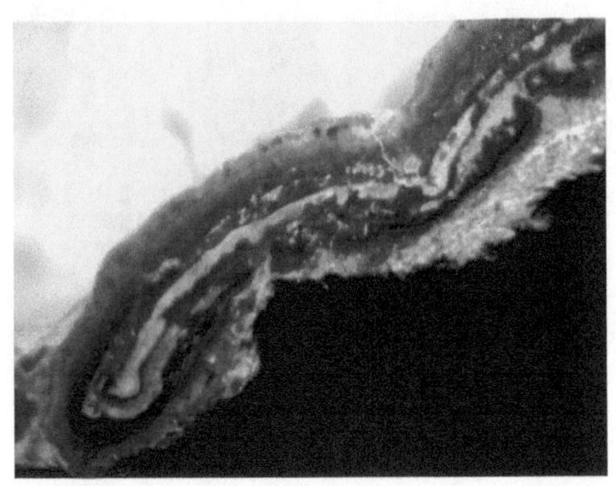

◀ 图 2.2.9 黄杨清莲体叶片切面光学显微照片；位置如图 2.2.8 红线所示

◀ Figure 2.2.9 Optical microscope photo of *Chinglian huangyang* 2009 Huang. The cut line is the red line as shown in Figure 2.2.8

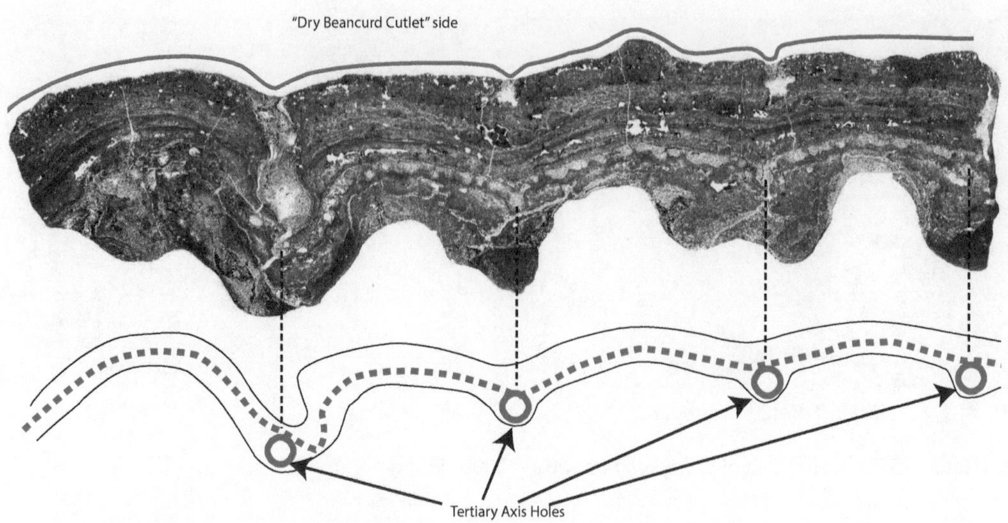

▲ 图 2.2.10　黄杨清莲体叶片另一张切片照片与解读示意图

▲ Figure 2.2.10　Another cross cut of the *Chinglian huangyang* and the interpretation

◄ 图 2.2.11　黄杨清莲体的另一端，显示类似叶片与叶梗和终端的形态

◄ Figure 2.2.11　The other terminal of *Chinglian huangyang*, showing the stem and termination

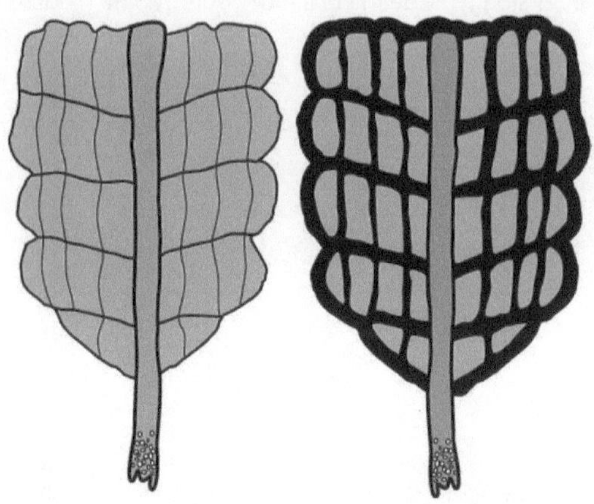

► 图 2.2.12　黄杨清莲体的想象形态结构示意图

► Figure 2.2.12　Illustration of the *Chinglian huangyang*

的，就像尖头树叶，而黄杨清莲体的终端（或说整体），却圆钝像芭蕉扇；②已知埃迪卡拉纪叶状生物叶片下面有根管子，如树叶的叶梗（茎）接着叶片，管子最终端，有个吸盘，吸附固定在海底，而黄杨清莲体的另一个终端，虽然有管子从叶片上延伸出来，却看不到有吸盘的结构，在化石现场，也没找到可能是吸盘的物体，反而，如果正对着叶梗主轴管终端横切面来看的话，有一点像兔唇，有几个裂沟；③和已知其他埃迪卡拉纪叶状生物最大的不同，在于整体的构造，可以看到有很多相互连接的管子，最中央的主轴管径最粗，从这根主轴管往左右两边，分出管径较小的二级管，再从二级管又分出管径更小的三级管，甚至有可能还有更细的四级管。后面这两点差异指出（从这两点差异我们可以推测），相对于已知的埃迪卡拉纪叶状生物固定于海底，有可能比较像"植物"，但是，黄杨清莲体，很可能是在海洋中自由游泳的生物，有可能它从管子一端吸入海水，利用体内的各级管状络，过滤取得食物，因此也就比较有可能像是趴趴走的"动物"。当然啦，在没有进一步深入研究、彻底揭开面纱之前，谁也说不准，目前只能从证据中做"有教育的猜测（educated guess）"，暂时做此认定。

在这些卡油和自己土法炼钢的研究当中，一个我很希望能做的，就是用我所提出来非破坏性的研究方法"看进骨头内"，探讨这些全球首次发现、到现在也是唯一的埃迪卡拉纪实体化

ing the specimens are unavoidable. Of course, the beautiful death look specimens cannot and shall not be used for sacrifice. Fragments shall be used instead. Devils are hidden in these details.

This hardy banana fan (*Musa basjoo*) shape Ediacaran creature *Huangyang chinglian* had some major differences from the known frond creature, such as *Charniodiscus*: ① The tip of the known Ediacaran frond creatures was pointed, likes the tip of a tree leaf. However, for the *Huangyang chinglian*, the termination (or the whole) was dull and blunt as a hardy banana fan. ② There is a tube under the frond of known Ediacaran creatures like the stem connected to tree leaf. At the end of the tube, it has a holdfast to suck on the ocean floor. However, the other terminal extended from the frond portion did not have such a holdfast. We could not find anything similar to such a suction at the fossil site. Instead, if it is looked from the main axial into the terminal tube, it looks more like the rabbit lips, several split ditches. ③ The biggest difference from the known Ediacaran frond creatures is that in the whole body, there were many interconnected tubes. The central tube is the biggest, then, from this to both sides smaller, secondary tubes, and from the secondaries, the tertiaries, and even finer quaternaries. The last two differences pointed out that those fixed on the ocean floor Ediacaran creatures maybe more like "plant". However, *Huangyang chinglian* was more like swimmers in the ocean. It could suck water from one end into the body tubular networks and filter the foods. Therefore, they were more like moving 'animal.' Of course, before further study to remove the veil, nobody

石 (body fossil)。2011 年春天以后，我曾经在台湾同步辐射研究中心 (Taiwan Synchrotron Radiation Research Center)，用透射X射线显微镜 (transmission X-ray microscopy，TXM) 做了一些扫描（图2.2.13），有看到似乎像细胞的椭圆形物体，同时本人外甥女也用扫描电子显微镜 (scanning electronic microscopy, SEM) 帮我扫了一些（图2.2.14），似乎也验证透射X射线显微镜所看到的。不过，后来透射X射线显微镜那边，一方面碰到有人贪得无厌想整碗端过去，另一方面又需要做破坏性的试片，磨到15 μm以下的薄片，所以也就停止下来。扫描电子显微镜，本人外甥女后来离开了原本

can be sure about this. We can do some educated guesses.

In all of these endeavors of my effort, one thing I like to do is to conduct my thoughts of "Look into the Bone" nondestructively to explore this first discovery and only Ediacaran body fossils. After the Spring of 2011, I did some Transmission X-ray Microscopy (TXM) scans at the Taiwan Synchrotron Radiation Research Center (SRRC). Figure 2.2.13. Some oval things, similar to cells were seen. My niece also did some Scanning Electronic Microscopy (SEM), Figure 2.2.14 for me re-enforce the TXM observation. Figure 2.2.14. Unfortunately, on the TXM side, someone tried to take over the whole bowl and, TXM is a destructive examination. Less than 15 μm thin slice had to be made. So, it was

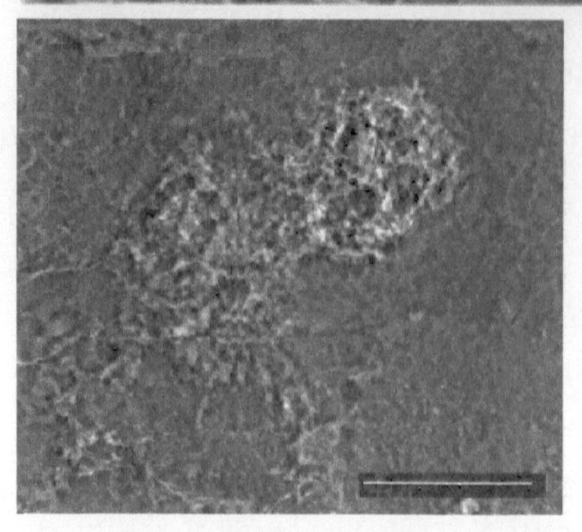

◀ 图 2.2.13　透射 X 射线显微组合影像，可看到一团团椭圆形的结构，尺寸与细胞大小接近，怀疑是细胞的结构

◀ Figure 2.2.13　Mosaics of TXM images. Oval structure can be seen, with size close to a cell. Wondering if this is a cellular structure

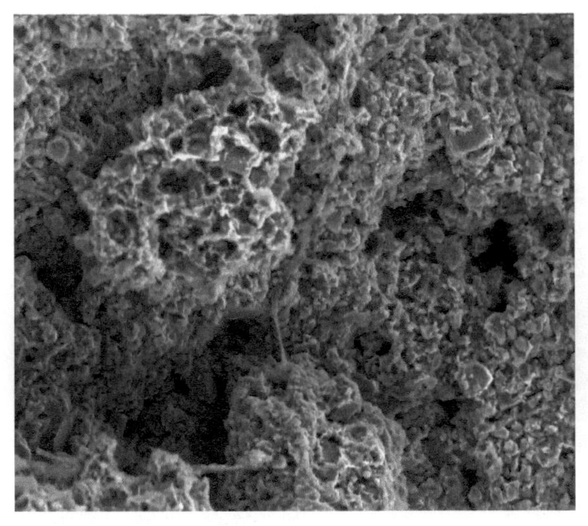

◀ 图 2.2.14 扫描电子显微镜下埃迪卡拉纪实体化石的内部结构，说如此精致的构造来自于矿物结晶，很难说得通，应该是生物体内的结构形成的化石

◀ Figure 2.2.14 SEM of an internal structure of Ediacaran body fossil. It very hard to say that these delicate structures were from minerals. They should be the fossilized biological structures

的研究室到英国去念博士了，也就无法多做下去了，我这到处沿门托钵的苦行僧，人家丢什么到碗里，只能说谢谢，没有挑食的空间。

此外，在这期间内，我也在台湾大学卡油做了一些多频显微 (multiple harmonic generation microscopy, MHGM) 的扫描，这是一种新的完全没有破坏性的三维影像扫描（图2.2.15）技术，我很喜欢。虽然从表面深入到化石样本内部的深度有限，大约只限于从表面往内100 μm左右，但是分辨率可达到0.47 μm，而且是完全非破坏性，扫描完成后，很快就可以做出三维的影像模型，对于研究古生物内部构造与组成，提供了其他仪器和方法没有的信息。无奈，这也是卡油研究，台大研究生毕业了，后续也就断了，真想，如果能找到钱，搞一台多频显微来玩玩，至少从我乞求人家恩惠施舍所做的化石扫描，我觉得是个非常有用的好工具，应该会对古生

stopped. Also my niece went to London for her Ph.D. and stopped too. As a bagger cannot be a chooser.

During this period, I did some Multiple Harmonic Generation Microscopy (MHGM) scans. This is a complete nondestructive 3D scan, Figure 2.2.15. I like it very much. Although the depth is limited to about 100 μm from the surface, the resolution can reach 0.47 μm and nondestructive. After the scan, a 3D model can be done very quickly. This provides the internal structure and composition information that many other instruments can not. However, this is also a favor given to me. The graduate student finished his degree, and my work stopped. I wish if I can find funding, I shall have one such instrument built for me. At least from my situation, as a bagger, be able to do MHGM fossil scans, this is a very useful tool. It will provide enormous impacts on paleontology studies. I already had a good title for a Ph.D. dissertation: MHGM for Paleontological Studies. This shall be in the level above the Master of Science de-

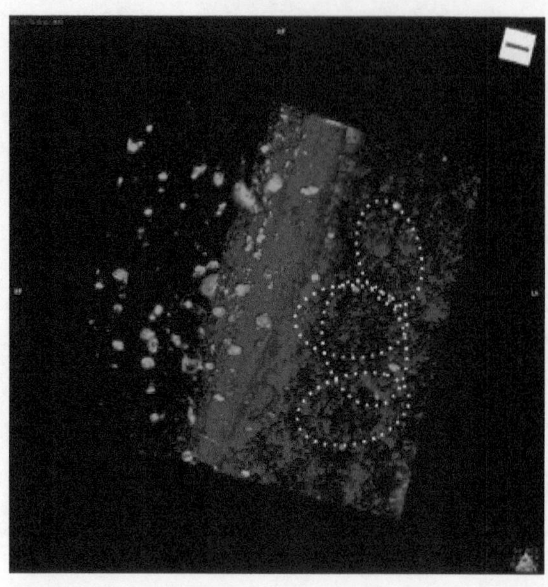

◀ 图 2.2.15　多频显微也可看到类似（图 2.2.13 上方）细胞样的椭圆形构造

◀ Figure 2.2.15　MHGM also shows the similar oval cell-like structures as the top of Figure 2.2.13

物研究，产生并带来重大的冲击影响。其实，我都想好了论文的题目："多频显微之应用于化石研究"，应该是硕士以上的论文，如果有研究生愿意好好地做，应该可当博士论文，发表到重量级的学术期刊去。没想到台湾的学术圈真的很有趣，在多频显微方面，搞了四年多，才终于在某分量不是很高的期刊上，发表了两小篇多频显微研究化石的论文，其中一篇 SCI 三点多的论文，简直是小咖中的小咖，根本没有把它最厉害的二光子、二倍频和三倍频整合起来，只说了现生和化石鳄鱼牙齿，可以用三倍频来看到普通光学显微看不到的牙小管构造，个人觉得有些暴殄天物，没有使出它最厉害的武功，很无奈啊！唷！我的问题，还是老问题，乞丐没有挑食的余地。

对了，还应该提出来的是：这篇记录里的那些实体照片或各种"看进骨头

gree. If anyone is interested and willing to do it, high caliber paper(s) can be published. Well, my great adventure in this regard ended with one paper on not so high caliber journal after more than four years of struggling with the hellish Taiwan academic circle. In comparison to what I had planned, this SCI 3+ paper is a small one. It did not shine on their most potent combination of Two-Photon, Second Harmonic Generation, and Third Harmonic Generation, and only used extinct and extant crocodile teeth to show the Third Harmonic Generation is good for dentin tubules.I felt this is a waste, not using the most powerful gun. What can I do? Same problem for me: Bagger cannot be a chooser.

Yes, I have to point out that the photos in this record, and all "Look into the Bone" images are never seen by most of the people worldwide. Even though I do not fully comprehend what these images really mean that I saw the cells inside the Ediacaran creatures. I am very sure that

内”所显示的诸多影像，都是全球以前没人看过的。虽然目前还不是很清楚到底这些影像的含义究竟是什么，是否真的看到了埃迪卡拉纪生物体内的细胞。可是，可以肯定的是，从未有人看过这些，你我是全球首先看到这些的人们，为此荣幸，可小小庆祝一下。

那么剩下来，还可行的路，就是利用微型计算机断层扫描 (μ-CT) 技术了，这是靠不同物质对于X射线有不同吸收程度的非破坏性检测。不过，在2014年之前，台湾所有的微断层扫描分辨率，最高只达到9 μm而已，对于我想进一步看看那些更细微的结构，实在是无能为力。9 μm的分辨率，对于地球生命演化有绝对关键作用的埃迪卡拉纪实体化石里面的结构，实在是不够用。9 μm的微断层扫描，每个像素等于9 μm^3，细胞大小约在10~20 μm，一两个像素无法显示细胞，根本解析不出来以前用其他方法（如非常破坏性的透射X射线显微镜制片和非破坏性的多频显微）所看到的细胞长什么样子，更不用说细胞里面了。

幸运的是，2014年台湾一家代理商，进口了一台全球最高分辨率的微断层扫描仪，分辨率可达 0.35 μm，真的很厉害啊！以前显示为一团的模糊结构，如今可以清楚看得见了，真好！因此，我就和这家公司合作，也是用惯用的卡油方式，我帮他们抓背，他们帮我抓背，双赢局面，他们的网站想要介绍所经销的最新微型断层扫描仪，展示最先进仪器超高的分辨率，要我提供一些有趣的

nobody ever saw these before. You and I are very lucky first. Celebrate.

The only road left is to use the μ-CT technique. This is a nondestructive examination based on the X-ray density of different matter. However, before 2014, all the μ-CT scanners in Taiwan can only reach the 9 μm resolution. For me wanting to see these unidentified but could be very significant for the life evolution on Earth, 9 μm is too rough. Each volume pixel is 9 μm^3. The average size of a cell is between 10–20 μm. So, one or two volume pixels just will not cut it as what TXM and MHGM can, not to mention I want to see inside the cells.

Luckily, in 2014, the distributor of the micro-CT imported the highest resolution μ-CT to Taiwan. The resolution can be as high as 0.35 μm! Wow! The blobby structures in the past can be seen clearly now. Excellent. Therefore we worked out a collaboration deal. Win & win. They want to introduce their new product to market to show the high resolution of this instrument and asked me to provide some fossils so that they can show excellent results on their website. After some considerations, I picked several potential good ones for them to scan. Of course, Ediacaran frond and terminal specimen were included. In the beginning, 6μm was used as a screening for selection of the best scan areas. Then, they repeat the scans with the best resolution for the selected areas. Nobody ever saw such fossil internal details before.

Looking at the 6 μm scan result, Figure 2.2.16, after our discussions, we decided to scan the highest possible resolution on a small "bean curd". They gave me the digital cut of the "bean curd" to play at home.

好化石样本，以便把扫描的结果秀在他们网站上。我评估了一下，找了几个看起来会有吓人潜力的样本给他们去扫描玩玩看，这其中当然包括了埃迪卡拉黄杨清莲体的叶状和终端部分，初步先用6 μm来做筛选扫描给我看，我才能告诉他们该样本高分辨率的扫描区域应该在哪里，再用该样本可能的最高分辨率来重新扫描，显示出从来没有人看见过的化石内部细微构造。

看着6 μm的扫描结果（图2.2.16），在我们讨论之后，决定接下来最高分辨率的扫描区域，大致上来说，就是一小块“豆干”，当场先把这筛选扫描的“豆干”数位性“切”下来，我带回家玩玩，很有意思哦！

不过，天下也没白吃的午餐，我不是指卡他们油做最高分辨率扫描这件事啦！那是相互抓背搔痒，而是指，随着分辨率越走越高，后续处理的计算机能力（几核、内存、储存空间、显示器大小等等计算机系统），也随之水涨船高，一般的计算机，恐怕吃不消。就举一个

There is no free lunch. I do not mean receiving highest resolution scan favor from them. That is a mutually beneficial deal. What I am saying is this. As the resolution goes higher and higher, the demand for post computer processing power (number of kernels, memory size, storage capacity, screen size, etc.) also went up significantly. The ordinary computer will not be able to hack the tasks. As an example: supports we want to scan 3mm in length with the 9 μm resolution, each image will be several Mb or not even 1 Mb. Three hundred images will be several Gb in total. My computer and the OsiriX software I am using can handle the job. Perhaps, it will take a little bit of time to accomplish. I can get good 3D QuickTime-VirtureReality (QTVR) output to let me and my teammates look at the object by QuickTime Player 7.x from any desired angle. Of course, this does not mean I do not have my complaints. For one, OsiriX was designed for medical CT, relatively very low-resolution computerized tomography and NMR. Do not know why after version 5, it can not export QTVR file anymore. So, I am stuck with version 4. Furthermore, QTVR let the viewer adjust to any

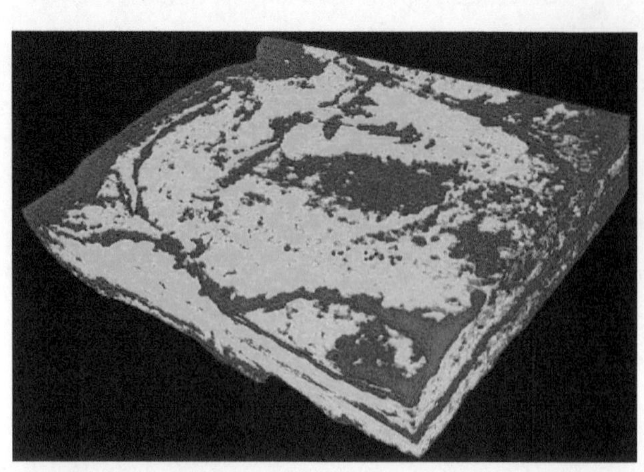

◄ 图 2.2.16　以 6 μm 分辨率扫描黄杨清莲体的一块“豆干”，从三轴切入，所看到的内部构造，和以前靠实体切片做的二维光学显微照片有些不同，可能需要修正以前对于黄杨清莲体内部构造的推论

◄ Figure 2.2.16　A bean curd of *Chinglian huangyang* scanned at 6 μm resolution. Look from 3 axial reveals the internal structure differs from the optical microscope. Perhaps, the previous deduction of this creature has to be updated

现实的例子来说，假设所要扫描的长度为3 mm，用9 μm分辨率来扫描，每张影像的大小，大约只要几Mb就够了，或甚至1 Mb都不到，三百多张影像下来，总共需要几吉(G)而已，我的计算机和所用做影像后处理OsiriX程序，大概还负担得了，花一点时间，可以做出不错的“三维虚拟QuickTime (3D QTVR)”文档，让我和团队成员，随意从任何角度来观看。苹果提供免费的QuickTime Player 7，给苹果和微软操作系统使用者去下载，所以没有所谓的携带性问题。当然啦！对于这套系统，我不是完全没有意见，比方说，OsiriX是设计给处理医学影像用的（相对来说，分辨率很低的计算机断层扫描、核磁共振等），可是从第五版之后，就无法输出三维虚拟QuickTime文档，所以我还是被绑死在第四版。再者，虽然三维虚拟QuickTime文档，可让观看者抓着鼠标转动立体模型到任意角度观察，可是，却无法调整如背景不透明度和设定调整黑白灰阶所对应的彩色等等。若要做这些更改，就必须回到OsiriX去慢慢小心调整到所要的亮度、对比、不透明度和彩色设定等等，重新输出三维虚拟QuickTime结果文档。

现在用到可说是最高分辨率的微型断层扫描，因为分辨率大幅提高，同时影像文件的大小，每一张从一两兆（或不到）跳到几十吉，整套扫描下来有几千张影像。如以1 μm分辨率来扫描每1 mm样本物件的结果，就有1000张影像，这些“天文”数字很快就叠乘出一套影

angle to view the object. However, no adjustment of background and opacity and the color lookup. If any of these adjustments are needed, then go back to OsiriX to accomplish the desired result of brightness, contrast, opacity, and color assignment, and re-export the 3D QTVR.

Now, the highest possible resolution μ-CT is used. the image size increased from one or two Mb (or less) to several deca Gb. Since there are possibly several thousand images produced. If the resolution is set at 1μm for 1 mm length scan, it will produce 1,000 images! The astronomic numbers can pile up very quickly. Each specimen will need several hundred Mb. A two Tb external hard disk can hold just several scans. In other words, Raid hard disks shall be considered. Hope to have someone kind enough to give me such.

The higher resolution scan results can be processed with their post-processing software, CTvox to make a 3D model with adjustable opacity, brightness, three color adjustments, cropping, and export video format. These fulfill my basic need. The complaints I have are ① The instrument manufacturer neglects the fact that many users are using Apple's Macintosh Operating Environment, not the broken Window. For them to provide a Mac version of such software shall not be a problem. They already provide a free iPad version. ② It can not export the QTVR file format. Although it can output cross-platform video, the output video cannot be adjusted by the viewer to any desired angle and limited to whatever done during the post-processing phase. Sorry.

Of course, I would like to show the 3D video in this book. However, limited by the 2D

像需要好几百吉(G)，一个2Tb的外挂硬盘，也只能收容几套扫描文档而已！也就是说，光是储存这些文档，那就要考虑列阵式硬盘了！对于我这到处沿门托钵受苦辛的大和尚来说，还期望着能早日碰到一个可施舍列阵硬盘的大德施主。

新的较高分辨率扫描的结果，在该仪器公司所提供的后续三维影像处理程序CTvox，可以做成三维模型，而且能提供调整不透明度、光度、三种基本颜色、裁切面、输出视像等等，基本满足了我的使用需要。可是，我的抱怨是：①该公司忽略了很多科研人员习惯使用苹果Mac操作环境；其实，提供一个可在Mac操作环境的软件版本，并非难到什么地步，因为该公司已经提供iPad可用的版本了；②无法或者没有提供三维虚拟QuickTime文档，虽然可以输出跨平台的视像，但是所输出的视像，对于终端使用者来说，无法随心所欲地调整到所想要观看的旋转角度，只能受限看到

publication media, what I can is to show one snap screen, Figure 2.2.17. Please compare with Figure 2.2.10 the optical image (upper right). The round beads curved line in the 'beancurd' looks like a series of neighboring tubes. However, from the μ-CT scan, it looks like a curve line. Moreover, if it is looked from the perpendicular direction cutting from front to the back, this curve forms a curved surface. Ha, this is interesting. It means within the "beancurd" there are two layers.

This is the Ediacaran Complaining of mine.

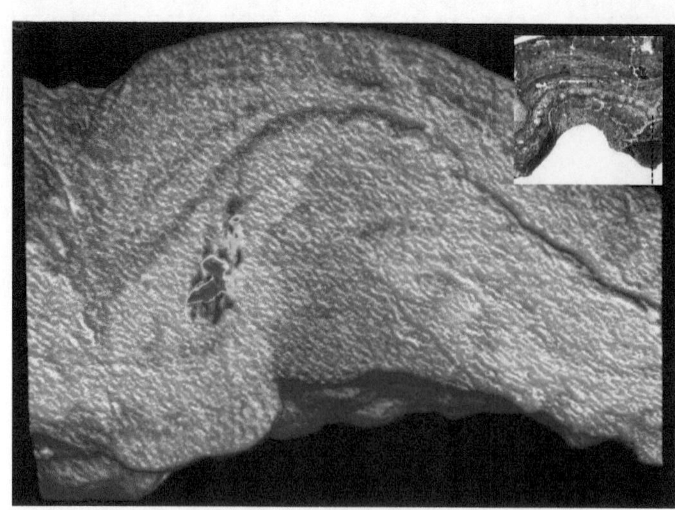

◀ 图 2.2.17 以 6 μm 筛选扫描黄杨清莲体的一块“豆干”

◀ Figure 2.2.17 A selected bean curd of *Chinglian huangyang* scanned at 6 μm

制作视像时所提供的控制（角度旋转、沿着某轴切入等等）画面，很遗憾。

当然啦！我很想在本文显示刚刚做出来的三维视像，不过，受限于平面媒体的二维展示，没有办法，只好抓出视像中某个画面（图2.2.17），请和图2.2.10切面光学影像（右上角插图）比较。图2.2.10“豆干”内那条往上鼓起来的串珠圆弧曲线，看来像是横切一系列相邻的管子。可是在 μ-CT扫描看来却是像一条曲线，而且若沿着垂直进入方向来前后切入的话，这条曲线，实际上形成一个曲面。哈，这就有意思啦！也就是说，“豆干”内部，其实有上下两层。

这就是我的“埃迪卡拉的怨叹”。

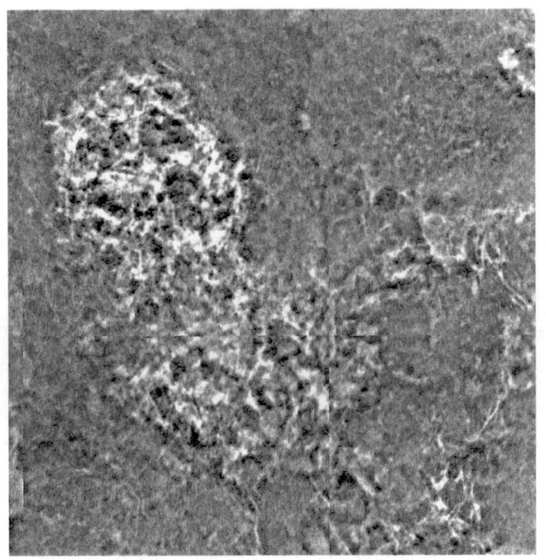

▲ 图 2.2.18　以透射 X 射线显微镜筛选扫描黄杨清莲体，类似于细胞分裂？

▲ Figure 2.2.18　Observed *Chinglian huangyang* TXM scan. Look like cellular division?

2.2.2　地质汪洋孤岛

这个地质“汪洋孤岛”的名称，是我这个不是搞地质工作的人自己想出来的，用来解释不止一处而是两个地方的独特地质情况。还请专业的地质学者批评指正，帮忙指点迷津。

我亲身碰到的两个非常困惑的情况是：①在云南楚雄市苍岭恐龙脚印地点；②在云南易门县埃迪卡拉纪化石点，依照云南地质图示，苍岭的地层是白垩纪地层,却出现侏罗纪恐龙脚印（详见第6章6.2节），而后者在地质图上标示为侏罗纪地层，却出土了埃迪卡拉纪化石，让我不知道该如何解释，只好提出这个“汪洋孤岛”的假说，让我能睡得着觉。

2.2.2　Geological Islet in Ocean

This geological “Islet in Ocean” was coined by this non-geologist author to explain not just one but two special geological situations. Hoping to have real geologist to help me and correct my ignorance.

I encountered two very puzzling occasions: ① the dinosaur track site at CanLing, ChuXiong City, YunNan, and ② the Ediacaran fossil site at YiMen, YunNan. According to the geological map, the dinosaur track site is Cretaceous. However, it has many Jurassic dinosaur footprints (see Chapter 6, section 6.2). The other place was marked as Jurassic but yield Ediacaran fossils. I don’t know how to explain these two oddballs.

我的论点是，任何地质图调查，即便高精度的地质图（如二万五分之一、五万分之一比例尺），不可能做到每寸土地都仔细调查确定其地层，只能做到较大面积的地层判断，这也是做工程的时候，通常需要进行工程地质调查，把更精细的实际情况搞清楚，否则靠着比较粗略的地质图就开干，工程和建筑肯定要出毛病的。记得，若以五万分之一的比例尺来说，地质图上的1厘米，相当于地表的500米。

我所谓的汪洋孤岛形成大致如此：图2.2.19有两组示意图，原先地层沉积堆叠（每组最左）老的在下面，新的在上层；经过造山运动（每组中间）地层鼓起，最后顶部被风化掉（每组右边），显露出相对来说很小块古老地层于广泛的年轻地层区块内，显露凸出来的古老地层被年轻地层包围，就如汪洋中的孤岛，故以称之为“汪洋孤岛”；再者，在这两个很小面积的化石点，都是在斜坡上，更进一步支持汪洋孤岛的可能性。

如图2.2.19左边苍岭“汪洋孤岛”示意图，白垩纪（绿色）和侏罗纪（青色）地层是整合地层；在云南楚雄市苍岭的

So I proposed this "Islet in Ocean" hypotheses so that I can sleep well.

My argument is this. Any geological map, even the high scale maps, such as 1/25,000 or 1/50,000 ones can NOT survey every square inch of the land. They did much bigger area. This is why when doing constructions, it usually needs to conduct the further engineering geological survey to get more precise and detail the real situation. Otherwise, some serious consequences could happen. Remember, if the 1/50,000 scale geographic map is used, each 1 cm on the map means 500 meters on the ground.

In short the Islet in Ocean I'm proposing is like Figure 2.2.19 There are two sets of the diagram. In each, the original strata are (on the left) the younger on top of the older stratum. Then the orogeny caused the strata to buckle upward (middle) and then weathering removed the very top (younger stratum, right), exposing the older stratum within the vast younger formation, as an isolated islet in the ocean. Thus, it's called "Islet in Ocean". Furthermore, these two fossil sites are on upward slopes. This further support the possibility of "Islet in Ocean" hypothesis.

As shown to the left of Islet in Ocean at the Cangling, the green Cretaceous on top of cyan Jurassic formation conformity. The dinosaur

▲ 图 2.2.19 汪洋孤岛示意图。（左）云南楚雄市苍岭恐龙脚印侏罗纪汪洋孤岛；（右）云南易门埃迪卡拉纪汪洋孤岛

▲ Figure 2.2.19 Diagram of Islet in Ocean. Left: Islet in Ocean of Dinosaur Tracks at CangLing, ChuXiong, YunNan; Right: Islet in Ocean of Ediacaran at YiMen, YunNan

恐龙脚印，若用云南地质图是说不通的，白垩纪的地层，怎么可能出现侏罗纪的恐龙脚印呢？如果是白垩纪恐龙脚印，那么，为何在楚雄州内甚至昆明玉溪等地，都没有发现白垩纪恐龙？

在埃迪卡拉纪化石点（图2.2.19右边）的情况是：上面的地层为侏罗纪（青色），下面是前寒武纪（亦即埃迪卡拉纪，棕色）的“不整合地层(unconformity)”。两地层中间缺少了古生代地层。这种不整合地层，在云南并非罕见。类似于上述苍岭恐龙脚印的情况，怎么会在侏罗纪的地层，跑出四亿多年前的埃迪卡拉纪生物呢？即便假设埃迪卡拉纪生物，存活进入寒武纪之前的大灭绝，没有死光光，那么，在这四亿年间，为何没有发现类似或中间过渡的化石？这一点难以生物学的角度解释得通。

2.3　寒武纪生命再爆发*

寒武纪大爆发（亦称寒武纪生命大爆发，Cambrian Explosion），是相对短时期的演化事件，开始于距今5.42亿年前的寒武纪时期，化石记录显示绝大多数的动物“门”都在这一时期出现了。它持续了接下来的大约2000万~2500万年，导致了大多数现代动物门的发散。因出现大量的较高等生物以及物种多样性，于是这一情形被形象地称为生命大爆发，这也是显生宙的开始。不过，最

* 本节资料，部分取自维基百科。
Information in this Section basically are from Wiki.

track at this site can NOT be explained well with the Yunnan Geological Map. How possible the Jurassic dinosaur footprints showed up in the Cretaceous formation? If these footprints belong to Cretaceous dinosaurs, then, why no Cretaceous dinosaurs ever found within the whole ChuXiong Prefecture, or even enlarged to the neighboring KunMing and YiXi areas?

At the right side of Figure 2.2.19, the Ediacaran fossil site, the top layer is the Jurassic (cyan) formation, and the brown Precambrian underneath unconformity, no Paleozoic formation in between. This kind of unconformity is not uncommon in Yunnan. Similar to the case at the Cangling dinosaur track site, how could it possible to find about 400 million years older Ediacaran fossils in the Jurassic formation? Even if the assumption that some of the Ediacaran creatures survived thru the big extinction before the Cambrian Period, then, within this 400 million years period, why no similar or transitional fossils were found? This can not be explained easily from the biology point of view.

2.3　Cambrian Life Explosion again

Cambrian Explosion, also called Cambrian Life Explosion is a relatively short period event, starting from 542 millions year ago of Cambrian Period. Fossil records showed the most phylum showed in this period of geological time. It lasts for the next 20–25 million years and diverse to the modern animal phylum. Since the appearance of the higher order of organisms and taxa

后这一句话，笔者有意见，因为多细胞大型生物的出现，最早是埃迪卡拉纪，显生宙的开始，应该从那时候算起，而非寒武纪。

在世界各地发现的化石群共同印证了这一生命演化史上的壮观景象，例如在加拿大的伯吉斯页岩 (Burgess shale) 和在中国云南省澄江化石地（图2.3.1至图2.3.3）等。这一时期的化石群相当典型，非常多不同种类的生物，几乎同时在这一时期出现。

寒武纪大爆发的事实证据，也曾让达尔文非常困惑，他在《物种起源》中写道：“这件事情到现在为止都还没办法解释。所以，或许有些人刚好就可以用这个案例，来驳斥我提出的演化观点。”但即使到达尔文死后一百多年的今日，寒武纪大爆发依旧是科学界的一大谜题，尚待更多的古生物证据出土，也许就能窥见当时的实际情况，找出真正的原因。

可能的原因

针对寒武纪大爆发，学界曾经提出过几种假设：

（1）地球在寒武纪之后，才出现足以保存化石的稳定岩层，而前寒武纪的沉积物，毁于地热和压力无法形成化石。

（2）生物到了寒武纪才演化出能够形成化石的坚硬躯体。

（3）大气中累积足够的氧气量，足以使大量动物短时间演化，并且形成臭

diversity, this is described as life explosion and the beginning of Phanerozoic eon. However, the author has a different opinion on the very last sentence. The earliest appearance of multi-cellular big life forms is the Ediacaran Period. Thus, Phanerozoic Eon started from that time, not Cambrian.

The fossils discovered in various places in the world witness this spectacular great life evolution, such as the Burgess Shale of Canada and ChengJiang fossil of Yunnan, China (Figure 2.3.1–2.3.3). The fossils of this time period are very typical containing many different organisms of the same time.

The Cambrian Explosion confused Darwin a great deal. In his *Origin of Species* he wrote: "Until now, we can not explain. So, perhaps somebody can use this event to rebut my evolution point of view." After more than one hundred years after his death, the Cambrian Explosion is still a myth waiting for more evidence to be unearthed. Perhaps by that, the actual scene can be seen and find the real reason.

Possible Reasons

There are some hypotheses for the Cambrian Explosion proposed.

(1) The stable rock formations appear after the beginning of Cambrian. Sediments of Precambrian were destroyed and no fossil formed by the heat and pressure.

(2) Hard bodies for forming fossils were not evolved until Cambrian.

(3) Enough oxygen concentration in the atmosphere so that mass of animals can evolve in short period of time, and form ozone layer to block out UV light.

氧层隔离紫外光。

（4）某些掠食性动物侵入物种稳定平衡的地区，减少原先占优势的物种，释放生态栖位给其余物种，进而促进大量物种歧异度的增加。

近年来研究认为，寒武纪大爆发跟埃迪卡拉纪末期灭绝事件有关：该次菌藻等单细胞生物灭绝事件结束了隐生宙，为多细胞生命在显生宙的发展铺平了道路。

澄江化石地是中国西南部云南澄江县境内40多处散存的寒武纪多门类古生物群遗址，距今约5.20亿到5.25亿年，主要集中于帽天山的页岩，开始发掘于1984年。由于化石埋藏地质条件特殊，不但保存了生物硬体化石，而且保存了非常罕见的清晰生物软体印痕化石。它的特点是动物体内没有矿物质的软组织部分保存得非常好，例如表皮、纤毛、眼睛、肠胃、消化道、口腔、神经

(4) Some predators invaded the species balanced areas, reduce the number of advantage species and release the spaces for other species to increase massive species diversion.

Recent studies showed there is a relationship between the Cambrian Explosion and the extinction at the end of Ediacaran Period. The extinction of single cells terminated the Cryptozoic Eon and pave the way for Phanerozoic Eon.

More than 40 Chengjiang Fossil areas located inside Chengjiang County preserved many phyla of ancient life from 520 to 525 million years ago. The main site is at MaoTenShan Shale. The excavation started in 1984. Due to the special preservation condition, not only the hard parts of the creatures but also traces of very clear and rare soft body creatures. The unique point is that the non-mineralized soft tissues were preserved very well, such as skin, cilia, eye, intestine, digestion tract, mouth, and nerve. In which the *Myllokunmingia fengjiaoa* is the oldest vertebrae known, found in the Precambrian forma-

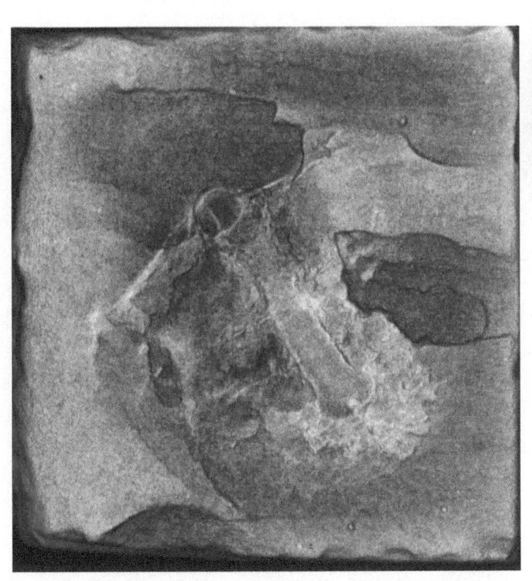

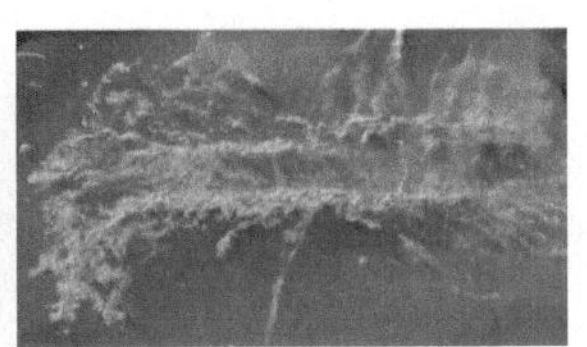

◀ 图 2.3.1 澄江动物群的纳罗虫 (*Naraoia*)，微型计算机断层扫描重建（左），区段化（右）

◀ Figure 2.3.1 *Naraoia* of ChengJiang fauna, μ-CT reconstruction (left) and segmentation (right)

◀ 图 2.3.2 澄江动物群的等刺虫 (*Isoxys*)

◀ Figure 2.3.2 *Isoxys* of ChernJiang fauna

等。其中，丰娇昆明鱼 (*Myllokunmingia fengjiaoa*) 是已知最古老的脊椎动物，化石在5.3亿年前寒武纪的地层被发现。它似乎是有由软骨形成的头颅骨及骨骼，澄江动物群比加拿大的伯吉斯动物群丰富，而且年代方面更早一千万年，使我们得以窥见寒武纪生命大爆发时代的海洋景观。

寒武纪大爆发的事实证据，产生了很多科学争论，1859 年达尔文也曾说到，寒武纪生命爆发是他的物竞天择演化最大的挑战。寒武纪动物群突然间出现的长期困惑集中在三个要点：是否在寒武纪早期相对短时间内有大量复杂生物的发散？是什么可能造成如此快速的变化？它所暗指的动物生命起源是什么？受到有限证据的限制，主要是寒武纪岩石内的化石纪录和化学资料不完整，解读有困难。

tion of 530 million years ago. It seems to have cartilaginous skull and bones. The ChengJiang Biota is about 10 million years older than that of Burgess Shale, enabled us to see the ocean view of Cambrian Explosion.

The Cambrian explosion has generated extensive scientific debate. In 1859 Charles Darwin discussed it as one of the main objections that could be made against the theory of evolution by natural selection. The long-running puzzlement about the appearance of the Cambrian fauna, seemingly abruptly, without precursor, centers on three key points: whether there really was a mass diversification of complex organisms over a relatively short period of time during the early Cambrian; what might have caused such rapid change; and what it would imply about the origin of animal life. Interpretation is difficult due to a limited supply of evidence, based mainly on an incomplete fossil record and chemical signatures remaining in Cambrian rocks.

◀ 图 2.3.3 澄江动物群的海口虫(*Haikouella*)

◀ Figure 2.3.3 *Haikouella* of ChernJiang fauna

地质化学观察

几个化学标志指出，在寒武纪早期发生了环境大改变，这些标志和大灭绝或甲烷冰释放造成大温暖是一贯的。这些变化可能是寒武纪爆发的原因，但也可能是生物活动增加的结果，造成了爆发。虽然这些不确定，地质化学证据让科学家专注到此事件的发生，是因为至少有一种可能环境变化的理论。

2.4 古生物化学的意义

古生物学作为一个整体的基础科学的领域，其主要目的为何？为何要去花那么多时间精力探讨一些已经死亡、灭绝很久的生物？这门学问又不能直接赚钱，多发现一个"新"品种的三叶虫（或任何其他古生物），那又如何？对于现代社会来说，古生物学有什么直接的效益？

Geochemical Observations

Several chemical markers indicate a drastic change in the environment around the start of the Cambrian. The markers are consistent with a mass extinction, or with a massive warming resulting from the release of methane ice. Such changes may reflect a cause of the Cambrian explosion, although they may also have resulted from an increased level of biological activity – a possible result of the explosion. Despite these uncertainties, the geochemical evidence helps by making scientists focus on theories that are consistent with at least one of the likely environmental changes.

2.4 The Meaning of PaleoChemistry

Paleontology as one of the fundamental sciences, what's the main purpose of it? Why so many times and efforts shall be spent on those long dead creatures? This is not a money-mak-

台湾光复后早期，曾经发生过吴大猷和李国鼎的路线争议：李国鼎说，台湾刚光复，百废待兴，先顾好老百姓的肚子要紧，才能从废墟中站立起来，因此，首要投注可以赚钱的项目；吴大猷说，基础科学不能废啊！虽然基础科学无法马上赚钱，但是，谁也不知道来日经济转型会用到什么基础科技，我们现在不培养人才，到时候需要而没人可用，如何转型？从后来的发展来看，李国鼎派占了上风，不赚钱的基础科学，如古生物学被荒废了，在我们搞恐龙胚胎科研过程中，在台湾就是找不到稍有分量的古生物学者，只好“礼失求诸野”，找外国专家组成联合团队。

然而，塞翁失马，焉知非福？正是因为台湾客观环境和个人的遭遇，因缘际会，误打正着，化危机为转机，从而开启了古生物学的一个新领域：古生物化学。古生物学是一门基础科学，让我们了解地球无数生命的演化，也因为多有了解，也会多费心思照顾爱护这个浩瀚宇宙中的一个小蓝点地球。不过，对于古生物的了解，传统的古生物学有其限制，因为无法回到从前，只能从手中的化石形态去探讨，所以有人说，古生物学不是实验室科学，没有办法一次做实验失败重来，因而未能深入到化石里面，从里到外彻底探讨这些有趣古代生物的奥秘。相对地，古生物化学提供了从最基础的化学层面探讨的一扇门，以窥究竟。

就以本章所说的地球早期多细胞巨

ing course. Discovery of a “new” trilobite or any other ancient organism, so what? To the modern society, what’s the direct benefit of paleontology?

During the early time of Taiwan Recovery, a debate between Dr. DaYou Wu and Mr. KaoTing Lee took place. Lee said that Taiwan just recovered, many many things had to be done. Taking care of people’s stomach is the most important thing so that we can stand up from the ruins. Therefore, we have to invest in the money-making issues. Wu said basic sciences should not be neglected. Basic sciences won’t make money immediately. However, nobody can be sure when and what will be needed for economic transformation. If we don’t educate people on these, we won’t have people to use when the time comes. From what took placed in Taiwan, it seems Lee got the upper hand. Those, not money-making courses, such as paleontology were deserted. During our dinosaur embryology project, we couldn’t find any competent paleontologist to participate and have to form an international team.

However, Every cloud has a silver lining. It was the tough Taiwan society and personal hardship gave me the chance to turn the crisis into favorable, and opened up a new domain in paleontology: PaleoChemistry. Paleontology is one of the fundamental sciences for us to understand the evolution of countless live evolutions. With more understanding, more cares and loves to this tiny blue dot we called Earth in the vast universe can be expected. But, for this understanding, it is limited by the conventional paleontology, because we can not go back to the geological time and bound by the fossils in hand. It is said that paleontology is not a laboratory science. Failure once cannot be repeat-

观生物（埃迪卡拉和寒武纪爆发）的演化来说，除了传统的形态系统描述之外，最缺乏的一块，就是相关的古生物化学资料非常欠缺，导致这些美妙生命演化的诸多奥秘，仍然深锁阁楼，比方说，为何在埃迪卡拉纪会出现好些可以大到一两米的生物？它们到底是动物界还是植物界的生物？如何生存繁殖？然后，到与寒武纪交界的时候，又全部死光光，没有存活演化成后代的生物？难道上帝第一次做大型生物实验失败，到寒武纪的时候重头来才实验成功？照目前的证据显示，埃迪卡拉纪生物，没有猎食者，不必打地洞躲藏身免得被吃掉，大家生活在一片祥和的伊甸园里，狮子和绵羊一起唱赞美歌（圣经描述的天堂），诸多这时代的生物，看来并没有口器，那它们怎么吃东西？靠吃什么维生？为什么这两个地质纪的诸多生物长得那么奇怪？如寒武纪爆发时期身上长有九个像眼睛的微网虫 (*Microdictyon*) 或搞不清楚上下的怪诞虫 (*Hallucigenia*)？生物界的第一只眼睛出现之后，整个生物法则改变了，不是猎人，就是别人的猎物 (hunter and hunted)，生物军武竞赛从此开始，无法回头。三叶虫复眼的镜头——就是开启猎人猎物的罪魁祸首——竟然是简单的方解石？诸如此类的问题一牛车。接着下来的五亿多年到现在，各种生物的演化还有更多迷人的问题，比方说，恐龙胚胎等等，等待着人们去解答。

透过古生物化学，无法回答所有生物演化的问题，但是如果能解开其中之

ed in the lab again. Thus, it cannot penetrate deep into the fossils to explore the mysteries of these interesting ancient creatures. On the other hand, PaleoChemistry opens up the door from the most fundamental chemistry level to explore further.

Using the evolution of the Earth early multi-cellular macro-organisms (Ediacaran and Cambrian), discussed in this chapter as examples, among the conventional morphological descriptions, one thing missing the most is the lack of PaleoChemitry information, which leads to locking so many mysteries of wonderful life evolution in the attic. For example, why in the Ediacaran Period there are one or two meters creatures? Were they belong to the Animal Kingdom or the Plant Kingdom? How do they reproduce? And to the boundary of early Cambrian, all of them vanished. None of them survive and evolved to decedents? Is that possible that God failed the first experiment of creating macro-organism in the attempt, so restarted in the Cambrian? From the current evidence to say that during this Period, there was no predator. They didn't dig holes to hide to avoid being eaten. Every member lived in a peaceful garden of Eden. Lion and sheep song together as the Bible said. Many of these organisms did not have a mouth. How did they eat? Surviving on what? Why were so many creatures of these two geological Periods were so odd looking? For example, the *Microdictyon* seems to have 9 eye-like things on its body, and the hard to tell upside-down *Hallucigenia*? After the first eye evolved, the biological rules changed. Hunter and hunted initiated the biological arms race, and can not turn back. The compound eyes of the trilobite were the very first smoking gun. But how could it made of simple

一二，回答其中几个小问题，如为何有机物能保存好几亿年？一辈子的投入，也就没有人生白走一趟，不是吗？让我们走出舒适的窝，在“化”石和“化”学两个共同的“化”字上，一起大步航向未被探索过的领域吧！

人都将去，化作灰尘，随风飘散。
人在浩瀚宇宙洪流中，只不过是无数轮回
　的几粒星尘，排列组合，生生灭灭。
人的出生，嘴巴里面没有含着金汤匙，
人生日子，只能努力活着不浪费时光，
人的尘化，但愿能留下些许飞鸿爪痕。

calcite? ··· Questions like these are endless. In the following 500 million years, more interesting evolution questions popped up, such as dinosaur embryology, waiting to be answered.

Thru PaleoChemistry, we can not answer all the life evolution questions. However, if just one or two can be solved to answer just a few small questions, such as why organic matter can be preserved for hundreds of million years, our life won't be void. Right? So, let's based on the common character "Hua（化）Change" of "Hua（化）" of fossil and the "Hua（化）" of chemistry, get out from our comfort zone and sail to the uncharted territory with the big strike!

People will be gone with the wind as dust.
People in the stream of the vast universe are just
　numerous reincarnated star dust, permutation,
　and combination, alive and die.
People born without golden spoon in the mouth,
People's days could be only doing the best we can,
People will become dust and hope to leave behind
　some traces.

第3章　操作背景

Chapter 3　Operation Background

本章介绍进行古生物化学所需要的特性、方法、原则与期望。

This chapter states the characteristics, method, guideline, and expectation of conducting PaleoChemistry.

3.1　从活着到化石

3.1　From Life to Fossil

图3.1.1显示某生物从活着到变成化石可能的阴间路途，这其中的所有因素都可能影响到最终我们所找到化石的状况，特别从古生物化学的角度来说，几乎每个因素，都可能造成我们检验化石相关的化学差异，因此，稍微来思考一下从活着到化石漫长过程中可能的地质与化学变化。

从最上面的“活着”说起，我们必须记得任何生物，即便是最简单的生物，都含有很多很多的生物化学成分，而非如化学实验室内的单纯试剂。因此，从最开始出发点来说，古生物化学从所面临化合物数量来说，绝非单纯某化合物，而是一大堆混合物。举个实际的例子来说，本人的硕士论文，A Phytochemical Investigation of Liverwort *Frullania fran-*

Figure 3.1.1 shows the possible underground variables from a living organism to become a fossil. All of these factors can influence the final condition of the fossil we found. Particularly, from the viewpoint of PaleoChemistry, each of these factors could cause the chemical differences in fossil we examine. Thus, let us take a long journey from life to fossil to see what possible geological and chemical changes.

On the very top box "Life", we have to keep in mind that any living organism, even the simplest form of life, contain so many biochemical components, not as the pure reagents in a chemical lab. Therefore, from the very starting point, the number of chemical compounds that PaleoChemistry is facing is not just a single chemical compound, but a great mixture of compounds. As a real example, the author's Master degree

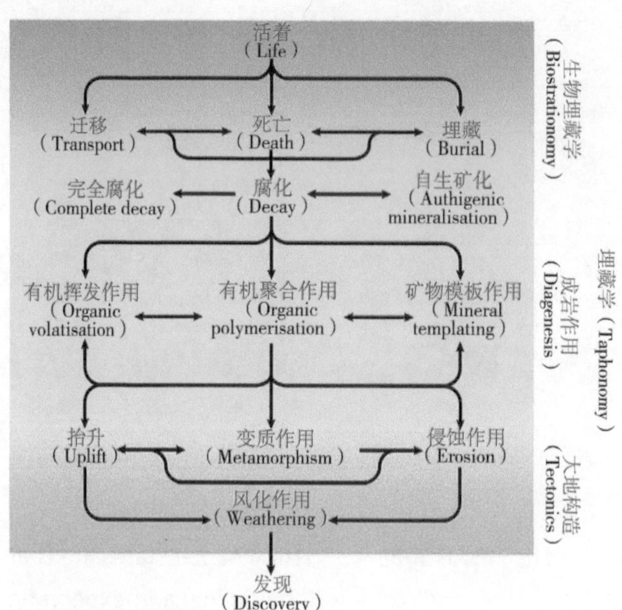

▲ 图 3.1.1　从活着到化石的可能变化历程

▲ Figure 3.1.1　Possible changes from Life to Fossil
(From Soft-Bodied Fossils Are Not Simply Rotten Carcasses – Toward a Holistic Understanding of Exceptional Fossil Preservation; Luke A. Parry, Fiann Smithwick, Klara K. Norden, Evan T. Saitta, Jesus Lozano-Fernandez, Alastair R. Tanner, Jean-Bernard Caron, Gregory D. Edgecombe, Derek E. G. Briggs, and Jakob Vinther)

ciscana Howe Species，分析了很低等的叶苔萃取物，发现里面含有超过八十种不同的化合物成分！何况更复杂的生物，可能在初始就有成千上万种不同的化合物混在一起！

从“活着”到“死亡”之后，基本上可能有两种情况：一是被迁移，另一种情况是快速埋藏，甚至如被山崩土石流等埋藏死亡。这三者之间，都是双向道，也就是说死亡、迁移与埋藏，有相互的关联。整体来说，这一段属于生物埋藏学 (Biostratinomy) 的领域，探讨生物怎么死的，刚死亡和死亡之后，又发生了什么。曾经看过报道,美国有个“尸

theses: A Phytochemical Investigation of Liverwort *Frullania franciscana* Howe Species, I analyzed the extract of the very low-level liverwort and found that more than 80 different chemical component compounds! Not to mention for the higher level life can have many thousands of different chemical compounds mixed.

From “Life” to “Death”, there are two possibilities. One is “Transport” and the other one quick “Burial”, such as buried by a mudslide. Between these three, they are bidirectional pathways, meaning they have interrelationship. This section belongs to Biostratinomy domain, which studies how the living organism dies, what immediately happened after they died. There was a documentary on “Body Farm” in the USA. This

体农场 (Body Farm)”，专门研究人的尸体腐化分解进展，把各种男女老少的尸体，放在不同环境之下，观察尸体如何腐化分解。此项研究，对于刑事鉴定，有非常重大的贡献，当司法单位在某处发现一堆人骨，可能只有某树林里的一堆炭火残渣里的一小段骨头，总得进行各种判断，死者性别、年龄、死因、时间……才有办法进行必要的刑事鉴定。研究古生物，也得从有限的环境信息，了解探讨古代生物相关的信息。比方说，笔者的第一篇《自然》期刊论文，描述辽西的一窝鹦鹉嘴龙，一个成年的头颅，旁边围绕着 34 个幼仔，见图 3.1.2。每个幼仔的“死相”都栩栩如生，像很可爱的小狗趴在地上，毫无死亡前的挣扎迹象，那么，它们是怎么死的？会像是如同当年意大利庞贝城那样吗？

接下来的“腐化”，如果是完全腐化，那也与化石无缘，死路一条。但也有可能在沉积岩内“自生矿化 (Authigenic mineralization)”作用。沉积物的颗粒、岩石碎片和化石，可以在形成岩石（成岩作用）过程中被地层或河海水所溶解的其他矿物取代。请注意，“腐化”和“自生矿化”是个双向道。

再接下来，还遗留在生物遗骸内的有机物，较小分子有机物质有可能会挥发掉或被水洗掉，留在躯体内被降解的小分子也可能产生重新聚合作用 (polymerization)，形成原本生物体内没有的“新”聚合物。请注意，这个过程是双向的。此外还有可能透过“矿

facility specialized in the study the decomposition and decay of human corpse. They place various human bodies of young to old, male, and female in various environments and observe how the corpse decompose and decay. This is very important for forensic investigation. The law enforcement occasionally would encounter a pile of white bones or a small fragment of bone in a charcoal pit. They have to identify such as gender, age, cause of death, duration, etc. before their investigation can go further. For paleontology, we have to gain and understand the relatively limited information that the environment can provide. For example, the author has a Nature paper describing a cluster of *Psittacosaurus* containing one adult and 34 hatchlings as in Figure 3.1.2. The death posture of each of the little guys is so cute as a bunch of little puppies. No sign of death-struggle can be seen. So, how they died? Like the famous Pompeii of Italy?

Then the next step is "Decay". If the corpse decay completely, no fossil can be found. This is a dead end. On the other path is going thru Authigenic Mineralization bidirectionally. The sedimentary particles, including rock fragments and fossils, can be replaced by other minerals by water during Diagenesis.

Then, the smaller remaining organic molecules inside the corpse could be volatilized or been washed out. The smaller molecules from degradation could also re-polymerized to "new" polymers not initially inside the body. Please note that these are also bi-directional. Alternatively, it went thru the Mineral Templating minerals accumulated on the surface of the organism organic components, such as pyrite on the surface of Crinoidea fossils. The changes in this

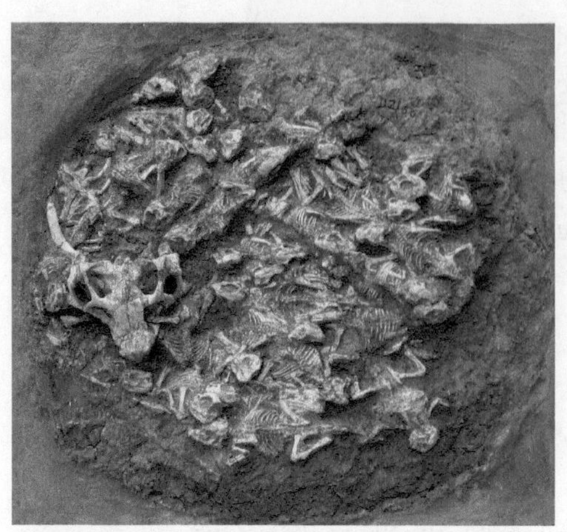

▲ 图 3.1.2 大连博物馆一窝鹦鹉嘴龙，一个成年，三十四个幼仔，怎么死的？

▲ Figure 3.1.2 A cluster of *Psittacosaurus* containing one adult and 34 hatchlings. How they died?

物模板作用(mineral templating)”，依附着有机成分表面堆积了某些矿物，如某些海百合化石表面的黄铁矿，这也是个双向道。这一段的反应称之为成岩作用(diagenesis)，生物尸体转换成化石。

再接下来透过地质现象的抬升或侵蚀的“变质作用(matamorphism)”，把埋藏在地里深处的化石带到地表或接近地表。再请注意，变质作用的两边“抬升”和“侵蚀作用”，也都是双向道，而且和上一层次的“有机挥发作用”、“有机聚合作用”和“矿物模板作用”有密切的关系。最后透过“风化作用”，终于让我们从（接近）地表发现采集到化石。

在这漫长的路途中，从出发点就是个含有很多生物化学成分的复杂生物体开始，加上上述埋藏过程中诸多的因素，都会影响到化石和紧邻围岩的化学

section are called Diagenesis, where the corpse transformed into fossil (rock).

Then, after the corpse fossilized, they could be brought to the shallower layer from the earth surface by the geological Metamorphism phenomenon of Uplift or Erosion, both are bi-directional. These actions are closely related to the Organic Volatization, Organic Polymerization, and Mineral Templating of the Diagenesis stage. After this Tectonics stage, finally, the Weathering processing provides the chance for fossil collecting on the earth surface.

During this very long journey, starting from a very complex living organism containing numerous biochemical components to go thru so many factors underground (hell) many chemical reactions and changes of chemical compositions were influenced. Do not forget that the roads inside hell are not just from top to bottom. Many bi-directional arrows are meaning they could go both ways. Furthermore, some of these arrows pointed upward and formed three loops, which make the taphonomy more complicated. Think, during these millions or even hundreds of millions of years, how many geological changes with different pressure, temperature, and water. All of them interacted with the fossil and the immediate surrounding matrix; it is very difficult to be sure what changes took place. For the PaleoChemistry study, we can not use a very simple view trying to explain the very complex natural phenomenon. Otherwise, the wrong conclusions could be the results.

3.2 Multiple Disciplines

Along with the progress of science and

成分。再者，请别以为图内的路径，只有由上往下走而已，图内有很多双向的箭头，可以从这边到那边，也可能从那边又到这边。此外，这些箭头群，也有往上走的，形成了三个回路，让整个埋藏学更为复杂。试想，在百万千万甚至几亿年漫长的时间中，曾经有多少次地质变化、不同的压力温度水质变化等等，也都不停地与化石和邻近的围岩起化学作用，也就是说，从生到死到化石，实际上是个很复杂而且往往很难确定的变化系统。我们研究古生物化学，不能一厢情愿抱着很单纯的思维来解读大自然的现象，否则就可能产生错误的结论。

3.2　多领域交叉学科

随着科技的进步，传统古生物学所使用的工具仪器，也逐渐跟着潮流在转变当中，特别是集成电路 (integrated circuit, IC) 的神速发展，带来了计算机和摄影的平民化，手机就是最佳的结合证明，一滑天下通，对于整个人类社会产生了无法想象的冲击。年轻学子们可能很难想象，父执辈时代，计算机是象牙塔里的大祭司们才有机会看得到，照相也是非常奢侈的，就连家里装有电话，都是一种社会地位的表征；但是，这一切明日黄花，也只能到博物馆去凭吊了。

可是，俗语说，天下没有白吃的午餐，我们每人每天，也都得付出现代科技的代价，包括古生物学领域，以前相对来说，照相麻烦又昂贵，所以为了形

technology, the tools and instruments used by conventional paleontology are riding the tide and changing. In particular, the fast speed of integrated circuits brought the computer and photography down to the laymen level. The smartphone in everybody's hands is the best evidence for this combination, in touch with the world with one stroke, creating an unknown impact on humanity. Young students probably have a tough time to understand during their parent generation; the computer was locked in the ivory tower reserved for the holy priests. Photography was also luxury, and even home telephone was social status. However, all of these were the yellow flowers of yesterday in the museum now.

As it says, there is no free lunch. Each of us has to pay our dues to modern technologies, including paleontology field. In the past, photography is cumbersome and expensive. Thus, for morphological description, students have to learn and practice very well linear hand drawing of a specimen, as Figure 1.2.3. Otherwise, he would fail and will not make it. However, how about now? Using the phone camera to snap a shot and further process by image processing app, it will not take long to have a good, even better than the linear hand drawing. In the past during the field works, someone is assigned to do the excavation record detailing which bone pointed to which direction, etc. However, now, just a snap of a pocket camera, a better record was done. Moreover, the unmanned flying drone can do exact areal recording, while the LIDAR (Light detection and ranging) can produce 3D tomography quickly.

This means forgoing the paper, and pencil

态描述，学生们必须学习样本的素描，而且还得学得很好，要能画出精美的形态，如第1章图1.2.3，要不然肯定过不了关、成不了大器。但是如今呢？手机相机拍一下，影像处理应用软件接着处理，不必多久，也可得到好，甚至比手会更好的成果。以前到野地工作，总要有人画下发掘的记录，哪根骨头朝什么方向等等，如今口袋相机拿出来咔嚓一下，记录更详尽，甚至用小型无人机来做精准的现场记录，更甚至用雷达 (light detection and ranging, LIDAR) 快速做出三维空间模型。

这意味着，脱离过去纸笔迈入计算机数据化的时代，谁也无可抗逆这种潮流，因此，基本功夫也有所改变，现在使用现代化科技产品，需要了解计算机与摄影的基本常识，影像没拍好、采光不对、影像处理有问题等等，那是自己没有跨到这两个领域去，不能怪老师没有教好。

如果跟上时代研究古生物学，即便是描述古生物学，现代化的仪器设备，如扫描电子显微镜 (scanning electronic microscope, SEM)、计算机断层扫描和微型计算机断层扫描 (computerized tomography, CT; micro computerized tomography, μ-CT) 等等，都是不可或缺的威力科学仪器，它们所牵涉的理论背景，更为专精，使用者总该多少了解一下跟上时代吧。相对于过去的显微切片等方法，这些新的科技仪器，能提供什么以往做不到的信息？仪器设备的极

and into the computer age is the trend nobody can avoid. Thus, old basic skills have to be changed. Using modern scientific technologies have to know computer and basic photography common sense. If bad photos, wrong lighting, problems from image processing, etc. come up, these are not the teacher's responsibility; it is you did not learn well these two fields.

If following the current pace, even for the descriptive paleontology, the modern instruments and facilities, such as Scanning Electronic Microscope (SEM), Computerized Tomography (CT), Micro-Computerized Tomography (μ-CT) are essential. The theory and background are more specialized. Shouldn't the user have some basic understanding and catch up? Contrast to the conventional thin slice, what the information the new technology can provide which was not available from the old ways? What is the limitation of these instruments and equipment? What range can they do? What can we see? What can we see NOT? … All of these demands us to have some degrees of understanding (Figure 3.2.1). For example, SEM uses electrons to bombard the specimen to produce the image. The electron has very little penetration power. Only the very surface can be seen, not three-dimensional images. Furthermore, if the specimen was under the powerful electron bombardment for too long, the surface could be burned out. When trying to look back on that high magnification spot, it likely cannot be found again. Then, in order to operate the instrument well, it will take times and practices. Moreover, how to process the image produced? How to interpret them?

Talking about the radiology department's CT and μ-CT, both of them use X-ray as the radi-

限在哪里？某项先进设备的能耐范围如何？能让我们看到什么？有什么看不到的……这些都需要有某种程度的了解（图3.2.1）。比方说，扫描电子显微镜，利用电子打击样本，产生二维影像，电子的穿透力很薄，因而受限于样本的最表面，无法提供三维影像，而且如果样本在强烈电子“轰炸”太久的话，样本表层会被烧掉，回头想再找看那个点的高倍率放大照片，可能就找不到了，因为那层样本已经被烧掉了。此外，光是操作该设备，总需要一段时间才能摸熟悉，所得到的影像结果，要如何处理，该如何解读。

就拿属于放射科的计算机断层扫描和微型计算机断层扫描来说，两者都以X射线为光源，进行对样本 *Z* 轴的 360° 环绕扫描，然后经过计算机处理，产生一堆沿着 *Z* 轴的平行 *XY* 轴切片影像。计算机断层扫描一般用于工业和医学，分辨率不太高，通常在 0.5 毫米 (mm) 左右（已经算不错的了），所产生的影像堆叠文件大小有限，一般的计算机配合适当的软件（很多免费版本），大概就应付得了。但是为了求取高分辨率的影像，使用微型计算机断层扫描，分辨率可达 1 微米 (μm) 或甚至高到几百纳米 (nm)。因此，光是如此扫描出来的影像堆叠，就非常庞大：假设用 1 微米分辨率来扫描 1 毫米的长度，就会有 1000 张、每张超过 30 Mb 的影像，光是这一小小范围的扫描文档，就会超过 30 Gb！想要处理如此庞大的数据量，计算机的计算能

ation source to do a 360° scan along the *Z* axis and produce a set of parallel *XY* plane slice images. For the medical and industrial usages, the resolution does not have to be very high, usually less than 0.5 mm, which is considered good enough. The size of the image set is not that big and can be processed by the average computer and appropriate software; many of them are in the public domain. However, to have a higher resolution by using μ-CT, which can be 1 μm or even several hundred nanometers. Thus, the size of the resulting image set can be huge. Assuming using 1 μm resolution to scan 1 mm length, it will have 1,000 images of more than 30 Mb each. The total file size will be more than 30 Gb! In order to process such massive data, some knowledge of computer power, graphics power, storage space, transmission speed, ⋯, are unavoidable. We cannot always rely on the help of computer science students.

So, the next issue is the post-processing of these large image sets to do something like make 3D models or to 'dig' out the desired portion, or to segment out a particular area with different color, ⋯ etc. This is related to another specialized field called Medical Image Processing. The trained medical image processing people may be able to teach us how to do this or to do that for such effect. However, they do not know paleontology or whatever we want, and may not have enough times to help us or to do that for us. Only we know exactly what we want. Therefore, another thing from different discipline has to be mastered. Furthermore, no matter if the software is commercially available or public domain, each package has its idiosyncrasy. The way to use it and processing capability

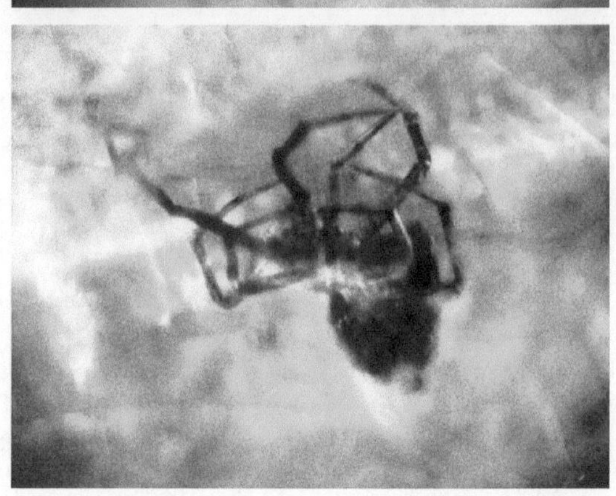

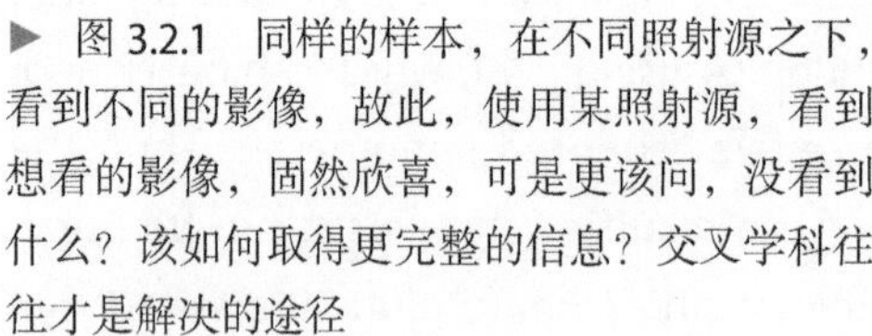

▶ 图 3.2.1　同样的样本，在不同照射源之下，看到不同的影像，故此，使用某照射源，看到想看的影像，固然欣喜，可是更该问，没看到什么？该如何取得更完整的信息？交叉学科往往才是解决的途径

▶ Figure 3.2.1　For the very same specimen under different radiation sources will yield different images. Thus, using one radiation source and seeing something shall be glad. But, the question of "What is not seen?" is more important. How to get more complete information? Multi-discipline is the answer

力、图形处理能力、储存空间、传输数度等等连带出一大堆计算机科学专业的领域，不懂行吗？总不能老是叫计算机系的学生帮忙处理啊！

接下来是这堆大影像的后制作处理，把扫描的结果做出三维的模型，或者只“挖”出所要的部位，或者将不同部位以不同颜色标示出来[称之为区段（segmentation）]等等诸多可“玩”的伎俩。这又牵扯到另外一门专业：“医学影像处理”，一般的医学影像处理专业人员，或可教你我一两招如何做这个动作得到那样的结果。但是，他们对古生物，特别是我们有兴趣的部分，不见得懂，

vary a great deal. The command to do the same thing may be different. Each one has its characteristics. Thus, in order to do your research well, you have to master several post-processing software packages. That is an unspoken homework assignment. Then, if you have to write your own plug-in, you also need to learn a computer language or two.

Besides the issue of post-processing, after the object scanned, there are other scanning scientific instruments can produce the 3D morphological structure. Each has its imaging principle. In order to have the best results we want, can we not to have some basic understanding of the background knowledge? For example,

也可能没那么多的时间来帮忙或替我们做，只有我们自己才知道我们所要的是什么。因此，又要多学一门不同领域的功课了！此外，不管是商业版本的后制作软件或是公开的免费软件，每套都有它自己的个性，处理方法与能力又不尽相同，甚至连某功能或基本名词都不尽相同，各吹各把号。因此，为了做好自己的科研课题，学会用几种不同的后制作软件，成为必须自己下功夫的家庭作业。如果碰到需要写个“外挂 (Plug-in)”的话，那么计算机编程也得懂一些。

除了上述扫描后影像后处理的议题之外，另外一些能产生三维形态构造的扫描仪器，各都有其成像的原理，为了得到我们所要的最佳结果，能够不对这些科仪的原理背景有些基本的认知吗？比方说，上述的计算机断层扫描和微型计算机断层扫描成像原理是什么？它们都使用X射线源，随着样本物质的原子序数增大，X 射线成像密度越高。因此，到医院照X射线的时候，看不到肌肉，而能看到骨头，因为肌肉的主要成分是由氢氧氮碳等比较轻的元素组成，对于X射线来说是透明的，而骨头是由比较重的磷钙等元素组成，有更高的X射线密度，因此，用计算机断层扫描或微型计算机断层扫描来做的三维扫描，也要记得这类光源是有它天生的盲点，有些东西它是无法看到的。相对来说，中子扫描，利用原子炉的中子作为照射源，而不同元素的中子成像密度，有其独特的方式。轻的元素，如氢氧氮碳等有机物主要组

what is the principle of CT/μ-CT? Both of them use X-ray as a radiation source; the X-ray density goes up along with the atomic number. Therefore, when taking an X-ray image in the hospital, muscles will not be seen, but bones are apparent. The reason is that the muscles are made of lighter elements, such as hydrogen, oxygen, nitrogen, and carbon. These elements are X-ray “transparent” and can not be seen. However, bones are made of heavier elements, such as phosphor and calcium. These elements have more X-ray density. Thus, using CT/μ-CT to do tomography, their nature blind points shall be kept in mind. Something cannot be seen by X-ray. On the other hand, using neutron as the radiation source for tomographic scanning, it is a different story. Each element has its unique neutron-image-density. Lighter elements, such as the main component of organic matter, hydrogen, oxygen, nitrogen, and carbon, have excellent image density, while aluminum is “transparent” to the neutron. Therefore, it can be said that CT/μ-CT and neutron tomography are complementary to each other. What will not be seen in this can be seen clearly on the other, and vice versa.

When I was in my Ph.D. graduate school, our professor wanted us to take a course: Glassware Workshop. Some of us students complaint. We were for our Ph.D., not for learning a handy craft. Why should we take this course? Professor said, yes, you guys are for your Ph.D. degree. Our school will provide the glassware beaker, condenser, etc., you need every day in the lab. However, along with your experiment progress, you may find a situation like this. If this glassware can be modified in a such and such way, it will

成元素，成像密度特别好，而铝元素却是透明的。因此可以概括来说，计算机断层扫描/微型计算机断层扫描和中子扫描，两者相辅相成，在这边看不到的，在另一边可以看得很清楚，反之亦然。

在美国念博士班的时候，教授指定我们修一门课程：玻璃工。有同学就抗议说，我们是来念博士学位的，不是来学手工艺的，为什么还一定要学这玩意？老师回答说：没错，各位同学都是博士生，你们做实验所会用到的玻璃器皿，烧杯试管冷却管等等，学校都会提供现有的，不过随着你们的实验进展，往往你们会发现，如果某个玻璃器皿能稍微改一下，如多个接口，或角度往上斜一点点，你的实验可以方便一些，可以快一点做出成果来；没错，你可以把想要的修改图形画出来交给库房，他们会到玻璃工厂帮你订制，不过，这一来回，可能需要等好几个月，如果还要再修改，又要等好几个月，你们的学位，能如此耽搁吗？因此，这门“玻璃工”，看来和博士课程无关，但是却是你们每个人都必须，而且需要把此功夫练好！科技是什么？科学的进步，需要有先进的技术支撑，两者相辅相成，没有望远镜、显微镜等技术，就看不到远处的星星和细胞。科学家有很多时间，必须花在科技仪器改进方面，做出自己最适用的工具！老是靠着买用现成的科仪，不只是学校的经济负担，往往还会牵扯到公家预算等等，最关键的问题在于，现成的科研设备，就像买成衣，总会有袖子太长裤

be easier to do the job. Of course, you can draw the modification and give to the stock clerk. They will send your request to the glass factory and made one for you. However, this may take months before you get it. If further modification is needed, another several month. Can your degree be delayed by things like this? So, the glassware workshop course may not look like directly associate with your Ph.D. courses, but it is a skill that each one of you has to master. What are science and technology? The advance of science needs technology to support. They go hand-in-hand. Without the technology of telescope, far away stars cannot be found. Scientists shall spend a great deal of time to improve and make their own tools. Depending on the commercially available instruments and equipment are not only a financial burden of school, but also like the mass-produced clothes. Maybe the sleeves are too long, the pants are too tight. The custom-made clothes fit you the best and most comfortable. So, go to the glassware workshop and start to practice your skill. At the end of this term, I want you guys to turn in what you did. Besides the glassware workshop, our department also has a mechanical workshop.

One similar situation: When I was near the end of my master graduate school, I applied for my Ph.D. graduate study. In one morning, I dropped my scholarship acceptance letter to Ohio State University into the mailbox. When I came back to my room, around 9 p.m., I got a telephone call from Chicago, which is 3 hours ahead of Pacific Time Zone. It was midnight in Chicago. The Dean of the graduate school called informing me that they decided to give me six years of scholarship for two degrees. When I

管太短等疑难杂症，相对来说，你自己量身定制的衣服，穿起来最合身舒服。现在就到玻璃工实验室去练功吧！学期终了，把你们的成果交上来，顺便带一句，系所除了玻璃间之外，还有机械车间，学生可以自己去车个铁工接环等。

类似如此的情况，还有一个例子：当快念完硕士班申请博士班研究所的时候，那天早上我刚把接受俄亥俄大学给我奖学金的回函丢进邮筒，晚上回到宿舍九点多了，接到时差三个小时芝加哥（半夜十二点多）那边某研究所所长亲自来电，说他们决定给我六年双学位的奖学金，毕业时同时取得化学和医学博士学位。我有点好奇，别校的博士班都只是四年一个学位，怎么这家要六年给两个学位，所长解释说，我校医学院这个课程博士班研究所，不收医学本科系的硕士学士生，全部招收其他科系背景的学生，如化学、物理、电子等等。原因很简单，他说，医生用药，有时候会发现，某药的副作用很不好，一般学医者，碰到这种情况，只好依赖药厂去做化学药品结构式的修改，那一来回，可能要十几年，病人不知已经死掉多少人了。但是学化学的，就有能力修改某化学构造，很快解决当下的问题。他继续说，他们招收学电子的学生，也是同样的道理，某种医疗仪器设备，得做某些修改，传统医学的医生，无法做到，但是懂得电子的医生，可能很快地修改一下某个电路，问题就解决了！

上面两个我个人实际经历的例子，

graduate, I would have both a Ph.D. in chemistry and medicine. I was curious. All other graduate schools have four year Ph.D. program and for just one degree. Why this school has six years and two degrees? The Dean explained that they would only accept new graduate students from others than the master degree of medical school. All new graduates were from a different background: chemistry, physics, electronic, … etc., for a simple reason. He said when the medical doctor was doing their job, it is often to encounter a situation like a given medicine may cause some adverse side effects. Usually, the doctor had to rely on the pharmaceutical company to do the chemical structure modification. This round around would take more than ten years. Many patients would die during this waiting period. However, for the chemistry trained doctor, this chemical structural modification can be done quickly to resolve the problem on hand. He continued to say the same principle for them to accept new graduate students of electronic major. When a medical instrument needs some circuitry modification, the conventional doctor will not be able to do it. However, a medical doctor with an electronic background repairs the circuitry quickly to resolve the problem.

These two personal experiences of mine are examples of multi-discipline. The American university already did this half-century ago. In their higher education to train new scholars, they do not just teach the students one thing only. They also taught them that during the pursuit of their study, if specific tools are needed, try to make the tools for themselves. Only you know exactly what your needs are. If there is no available tool,

就是跨领域交叉学科的典范，人家美国大约半世纪以前就如此搞的，在他们高等教育的管道中，训练一个某专业的学者，绝对不只给他一把刷子，还教导学生，科研过程中碰到自己需要的工具，自己想办法自己做！只有你自己才最了解你自己的需求，没有现成的仪器设备，自己想办法做一台，需要学习其他学科，就去好好学习！

临场考题：如果想看琥珀内的昆虫，该用什么设备？

3.3 在原位同点多重检测

在做样本各种探讨实验过程中，难免要先作样本的处理，例如，传统的形态描述古生物学，没有把化石清理，看不到某化石的原貌，如何进行呢？但是，样本的处理，需要花时间，很长的时间，甚至在立体显微镜下干好长一段时间的苦工。虽然现在可透过计算机断层扫描(CT)来协助进行“数位清理”，可是从费用的角度来说，不是每个样本都花得起扫描的费用，即便如此，有些样本对于某分析方法来说，还是需要进行某种程度的样本准备。

前面提过，即便煎个荷包蛋，其实也是因热而发生了化学反应，若拿煎好的荷包蛋来做生鸡蛋的分析，肯定不是鸡蛋的原汁原味成分，而是经过化学反应之后的结果。也就说，在样本准备过程中，每一道样本处理的过程，必须考虑一个严肃的问题：处理的过程中，会

then, make one for yourself. If a new subject had to be learned, then learn it.

Quiz: What facility will be needed to see the insect inside an amber?

3.3 *In Situ in Loco* Multiple Examinations

It is unavoidable before various experiments the specimen has to be pre-processed. For example, in the conventional morphological description paleontology, without removing the matrix and have proper preparation, how can the study go further? Sample preparation takes times, very very long time, and perhaps have to be done under a dissecting microscope for a long labor. As now, the CT can be used to conduct "digital preparation" to help out. But not every specimen can afford to spend the money for a CT scan. Even if it is scanned, some preparations may still be needed.

As mentioned before, even as simple as frying an egg, there are chemical reactions that took place by heat. If a fried egg is used for the analysis of raw egg composition, indeed, the results are not that of the original, but the chemically changed state. In other words, each step or action was taken must consider a serious question: During the preparation, did any chemical change/reaction took place, which can cause the wrong result?

In dinosaur study in the late 90 of last century, a shocking discovery was made by Mary Schweitzer team of Montana State University. They found soft tissues and heme from *T. rex* femur of 70 million years old. It was an astonishing discovery. Because in general believe that

不会把化石破坏掉？有没有进行了没有想到的化学变化，甚至导致后续的分析得到错误的结果？

在恐龙学方面，二十世纪九十年代末期曾经有个很震惊的发现，当时还在美国蒙大拿州立大学的Mary Schweitzer在七千万年前霸王龙的大腿骨内找到软组织、血红素 (Heme) 等，造成了轰动，因为从一般的观念和想法，有机物质不可能保存这么久，化石已经都变成无机物了，怎么还可能保存着有机质的软组织等呢？她后续发表的论文接着说，骨头化石内能保存有机质的软组织，是因为针铁矿 (goethite) 扮演了保护机制的催化剂 (catalyst) 等等云云。

可是，回头检验她团队所使用的样本准备方法，使用去钙法先把骨头内的磷灰石群和其他矿物以乙二胺四乙酸 (EDTA) 溶解掉取得胶原蛋白等软组织，如此一来，原本的有机软组织，已经离开了原本位置，也就是说，她们所观测到的有机软组织，无法确定是从骨头里面哪个位置来的，来源位置无法确定，总是个遗憾，此为缺失之一。再者，因为所采取的先导样本处理方法，使用去钙法，将骨头内硬质的磷灰石群矿物溶解，在此同时也“顺便”夹带着某些她们没注意到的化学反应，将赤铁矿 (hematite) 水解成针铁矿 (goethite)，因而导致把冯京当马凉的结论，此为缺失之二。

扫描电子显微镜 (scanning electron microscope, SEM) 和透射电子显微镜 (transmission electronic microscope, TEM)

organic matter cannot be preserved for such a long time. Fossils are ancient life changed to inorganic rock. How can organic soft tissue be preserved? In their follow up papers, they stated that goethite is the catalyst for the preservation.

Her team's preparation method was to dissolve the apatites and other minerals by Ethylenediaminetetraacetic acid (EDTA) to get the collagenous soft tissue. By doing so, all the organic soft tissues are no longer *in situ*. In other words, the exact location of the observed soft tissues cannot be specified where they came from. This is one unfortunate issue. Furthermore, since they use decalcification preparation to dissolve the apatite minerals, some chemical reactions took place unnoticed, hydrolyzed hematite into goethite. This caused the wrong conclusion. This is the second issue.

Scanning Electronic Microscope and Transmission Electronic Microscope (TEM) are two very important research instruments for many scientific fields. The magnification they can provide is far far more than the conventional optical microscopes, and can even reach nanoscale. Wonderful! Disregarding to the thin slicing sample preparation and the gold coating aside, the specimen now can be examined directly in the chamber. Both of these use high energy electrons to bombard the surface of specimens to produce images. Thus, the consideration is that under such a high energy electron bombardment, how long the specimen can stand without got burn out? If the high energy electron beam stops on a given spot too long, that area will burn out. When coming back, perhaps no image can be seen.

For the Transmission Electronic Microscope

可说是现代生物学与相关学科的重要科研利器，它们让研究者看到放大倍率远远高于普通光学显微镜的生物构造，甚至可达纳米的分辨率，真棒！不过，先不用说以前的检测，必须先把样本镀上一层金，以便导电，新型的这类设备，已经可以不必如此“破坏”样本，直接放进电子显微镜的检测腔就可以了。这两种电子显微镜都是利用高能量电子打击样本产生影像的，因此，必须考虑的是，在高能量电子的轰炸下，样本能耐多久而没被烧掉？如果让高能电子束停在某点过久，该处就会被烧灼，回过头来看，可能就看不到想要看的影像了。

透射电子显微镜，因为电子的穿透力很弱，导致必须把样本切成纳米级的厚度，在实际检测样本制备过程，这是一个很大的挑战，如何制作出可用的试片，是一门很大的功夫和技巧，很多科研人员做不出成果，就是因为做不出可用的试片，被卡在最基本的出发点，后续什么都不用说了，不是吗？好多博士生毕不了业，不就是卡在这一关吗？因此，各种尖端科技的切片设备，成为必备的科研设备。这其中有一个很厉害的家伙，称之为“聚焦离子束 (focus ion beam, FIB)”，它是一种样品制备工具，可以准确地制造出样品的横截面，聚焦离子束为透射电子显微镜样品提供革命化的样品制备方法，可以辨别亚微米 (sub-micron) 级的特点，以及精确地制作横截面。聚焦离子束制备样品被广泛使用在扫描电子显微镜，其中聚焦离子

(TEM), since the electron penetration power is very weak. The sample must be sliced down to the nanoscale. In reality, for the sample preparation, this is a great challenge. How to make such thin, suitable slice requires excellent skill and technic. Many types of researchers got hung on this issue from the beginning, just because they can not make a good thin slice. Thus, various thin slice preparation equipment was developed. One of which is called Focus Ion Beam (FIB). It is an excellent sample thin slice preparation tool. It uses the focus ion beam to precisely cut thru the cross plane. TEM provided a revolutionary sample preparation method and can do the job in ~μm scale precisely. So FIB thin slices are widely used by TEM. Some instrument even puts these two together into one machine.

Theoretically speaking, the FIB can make a very thin and usable slice. Real good idea. However, let us calmly think about this. For the living sample under such a powerful bombardment of an ion beam, could it preserve the morphology of the sample without the contents were cooked? Or their chemical composition changed? Do not forget the kitchen chemistry of frying an egg!

For the same reason, under the ordinary condition, when big strong power tools, such as laser cutter, are needed to remove the matrix, will the heat generated cook the chemical reaction? Thinking from the PaleoChemistry point of view, this is one thing that must be considered. Furthermore, the thin slicing of fossil material, it's common to use some glues, such as epoxy to encase the specimen. Then cut to a specific thickness with a diamond saw. Then transfer and glue that to a suitable carrier for the final

束制样，扫描电子显微镜成像和元素分析，可以发生在同一个多技术的机台中。

从理论和实用上来说，使用聚焦离子束设备来做试片，可以做到很薄可用的试片，非常理想啊！不过，稍微冷静地想一想，活体生物样本，在如此强大能量聚焦离子束的轰炸下，会不会在整个过程中，虽然保存了样本部分的形态，但是已经把样本内容（成分）烧焦了？或改变其化学成分，甚至细胞形态呢？请别忘记，煎个荷包蛋，也是（厨房）化学反应！

同样的道理，在一般的情况下，如果围岩需要动用大力的动力工具来切割清除，如激光切割，工具作用所产生的热量，是否也会产生煎荷包蛋的化学反应？从古生物化学的角度思考，就必须注意。再说，先进的古生物化石切片，通常先把化石以某种塑胶胶水（如环氧树脂）包裹起来，以便利用钻石锯片切成某种厚度，然后转到适当的载体（又用胶水黏合在载体），进行最后打薄到所需要的厚度，通常少于30微米，以便透光在透射显微镜下观看，在整个切割研磨过程中，使用水（包括蒸馏水）为冷却剂。

然而就以我们探索古生物化学的角度来说，将样本用塑胶胶水包裹起来，那就是个大“诺诺（No, No）”，很有可能胶水会渗透进入骨头里面，污染了样本，干扰检测结果，当然要想办法一定避免，改用如手工切割研磨等等高技巧性的苦功夫，期望能得到仪器所需要的

grinding to the desired thickness, usually less than 30 μm so that it can be observed under a transmission microscope. During the cutting and grinding, usually water (including distilled water) is used as the coolant.

From the PaleoChemistry point of view to see the encasing of the specimen in plastic glue is a big NO, NO. The glue may penetrate the bone and contaminate the specimen and interfere with the test result. This must be avoided by all means, such as do this totally by skillful hands to the desired thickness. For a real example, in our study of the dinosaur bones of almost 200 million years old for organic remains and collagen, we know that in order to prevent the contamination and interference from glue, no glue was used for the thin slice preparation. However, the results from sr-FTIR were very puzzling. Some showed up; some will not. After many reviews, we found it could be caused by wrong directional cross-cutting. When the bone was cut transversally, the size of the vascular canal areas is tiny. Thus, it was difficult to see the organic remains inside the vascular canal. We learned a lesson. So, we changed to the longitudinal cut. The spectra result improved, but still not ideal. Why? Then, we realized that during the thinning process, we used purified distilled water as the coolant as standard practice. We did not think that collagen is water soluble! So, along with cool water grinding, collagen was washed out. That caused irregular results. So, the coolant was changed to absolute alcohol (collagen will not dissolve in alcohol), and the annoyance vanished.

By carefully considering the details of each sample preparation step can avoid many un-

厚度。举个实际的例子来说，在我们团队研究二亿年前恐龙化石内有机残留物（第一类胶原蛋白）的样本制备过程中，事先知道为了要检测有机物，完全不用任何胶水包裹和黏着。可是，从同步辐射傅里叶红外光谱检测的结果，很让人困惑，一下子好像有，一下子又好像没有，经过反复分析，有可能是样本切割方向（横切）不对所造成的，因为在横切情况下，切过硬骨头里面的脉管渠道(vascular canal) 截面积很小，就很不容易看到脉管渠道内的有机残留物，因而学了一堂课程，改为纵切。可是改成纵切割方向，效果虽然有改善但还不是很理想，到底怎么回事？再度检讨之后才发现，在打薄研磨过程中，照传统的方法使用蒸馏水为钻石磨磐的冷却剂，却没想到胶原蛋白会溶解到水中！随着研磨同时也把胶原蛋白溶解掉，一面磨一面溶解掉，难怪检测结果不稳定要看运气！后来改用无水乙醇当冷却剂，胶原蛋白不溶解于乙醇，问题才解决。

透作仔细思考检讨样本制备的每一个细节，避免可能没想到的差错，确认没有化学污染，才可以得到所要的检测试片，原汁原味在原位的样本试片！就如前面所说的，我们团队所采用的方法，不会有也不可能如Schweitzer那样，把骨头脉管渠道内赤铁矿水解成针铁矿，导致错误结论的窘状。我们可以大声地说，我们的方法才真正是“在原位”，我们在二亿年前恐龙骨头内找到了被保存的第一类胶原蛋白（图 3.3.1）！

expected jumping hammers. Make sure no chemical contamination to ensure the desired specimen samples are in the original pristine state! As just mentioned, our preparation method would not and could not hydrolyze the hematite to goethite as the wrong conclusion Schweitzer did. We can say very loudly that our method is real *in situ*. We found the preserved collagen type I in the almost 200 million years old dinosaur bone. See Figure 3.3.1.

Please look back at Figure 1.1.7. This is an overlapped image of not one but four different radiation sources. It has very significant meaning. In the future, such as 2D Raman image and X-ray fluorescence, etc. can be added. What the proposed here, *in situ in loco* multiple examinations, will have a huge impact for paleontology or PaleoChemistry.

Why?

From the point of view of fossil composition to say, fossils are very very heterogeneous material. Two adjacent examination spots, which maybe just several μm apart, can have a very different chemical composition, which causes different observation result. Making microscopy slice is a common practice in paleontology. However, the thickness of the saw could be several hundred μm. Thus, two next to each other thin slices could be from several hundred μm apart. Internal and composition could be very different. They are two different specimens from the same sample. The critics could argue that different specimens were used, not the same thing.

This is one of the major problems with Mary Schweitzer's discovery of organic remains in a *T. rex* femur. Her method used decalcification to

请回头看第1章图1.1.7，能够在同一个样本同样位置，取得四种不同照射源的结果，有很重大的含义。来日，可再把二维拉曼光谱、X射线荧光等等加入测试，我所提出的这个在原位同位多重检测 (*in situ in loco* multiple examinations)，对于古生物学或说古生物化学，会有重大的影响。

原因何在？

从化石的成分角度来说，化石的成分非常不均匀，相邻两个检测点，可能只有几个微米的距离，化学成分就可能有很大的差异，导致观测结果的不同。把化石做成显微切片，光是锯片本身的厚度，至少就几百微米了，所以即便是相邻的两个试片，实际上的距离，可能就相距了好几百微米，内部结构成分当然很有可能有很大的差异，因为本来就是同一个样本的两个不同样本啊！批判者可以以此来说，你的样本不同，怎可说成同样的样本呢？

这也是Schweitzer最早发现霸王龙大腿骨内保存有机残留物的最大致命伤，她的实验做法，将骨头去钙溶解，导致她认为针铁矿是有机物保存的催化

remove the apatite bone and cause her to claim that goethite is the catalyst for organic preservation. In fact, in our paper, we showed it is the hematite *in situ*.

So, using this '*in situ in loco* multiple examinations,' I present here is not only nondestructive but also on the exact same specimen at the very same location to obtain multiple data pointing to the same conclusion. No room for critics. The specimen can be re-examined by such critics.

The illustration of Figure 3.3.2 is a small initial result. It overlapped four different examination images of (from top to bottom) IR, MHGM, microscopy, and neutron tomography. This very early and rough start proves that we can look at the exact same spot of the same fossil by multiple different methods to gain multiple data sets for further studies.

In the future, many nondestructive examination images can be overlapped, such as XRF (element distribution), μ-CT, Raman-CT, Laser Stimulated Fluorescence, etc. By doing so, we can combine multiple information of the very same spot of the very same specimens. This is a very compelling research methodology we can

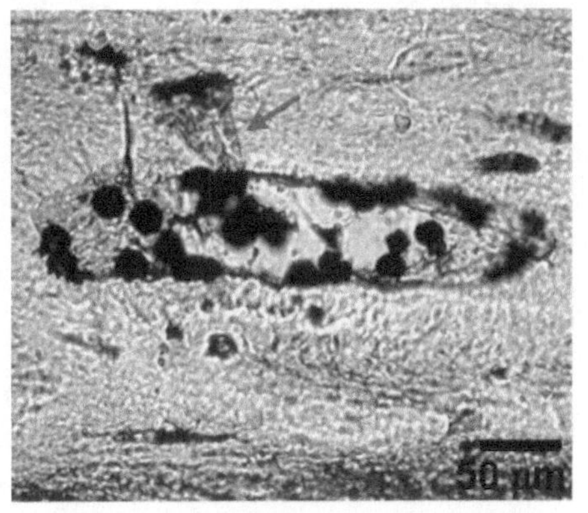

▶ 图 3.3.1　恐龙骨头纵切面。影像中压扁椭圆形，就是刚好切过脉管渠道的地方，其上左方红色箭头所指有一块三角形的东西，那就是胶原蛋白，而黑色的原点，就是赤铁矿

▶ Figure 3.3.1　Longitudinal cut of dinosaur bone. The squashed oval is the place cutting thru a vascular channel. The triangular thing pointed by the red arrow is the collagen. The black dots are hematite

剂，实际上是赤铁矿（我们在原位论文证明了这一点）。

话说回来，用我这套“在原位同点多重检测”，不只是非破坏性，而且就在同样同位的样本，取得多重数据，多从数据指着同样的结论，别人没得啰唆，某人不信的话，还可拿这个样本自己去做。

图 3.3.2 这张示意图，表示一点最初的小成果，把同一个样本同点的四种影像（四种影像检测方法）重叠在一起，从上到下为：红外线、多频显微、普通光学、中子扫描。这个粗浅的起步，表示我们可以在同一个样本的相同位置，使用多种不同的检测方式，取得同一个位置的多种信息，以供研究分析。

将来还可以把好多种非破坏性的检测影像叠加进来，如X射线荧光（看元素分布）、微型计算机断层扫描、拉曼断层扫描、

work hard on.

It's not paranoid, but we now can do and see 3D chemical composition distribution inside fossils.

Please look at Figure 3.3.3. This is a 3D chemical composition distribution image obtained from a Raman-CT scan. The blue is the mineral apatites (represented by phosphate peaks), green is the carbonates (represented by carbonate peaks), and red is the collagen type I (represented by Amide III peaks). From this image at least two important and exciting things can be seen: ① As known, apatites minerals were formed on the surface of collagen type I. This image clearly show *in situ in loco* formation of apatites during the embryonic stage. Furthermore, there are plenty of red collagen areas with blue apatites attached, showing high percentage primary tubular cavity. ② The green carbonates did not

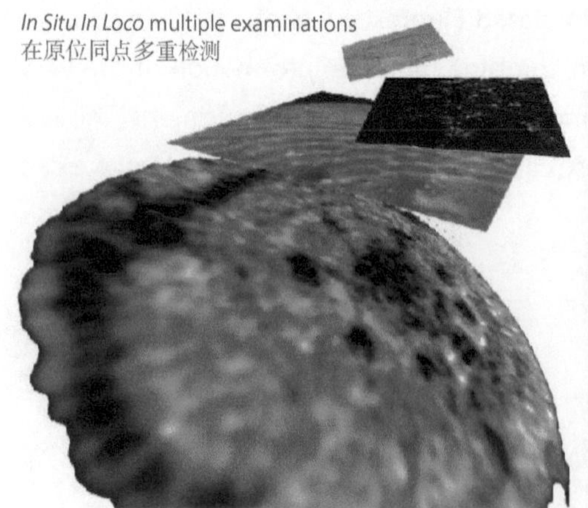

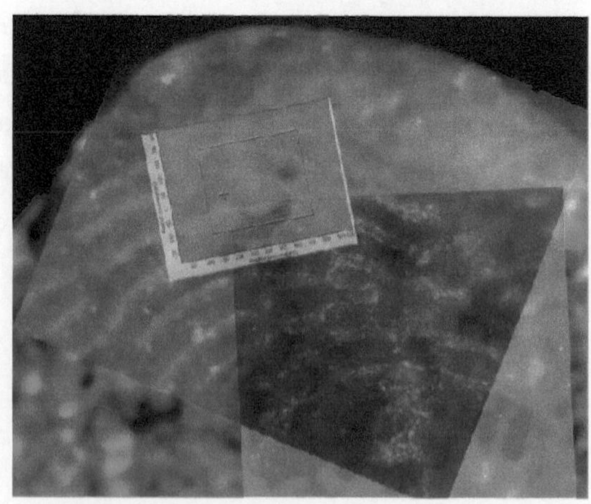

▲ 图 3.3.2　在原位同点多重检测示意与结果图。所检测样本的相片在左边示意图左下角，显示四种（由上到下：红外线、多频显微、普通光、中子扫描）非破坏性检测影像重叠，右边影像显示完美的重叠

▲ Figure 3.3.2　Illustration and result of *In situ In loco* multiple examinations. The specimen is at the lower left of the left illustration showing (from top to bottom) nondestructive IR, MHGM, microscopy, and neutron tomography overlapped images. The perfect matched result is shown at right

激光荧光……如此一来，我们可以综合同一块化石同一个点的这些多重资料，取得更完整的化石信息，此法是个非常有威力的研究方法，值得我们努力。

不是妄想，而是已经可做到并且看到化石内的三维化学成分分布。

化学形态学请看图 3.3.3，这是一张禄丰龙胚胎骨头的拉曼断层扫描 (Raman-CT) 做出来的骨头化石内化学成分的三维排列分布影像，此中蓝色的是化石骨头内磷灰石群（以磷酸根波峰代表），绿色的是碳酸盐（以碳酸根波峰代表），红色的是第一类胶原蛋白（以酰胺Ⅲ波峰代表）。从图片中可以清楚看到至少两件有趣而且重要的事：①依照已知，骨头内磷灰石群矿物的产出，是依附在第一类胶原蛋白表面，此图示很清楚地在原位在同点显示在胚胎阶段磷

penetrate too deeply, just about half way or less in this image. This agrees with the Carbonate Cementing in Chapter 7 Section 2. Carbonates are the main ingredient of cement. They formed an encasing surrounding layer outside the fossil without going too deep. We saw the same phenomenon in the dinosaur tooth fossil. Thus, is any relationship between this Carbonate Cementing phenomenon and the preservation of organic remains? Further investigations shall be conducted.

3.4 Shall Exhaust all Possible Nondestructive Methods before any Destructive Analyses

In the chemistry experiments, if failed one time, then take more reagents, adjust the ex-

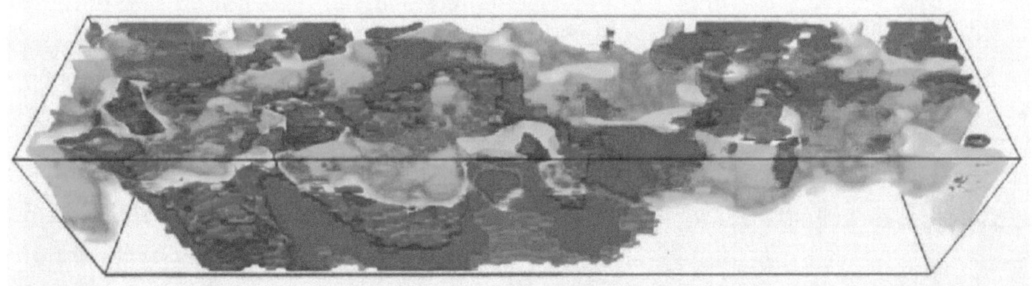

▲ 图 3.3.3　禄丰龙胚胎骨头三维拉曼扫描，显示骨头化石内化学成分的分布排列。蓝色的是磷灰石群矿物，绿色的是碳酸盐，红色的是胶原蛋白。这应该是世界首创化石内化学成分三维分布图

▲ Figure 3.3.3　3D Raman scan of a *Lufengosaurus* embryonic bone showing the chemical distribution inside the fossilized bone. Apatites are shown in blue, carbonates in green, and collagen in red. This is the first historical record of seeing 3D chemical composition distribution inside a fossil

灰石群的产生，也可以看到在胚胎阶段，原始管状空间比例相当高，很多红色的第一类胶原蛋白区域，附着着蓝色表示的磷灰石群矿物；②绿色的碳酸盐，并未深入太深，到了大约从表面深入不到一半深度的地方，就没有了，这和本书第7章7.2“碳酸盐胶合”所言吻合，碳酸盐（水泥主要成分）形成了一个胶合层，并没有非常的深入化石的内部，这种现象，我们在恐龙牙齿化石内，也观察到同样的情况，因此，此碳酸盐胶合现象，是否和有机残留物的保存有密切关系呢？很值得进一步深入探讨。

3.4 先进行所有可能非破坏性方法，不得已才用破坏性

相对于通常一般化学实验，如果有一次没做好，再拿些试剂重新调配、调整某些实验条件……往往还可以重做。不过，从古生物学的角度来说，它本身就不是“实验室科学 (Laboratory Science)”，它受到诸多的天然条件限制，老天爷给什么化石、给多少、品质如何……都不是我们可以控制挑选的。俗语说：“乞丐没有挑食的权利”，如果幸运一些，样本较多，进行破坏性的检测，或许有机会重来，然而也经常会碰到稀有、罕见、珍贵的样本，甚至是全世界唯一的样本，怎容得你我拿来切割研磨呢？在有些情况下，人家让你我来研究该样本，就算是很大的荣幸恩惠了，不是吗？

periment parameters, … etc., and you can re-do the experiment. On the contrast, from the point of view of paleontology, it is not a “Laboratory Science” and bond by many natural limitations. Whatever, how many, and the quality of specimen, … etc., provided to us are not in our control and for our pick. Bagger cannot be the chooser. If lucky, we may have more specimens so that we can conduct destructive analyses, and that may allow us to repeat a given analysis. However, its often we face rare and precious specimen, and even could be the only one in the world. Will it be possible for us to cut and grind such specimens? In most case, having a chance to study them shall be considered as a great honor and favor, isn’t it?

Therefore, when dealing with precious, limited number, or not so well preserved fossil specimen, before any experiment even starts, careful consideration on the destruction of the specimen must be given for each of the examinations. Any destructive analysis cannot be repeated if failed in the early stage. And other people can not repeat your experiment because the sample was used up. From the scientific point of view, not be able to repeat the experiment in other lab is not a convincing way. The basic scientific requirement is to be the same everywhere. It could not be only my lab can do, but not others for verification.

As a real example, dinosaur embryos, no matter what age they were. Perhaps more from late Cretaceous age. Only two known from Jurassic Period: ① *Massospondylus* from South Africa, and ② *Lufengosaurus* from LuFeng, Yunnan. The South African embryo was a cluster of eggs and embryo couldn’t be

因此，在面对珍贵、数量有限、保存（石化）条件不是很好的化石样本，从事任何科研课题之前，必须谨慎小心考量到某种检测对于化石的破坏性，因为任何破坏性的检测，如果实验没做好，无法重来，别人想要重做你的实验，也已经没有样本可用了。这一点，从科学的角度来说，无法在其他实验室重复验证（你的）检验结果，总是无法让别人心服口服，科学的基本要求之一，就是放诸四海皆准，不能只有在我实验室可以做到，而别的实验室无法复证。

举个实际的例子来说，恐龙胚胎化石，不管是哪个年代的，晚白垩纪的或许多一些，侏罗纪的恐龙胚胎，到目前为止，只有两个案例：①南非的大椎龙(*Massospondylus*)和②云南禄丰的禄丰龙(*Lufengosaurus*)。南非出土的样本在整窝恐龙蛋当中，从外表看不出来哪个是好蛋里面有恐龙胚胎化石，哪个是坏蛋里面没有恐龙胚胎的空包蛋，好好的整窝恐龙蛋，总不能把每个蛋都打开来看啊！因此，首先该做的就是先用计算机断层扫描，数位透视检查整窝样本，把好蛋找出来，再把那个有胚胎的蛋仔细小心地清理出来，显示出必要的形态结构，进行必要的系统和形态描述等等，如此珍贵罕见的样本，绝对不可能让你取下其中的骨头来进行破坏性的切割研磨；相对于南非的早侏罗纪恐龙胚胎完整样本，云南禄丰的禄丰龙胚胎样本，到目前为止，仅有我在 2003 年无意间捡到的那个样本，算是绞合 (articulated) 比较完

seen from outside. Before CT scan, it was no way to tell which one is a good egg and which is a bad egg. Once the embryo containing egg was identified by CT, then careful preparation of removing the top layers and finally the whole embryo was shown for morphological study and systematic descriptions. With such rare and precious specimens, it is impossible to remove any bone for any destructive analysis. On contrary to the South African embryo specimens, except the one articulated specimen I accidentally picked up in 2003 in Yun-Nan, all the rests are disarticulated embryonic bones (Figure 1.1.3), and we have many such bones. Thus, this allows us to be a little bit of 'luxury' to do some cutting and grinding to look into the bone. Even so, each bone being cut, one less in inventory. Thus, we have to choose very carefully with 'too much' wasting to prevent no rice for cooking in the future.

Luckily, along with the progress of science and technology, many non-destructive examination methods are available now. Besides such as μ-CT, Multiple Harmonic Generation Microscopy (MHGM), Neutron Tomography, and Laser Stimulated Fluorescence for morphological observations, on the chemical analysis side, there are several non-destructive examination instruments and methods, such as Raman-CT, LA-ICP-MS, Small Angle Neutron Scattering, XRF, sr-XRF, MS-DESI, and even LSFS and MHGM can be used for non-destructive chemical examinations to explore the elements of the chemical composition inside the fossil.

Of course, the term "non-destructive" is a relative term. People can pick up bones from

整的之外，后续所找到的诸多样本（第1章图1.1.3），都是非绞合 (disarticulated) 的零星死骨头，数量上也相对较多；因此，容许我们“比较奢侈”地进行切割研磨看进骨头内，不过，即便如此，每切掉一根，库存就少一根，也逼得我们必须做很谨慎地挑选，不容许“太浪费”，否则日后会碰上无米炊的窘状。

幸好，随着科技的进步，如今有很多非破坏性的检测方法，除了用微型计算机断层扫描、多频显微、中子体层成像 (neutron tomography, NT)、激光荧光等做细微的形态观测，在化学成分方面，也有一些可用的非破坏性检测方法和仪器，如拉曼断层扫描 (Raman-CT)、激光剥蚀电感耦合等离子体质谱 (LA-ICP-MS)、小角度中子散射 (small angle neutron scattering, SANS)、X射线荧光/同步辐射X射线荧光 (XRF, sr-XRF)、质谱解吸电喷雾电离 (DESI-MS)，甚至激光荧光光谱 (LSFS) 和多频显微 (MHGM) 等武器，都可拿来进行非破坏性的化学检测，探讨化石的成分组成等。

当然，在所谓的“非破坏性”这个旗帜之下，喜欢辩论者，总可在鸡蛋里挑骨头，找出理由或借口，甚至证明“完全没有破坏性”实际上是不可能的，好吧！从宇宙万物不停改变的角度来说，只要有观测者，就会影响到观测的结果。不过，也不必那么死脑筋，我们搞古生物学者，设计科研检测时，只要抓住一个原则就好了：能先进行所有可能非破坏性方法，就先用非破坏性方法，也该穷尽所有可用的

chicken eggs, and even prove that "completely non-destructive' is an impossible statement. Yes, from the universal point of view, everything changes all the time. As long as observer exists, the measurement will be influenced. Ha, we don't have to be that. For us as the paleontologists, when we design our examination, keep one principle in mind is enough: Do whatever possible non-destructively and exhaust all the non-destructive methods first, Then, when it is absolutely necessary, try the least destruction to the specimen. Figure 3.4.1.

3.5 Elevating Paleontology to a New Level, as well as Open up a New Field of Chemistry

Thru the combination of conventionally advanced paleontology with PaleoChemistry enable us to look at the details clearly not only from the inside but also outside the fossil and surrounding matrix to obtain more complete information. To paleontology, this is a systematic quantum jump to a new domain. At the same time, it opens up a new field of chemistry. By adding the efforts from the chemistry side, we will understand more mysteries of the life evolution of this Earth. This is an excellent win-win situation and is we what shall do for science advancement. Be able to understand from the most ground state up, and its power and influence are unlimited.

Therefore some suggestions for paleontology and chemistry major: This is an epoch-making historical good time. Do not hold on your

▶ 图 3.4.1 非破坏性看到石头里面去：同样的样本，上方是普通照片，下方是中子扫描，以非破坏性的检测，看到整块样本表面和埋在围岩里面肉眼看不到的骨头

▶ Figure 3.4.1 Non-destructively Look into the Bone. The top is a regular photo, and the bottom is a neutron scan of the same specimen. Neutron scan shows the surface and internal not eye visible bones inside the matrix

非破坏性方法之后，万不得已才用破坏性，在使用破坏性方法时，也力求对样本产生最少的破坏（图 3.4.1)。

3.5 提升古生物学到新水平，打开化学新领域

透过将古生物化学结合先进（传统）古生物学，把化石从里到外，包括邻近的围岩，全都看透透，取得更完整的整套资料，对于古生物学来说会是一个巨大的前进，将古生物学提升到新的境界。在此同时，也给化学打开了一片天地，透过化学的努力，让人们更彻底地了解地球生命演化的诸多奥秘。这是个双赢的好局面，也是我们所该追求的科学进

stereotype thought of interlaced like a mountain, walk out your self-set comfort boundary. Nobody was born of knowing it all. We all learn from mistakes and gain expertise. Right? So, to those trained paleontologists, go back to the basic chemistry you learned before. Explore instrumental spectroscopic analyses. Chemistry is not that horrible! To the trained chemists, those dead bones are not that scary as you thought. They are amusing to play with various fossils. Bring back some basic biology you learned in high school (or whenever). Meanwhile, enhance your basic knowledge about fossils. Suggestion for both: besides your own specialized training, walk into the other domains, not just one or two, but as many as you can, such as the post-processing of the scan images. In short, get out bravely from your own snail shell, and

展，能从最基层多来了解，它的威力和影响，无法限量。

因此，给予古生物专业者和化学专业者一些建议：这是一个划时代撰写历史的好时间，不要抱着“隔行如隔山”的陈腐思想，勇敢地跨出自我设限范围，没有任何人天生下来就是什么都懂的，我们都是靠从错误中学习而有所专精的，不是吗？所以，对于古生物专业者来说，回头去复习一下过去学过的一些基本化学，探索一下仪器光谱分析，化学没有那么恐怖难搞啦！对于学化学者来说，以前学过的生物学，也可拿来温习温习，还要加强一下古生物方面的基本常识，那些死骨头化石，没有你想象中那么恐怖，其实，它们蛮有趣好玩的。给予这两方面专业者共同的建议，在自己专精的本行之外，除了把脚步跨进另外的天地，还应该多跨一些领域，如计算机断层扫描之后所需要的影像处理等等。总之一句话，快快从你自己的蜗牛壳中走出来，你会发现另有一片天地等着你来彩绘，如果可以的话，不妨把另一专业的新闻报道学术期刊等，当武侠小说来看，一时没看懂，没关系，多看一些多看几遍，日积月累之后，小媳妇总有熬成婆的日子，不停止地弥补自己在另一领域的不足。

举个简单的实际例子来说：我们团队发现同时也证明了在快两亿年前的禄丰龙胚胎和成龙骨头内，都还保存着有机残留物第一类胶原蛋白，事实上我们已经证明了，实体化石内保存着有机残

you will find new heaven waiting for you to venture into. If possible, read some news reports, literature, or papers. Treat them as novels. If something that you do not understand at the beginning, it does not matter. Read more papers and more times. Sooner or later, you will metamorphose from daughter-in-law to mother-in-law. Keep learning what you do not know from the other side.

Using a simple real example here, our team discovered and proved the preservation of organic remains collagen type I in the almost 200 million years old *Lufengosaurus* embryonic and adult bones. We proved that in the body fossils, organic remains preservation is a common phenomenon (Figure 5.1.2), as long as you have the right strategy, find the right method, and use sensitive enough instruments, it is not difficult to find more or less amount of organic remains. This is the first step of PaleoChemistry.

Then, the next step of PaleoChemistry is to figure out how these organic remains were preserved? What is the preservation mechanism? If this can be resolved and apply to food preservation, how significant the impact will it be for the whole of humanity? In order to feed 7, 8, … or even10 billion population, if the food can be kept just a little bit longer, throw out less, fewer children suffer hunger to death. How great that will be. On the other side of the same coin. It is similar to if you have a lock, someone shall be able to make a key to unlock it. Since we know how to preserve food for a longer time, we can also reverse that and find a way to digest the ever-increasing amount of garbage. Nobody likes to have a garbage dump in his backyard. This reversal can help to reduce the pressure

留物是个普遍现象（第5章图5.1.2）。只要有对的研究思维，找出对的方法，使用够灵敏的仪器设备，都不难在实体化石内找到或多或少的有机残留物，这是古生物化学的第一步。

接下来古生物化学的第二步，试着进一步想想，这些有机残留物是怎么被保存下来的？这个保存的机制 (mechanism) 如何？若能解开这个谜题，应用于人类粮食的保存，对于人类的贡献会有多大？要养活七十、八十甚至百亿全球人口，能够保存食物久一点点、少丢掉一些粮食，少饿死一些孩童，不是功德无量吗？反过来说，既然知晓了有机物的长久保存机制，有个锁，就可以做出一把钥匙来打开这个锁，反过来用的功效也无可限量，举个简单的例子来说，没人希望垃圾掩埋场在自己家后院，如果能把垃圾的体积缩小，社会对于掩埋场的压力，不就减轻了好多吗？我们搞古生物和古生物化学，表面看来没有直接的民生经济价值，没法立即赚大钱，然而，我们的最上游研究成果，能够走下去，对于整个人类，会有很重大的贡献。

或者，不说那么庞大遥远拯救全人类的课题，仅就我们老中医的中药用了好几种化石来说，除了龙骨之外，还如石燕，为古生代腕足类鮍科动物中华弓石燕 (*Cyrtiopirifer sinensis, Graban*) 及弓石燕 (*Cyrtiopirifer* sp.) 等多种近缘动物的化石。分布于湖南、广西、四川、山西、江西，浙江亦产，具有除湿热，利小便，退目翳之功效，用于淋病，小便不通，带下，尿血，小儿疳积，肠风痔漏，眼

of the volume of the mountain piling garbage issue. We do PaleoChemistry may seem on the surface as no direct economic benefit, and can not make tons of money. However, our most upstream research can lead to a profound contribution to the whole of humanity.

Alternatively, let us not saying such a big idea of saving the whole world and shrink down to the fossils used in Chinese medicine for several thousand years. Besides the "Dragon Bone", there are "Rock Swallows (*Cyrtiopirifer; Cyrtiopirifer sinensis*, Graban; and *Cyrtiopirifer sp.*)" from Hunan, Guangxi, Sichuan, Shanxi, Jiangxi, and Zhejiang. There were used commonly by Chinese medical doctors for various illnesses. From the chemical compositional point of view, the basic ingredient is the carbonate. However, since there were buried for so long in the ground, usually more than 100 million years, they are different from the common carbonate. Would that be the reason for their pharmacognosy usefulness? Another example is the Rock Crab (Crab fossil). These are ancient Arthropods and related animals. Please note that both Rock Swallow and Rock Crab have the same ingredient of carbonates, but their pharmacognosy is very different, one for the bladder problem and the other for the eye problem. Why?

Is it true that the pharmacognosy of these fossils in Chinese medicine just inorganic minerals? Taking the Rock Swallow and Rock Crab (Figure 3.5.1) as examples, saying the inorganic calcite (carbonate) is the effective part does not work well. It must be something else. And that shall be the different organic remains preserved in the fossils. So they are used for different diseases. Same for the Dragon Bone. Why is it effective? What

目障翳。就其化学成分来说，基本上是碳酸钙，只是埋藏在地层内的时间很久，一般都有几亿年的历史，与普通碳酸钙有所不同，是否就因为这原因才能作医药之用？又如，中药石蟹（螃蟹化石），为古代节肢动物石蟹及其他近缘动物的化石，也常入药。请注意，石燕和石蟹的主要成分都是碳酸钙，可是它们的药效，有很大的差别，一个和膀胱病变有关，另一个用于治疗眼疾，这怎么说得通呢？为什么？

这些中药化石的药效，果真只是这类化石里面的无机矿物？就以石燕石蟹（图3.5.1）来说，如果把药理解说只是无机方解石（碳酸钙）是说不通的，肯定还有其他（微量）药效成分，应该就是这些化石里面不同的残留有机物，所以才能治疗不同的疾病。就以龙骨来说，为什么有药效？无机的方解石和磷灰石，有什么药效？我们认为，龙骨的药效，就是龙骨内的残留胶原蛋白。从生药学的角度来说，若能解开这个成分药理，延伸开发出新药，会造益多少人啊？

简单总结一下本章，我们透过古生物化学的新思维来探讨，把握着这几个原则，大胆地往前航行，数天下风流人物，还看今朝！

pharmaceutical effects of inorganic calcite and apatite have? In our opinion, the effective component of Dragon Bone is the preserved organic remains. From the point of view of Pharmacognosy, if we can figure out these pharmaceutical mysteries and develop new medicines, how many people will be benefited?

To make a summary of this chapter, we can use the new thoughts of PaleoChemistry for our explorations, holding these fundamental principles and sail to the new water. Count the number of world romantic figures, look at the present court!

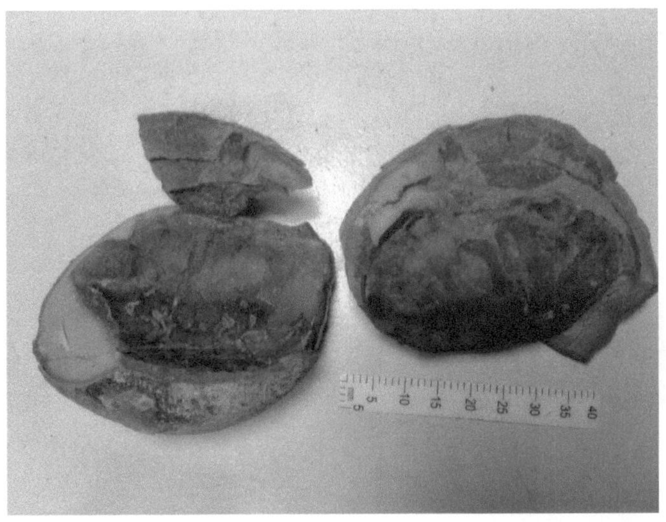

▲ 图 3.5.1 中药里面的石蟹（上）和石燕（下）

▲ Figure 3.5.1 Rock Crab (top) and Rock Swallow (bottom) in Chinese medicine

第4章　工具

Chapter 4　Tools

本章所介绍的各种古生物化学工具，只是现在比较常用的，虽然有一些可说是最先进的 (state of art)，但是并不完整，还有诸多化学分析方法，也可以运用到古生物化学的领域。再者，过几年之后，肯定会有更先进更好用的科学仪器出现，诸多形态和化学仪器分析设备，只要用得上，没有理由不使用。古语有云："工欲其善，必先利其器。"打仗之前，先了解弹药武器库内，有哪些可用的利器，针对某个课题，挑选最适当的科研分析方法和仪器设备，当然会做出较好的成果来。

因为近代科技的进展，诸多分析仪器设备的照射源和独立的某种仪器（如红外线光谱仪）相比较，都有很大进展，照射源的亮度和穿透力，远远超过独立仪器。这其中有两种大型的设施：同步辐射 (synchrotron radiation, sr-) 光源和中子散射 (neutron scattering, ns-)。前者可提供比太阳光强百万倍更亮的光源，波长从中红外线到硬X射线，后者利用核

The PaleoChemistry tools introduced in this chapter are just more commonly used as the writing time. Some of them are the most advanced state of the art, but not complete. There are many chemical analysis methods can be and shall be used in the PaleoChemistry field. In the coming years, more advanced scientific instruments will be available. There is no reason not to use various morphological and chemical analyses instruments. It says in order to get the job done well, sharpen your tools first. Before the battle, see what ammunition and weapons in your storage. For a given research topic, selects the most suitable method and instrument to achieve better results.

Due to the progress of modern science and technology, the radiation source of many analytical instruments, such as Infrared Spectroscopy have very significant improvements. They have much much more brightness and penetrating power than commercial instruments. Two main big facilities are Synchrotron Radiation (sr-) and Neutron Scattering (ns-). The synchrotron radiation source can provide one million brighter

反应炉或其他方法发射高穿透力的中子，进行各种光学和物理学的检测。因此，本章先通识性介绍这两种设施，比方说，傅里叶变换红外光谱仪 (Fourier transform infrared spectroscopy, FTIR)，在各化学分析实验室，是很普通常用的红外线光谱仪，但是这种独立仪器之红外光源强度，就无法和以同步辐射所提供的比较，因此分辨率就受到限制，也就是说，同步辐射傅里叶变换红外光谱仪 (sr-FTIR) 可提供更精准的扫描数据。同样的道理，小角度散射 (small angle scattering)，有独立的仪器，也有使用同步辐射X射线的设备 (small angle X-ray scattering, sr-SAXS)，但是照射源也可改成中子，变成小角度中子散射 (small angle neutron scattering, SANS)。

本章诸节中，有几个是用到同步辐射和中子散射设施的，特别提出来讨论，但这并不代表说这两种先进国家级的设施，只能做本章所提的几种实验，它们还有很多功能，等待大家从古生物化学的角度去发觉，去创造出更好的成果。

4.1 同步辐射设施*

同步加速器光源 (synchrotron radiation) 是二十世纪以来科技研究最重要的光源之一，已广泛应用在材料、生物、医药、物理、化学、化工、地质、

than our Sun ranging from mid-infrared to hard X-ray, while neutron facility provides high penetration neutron from a nuclear reactor or other means for various optical and physical measurements. Therefore, in this chapter, these two facilities will be generally introduced. For example, Fourier Transform Infrared Spectroscopy (FTIR) is commonly used in chemical analyses. However, the infrared source for stand-alone FTIR instrument cannot compare with that of synchrotron radiation. Thus, the resolution is limited. In other words, sr-FTIR can provide much better scan data. For the same reason, Small Angle Scattering can use synchrotron radiation X-ray as Small Angle X-ray Scattering (sr-SAXS), or with neutron source as radiation to Small Angle Neutron Scattering (SANS).

In the sections of this chapter, we point out that many facilities can be synchrotron radiation or neutron radiation source. However, this does not mean these two advance national-level facilities can only do the experiments discussed in this chapter. They can do much much more. You shall try to find out more from them to see how these can be used for our PaleoChemistry to obtain much better results.

4.1 Synchrotron Radiation Facility*

Synchrotron Radiation is the most critical light source for scientific research since the 20th century. It has been used widely in the material, biology, medicine, physics, chemistry, chemical

* 以下简介，取自于台湾同步辐射研究中心 (Taiwan Synchrotron Radiation Research Center, SRRC) 相关资料，特此感谢。
The following introduction is taken from the webpage of the Taiwan Synchrotron Radiation Research Center (SRRC). Gratitude expressed here.

古生物、考古、环保、能源、电子、微机械、纳米元件等等基础与应用科学研究，因而被称为现代的“科学神灯”。目前全球供实验用的同步加速器光源设施超过七十座，同步加速器光源设施的建造已成为各国高科技能力的指标之一。

什么是同步加速器光源？同步加速器光源也是电磁波的一分子，为一连续波段的电磁波，涵盖红外线、可见光、紫外线及X射线等，1947年首次在美国通用电器公司同步加速器上意外地被发现，因此命名为“同步辐射”或“同步加速器光源”（图4.1.1）。

二十世纪初，同步加速器是高能物理学家专门用来找寻基本粒子与探索宇宙本质的重要工具；自从同步加速器光源被发现后，一些物理和化学家们利用高能物理研究的空档，使用加速器所产

engineering, geology, environment, energy, electronic, micromachine, nanocomponents, etc., basic and applied sciences. Thus, it is called as the modern "Scientific Magic Lamp." There are more than 70 operational synchrotron radiation facilities for experimental usage worldwide. The building of synchrotron radiation light sources already become one of the indicators of the power of high technology.

What is the synchrotron radiation light source? The Synchrotron accelerator light source is a part of the electromagnetic wave. It is a continual section of the electromagnetic wave, including infrared, visible, ultraviolet and X-ray. In 1947 discovered accidentally on American General Electric synchrotron accelerator. Thus, it was named as Synchrotron Radiation or Synchrotron Accelerator Light Source (Figure 4.1.1).

During the early 20th century, synchrotron accelerator was an important tool for high ener-

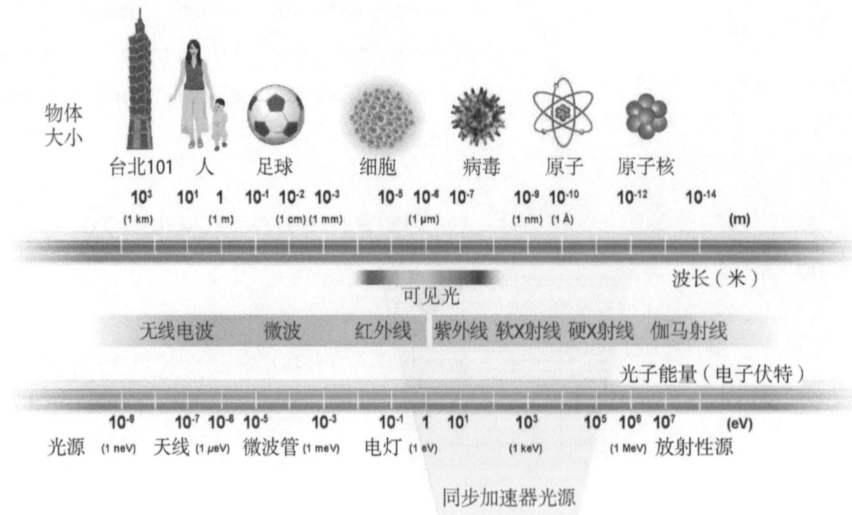

▲ 图 4.1.1　电磁波示意图

▲ Figure 4.1.1　Electromagnetic Wave

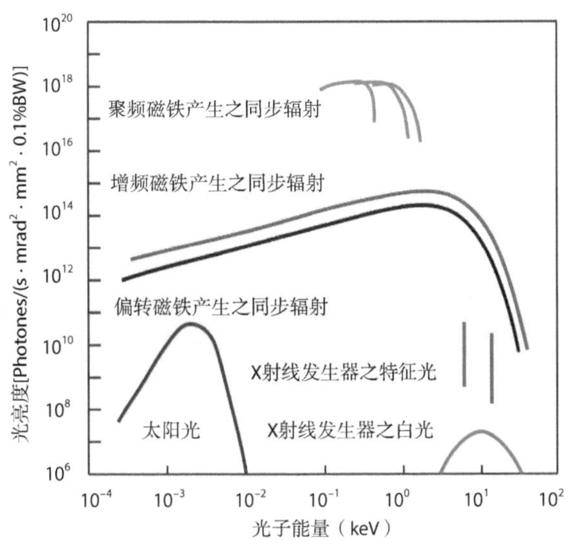

◀ 图 4.1.2　几种同步辐射与相关光源比较

◀ Figure 4.1.2　Comparison of several synchrotron radiations and related light sources

生的光做研究，后人便称此类与高能物理研究共用的加速器为“第一代同步加速器光源”。二十世纪七十年代，科学家们逐渐体认到同步加速器光源的优异性，于是纷纷开始兴建专门为产生同步辐射光的加速器，这就是所谓的“第二代同步加速器光源”。八十年代，科学家们提出一个构想，在储存环中装入特别的插件磁铁，例如增频磁铁或聚频磁铁，借此使电子由偏转一次变成多次偏转，同步加速器光源的亮度则可提高一千倍以上，这便是第三代同步加速器光源（图4.1.2）；目前全世界约有七十座实验用的同步加速器，其中第三代加速器于 1990 年后陆续建造完成，而台湾的同步加速器是在 1993 年完工启用，成为少数最先完成的第三代同步加速器光源设施之一。

同步加速器光源具有以下特性：

- 强度极高；

gy physicists for the search of fundamental particles and exploring the nature of the universe. Since the discovery of synchrotron accelerator light source was found, some physicists and chemists used the light generated during the spare times of high energy physicists for their researches. So, it was called the co-usage of high energy physic accelerator as the First Generation Synchrotron Accelerator Light Source. In the 1970s, scientists realized the superiority of synchrotron accelerator light source and started to build the specialized accelerator to produce synchrotron radiations. This is the so-called Second Generation Synchrotron Accelerator Light Source. In the 1980s, scientists proposed a thought to install specialized plugin magnets, such as frequency increasing magnet or frequency concentration magnet that causes the electron to deflect from once to many times, the brightness of synchrotron accelerator light source can be increased to more than one thousand times (Figure 4.1.2). This is the Third Genera-

- 波长连续；
- 准直性佳；
- 光束截面积小；
- 具有时间脉波性与偏振性。

若以X射线为例，同步加速器光源在这个波段的亮度，比传统X射线机还要强百万倍以上！过去需要几个月才能完成的实验，现在只需几分钟便能得到结果；以往因实验光源亮度不够而无法探测的结构，现在借由同步加速器光源，都可分析得一清二楚，也因此于近年内许多新的研究领域得以开发。

根据电磁学的理论，带电粒子的运动速度或方向改变时，会放射出电磁波；当电子以接近光速飞行，受到磁场的作用而发生偏转时，便会因相对论效应沿着偏转的切线方向，放射出薄片状的电磁波，这就是“同步加速器光源”（图4.1.3）。

以台湾同步辐射研究中心加速器为例（图4.1.4），由注射器（1）产生的电子，经由传输线（2）进入储存环（3），电子在环中经过偏转磁铁或插件磁铁（4），会产生同步加速器光源，经过光束线（5）

tion Synchrotron Accelerator Light Source. Among the more than 70 experimental synchrotron accelerators worldwide, the third-generation accelerators were built after 1990. The Taiwan synchrotron accelerator was done building and started to use in 1993. This is one of few operational third-generation synchrotron accelerator light source facilities.

Characteristic of synchrotron accelerator light source:

Very strong intensity;

Continuous wavelength;

Excellent collimation;

Small cross section area;

Have time pause and polarization.

Using X-ray as an example, the brightness of this section of wave band from synchrotron accelerator light source is more than one million times brighter than the conventional X-ray machine. The experiment requires many months to accomplish in the past can be done in several minutes. In the past, due to the light source is not bright enough, the internal structure cannot be explored. Now it can be seen clearly by using synchrotron accelerator. So because of this, many new research fields can be done (Figure 4.1.3).

According to the theory of electromagnetic,

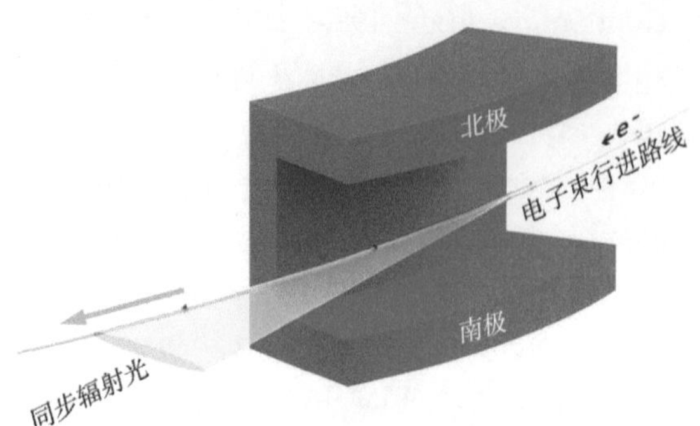

◀ 图 4.1.3 同步加速器光源的产生

◀ Figure 4.1.3 Generation of synchrotron accelerator light source

导引到达实验站（6），研究人员便可使用光源进行实验。

（1）注射器

电子束由电子枪（图4.1.5）产生后，经过直线加速器加速至能量为5000万电子伏特，电子束进入周长为72米的增能环后，继续增加能量至15亿电子伏特，速度非常接近光速 (99.999995%)。

注射器规格

注射能量：1.5 GeV；

注射频率：10 Hz；

线型加速器能量：50 MeV；

注射环周长：72 m；

增能环高频：499.654 MHz。

（2）传输线

注射器产生的电子束经由传输线进入储存环，传输线的长度为70米。

（3）储存环

电子束进入六边形设计、周长为120米的储存环后，环内一系列磁铁导引电子束偏转并维持在轨道上，如此一

when a change of speed or direction of charged particles will release electromagnetic wave. When electrons are flying with near light speed and influenced by the deflection caused by the magnetic field, according to the relativity theory along the tangent direction of deflection, thin sheet-shape electromagnetic waves will be released. This is the synchrotron accelerator light source.

Using SRRC's accelerator as an example (Figure 4.1.4), the electron generated by the injector (1) go thru conducting line (2) into the storage ring (3). Electrons in the ring passing thru deflection magnet or plug-in magnet (4) will produce synchrotron accelerator light source. Then the light beam (5) is directed to the workstation (6).

(1) Injector

Electrons were generated by the electron gun (Figure 4.1.5). They go thru linear accelerator to the speed of 5 K electronic volts. Electron beam enters to an energy enhancing ring of a diameter of 72 meters to increase the energy to 1.5 billion EV, and the speed is very close to the

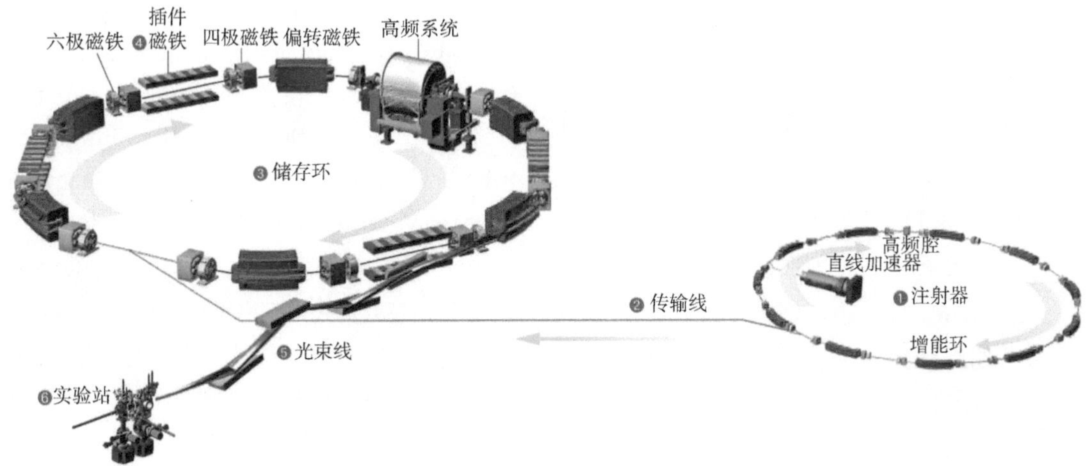

▲ 图 4.1.4 同步辐射光源的产生

▲ Figure 4.1.4 Generation of synchrotron radiation light source

电子枪

直线加速器

增能环

▲ 图 4.1.5 几个关键部件

▲ Figure 4.1.5 Several key components

来，电子束便能于每一圈的运行中在偏转磁铁切线方向或插件磁铁下游放出同步辐射光；由于电子会因辐射而损失能量，因此环内装置高频系统，用来补充电子的能量（图4.1.6）。

（4）插件磁铁（增频磁铁/聚频磁铁）

插件磁铁为一系列极性交错排列的磁铁，当电子束经过时，会产生多次偏转；若将磁场强度提高，可使所放射出之同步加速器光谱提升至更高能量，如软X射线甚至硬X射线，此类插件磁铁称为增频磁铁；若将磁场交替的空间周期变短，使带电的电子摆动幅度变小，放出的同步光可在特定光谱形成建设性干涉，大大提升光亮度，此类插件磁铁称为聚频磁铁。

（5）光束线

光束线是同步加速器光源与实验站之间的一座桥梁；理论上，在每一处电子偏转的地方或插件磁铁的直线下游，都可以打开一个窗口，利用光束线将同步加速器光源导引出来，最后到达实验站（图4.1.7）。

speed of light (0.99999995).

Specification of Injector

Injection energy: 1.5 GeV;

Injection frequency: 10 Hz;

Power of linear accelerator: 50 MeV;

Circumference length: 72 m;

Energy enhancing ring frequency: 499.654 MHz.

(2) Conducting line

The electron beam generated by the injector goes thru 70-meter long conducting line into the storage.

(3) Storage Ring

Electron beam enters into a hexagonal design of Circumference length of 120 meters. In the ring, there are many magnets directing electron beam deflection and maintains on the track. By doing so, the electron beam will release synchrotron radiation light along the tangent direction of the deflection magnet or plugin magnet downstream. Since electron will lose energy when radiation, therefore high frequency-system is equipped to supplement electron energy (Figure 4.1.6).

(4) Plugin magnet (Frequency increase magnet/frequency concentration magnet

▲ 图 4.1.6　储存环

▲ Figure 4.1.6　Storage ring

（6）实验站

同步加速器光源经由光束线的导引照射到实验站的试样后，研究人员借由量测反射、绕射、散射及穿透试样的光强度、能量及试样被光子激发出的电子及离子，可以进一步推断物质几何、电子、化学或磁性结构（图4.1.8）。

台湾同步辐射研究中心除了上述的台湾光源（Taiwan Light Source, TLS）之外，在 2014 年底，花了七十多亿新台币十多年时间新建设的五百多米台湾光子源（Taiwan Photon Source, TPS），也已经开放使用，先期开放七条光束线；以后拉出每一条光束线，终端工作站的设备还不算在内，至少还要花一两亿新台币！

从我们实用经验来说，在台湾的这

The plugin magnet is a series of pole crossing magnets. When electron beam passing by, it will cause many deflections. If the magnetic field is increased, it will cause the spectra released by the synchrotron accelerator to increase to even higher energy, such as soft X-ray or even hard X-ray. This type of plugin magnets is called Frequency increase magnets. If the space cycle of the magnetic field is shorten, causing the vibration of the electron to reduce, the related synchrotron light can produce constructive interference at specific spectra, and increase the brightness very significantly. This type of plugin magnet is called frequency concentration magnet.

(5) Light beam

The light beam is the bridge between synchrotron accelerator light source and workstation. In theory, at the direct line downstream of every electron deflection place or plugin magnet, a window can be opened using a light beam to bring out the synchrotron accelerator light source and reach the workstation (Figure 4.1.7).

(6) Workstation

The light from the synchrotron accelerator light source and thru the light beam to the sample of a workstation, the researcher can measure the light strength of reflection, diffraction, scattering, and penetration, and the electrons and ions excited by photons. That can further be used to deduct the geometry, electron, chemi-

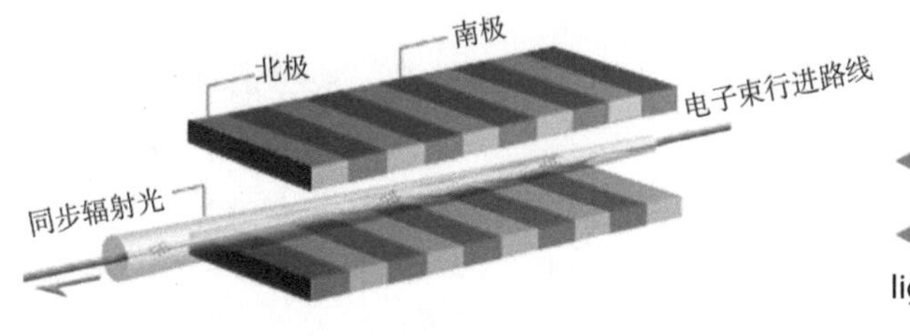

◀ 图 4.1.7　同步辐射光

◀ Figure 4.1.7　Synchrotron Radiation light

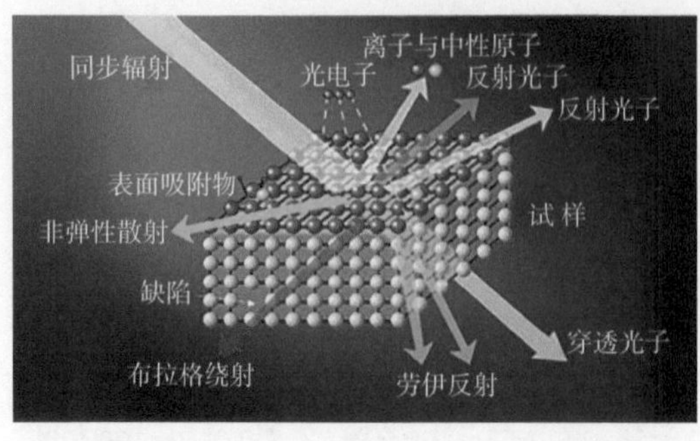

◀ 图 4.1.8 同步辐射光源能产生的各种光电反应

◀ Figure 4.1.8 Various photonics and electronic reactions by synchrotron radiation

个同步辐射设施，我们用过同步辐射傅里叶变换红外光谱 (sr-FTIR)、同步辐射透射X射线显微 (sr-TXM) 和X射线散射 (X-ray deflection) 等，都有惊人的效果，如第1章图1.2.6等。

全球各处的同步辐射设施，都对外开放，学术界免费使用，学者只要照各设施的规定申请，取得使用者资格和光束时间，都可利用这些设备。至于全球每个同步辐射有哪些工作站，可以进行什么测试，可上到他们的网站去搜寻。

4.2 核科技设施*

核科技是一门多领域、很广泛的学科，无法在本节中详述。这一节介绍将以澳大利亚核科技组织 (Australian Nuclear Science and Technology Organization, ANSTO) 的OPAL核子反应炉相关设施为主，本节资料也都取自该组织相关网页。相关更进一步的资料，请上网查询。

澳大利亚核科技组织是澳大利亚国

* 取自 ANSTO 网站。
From ANSTO Web Pages.

cal, or magnetic structure of the specimen under examination (Figure 4.1.8).

Besides the Taiwan Light Source (TLS), SRRC also spent more than 7 billion New Taiwan Dollars and more than ten years on building the more than 500-meter circumcircle Taiwan Photon Source (TPS) and in operation at the end of 2014. In the early phase, seven light beams are available for users. Each beamline costs more than 100–200 million NTD, not including the cost of a workstation.

Speaking from our experience, this synchrotron radiation facility provided us sr-FTIR, sr-TXM, and X-ray Deflection. The results are very astonishing such as Figure 1.2.6 of Chapter 1.

All the synchrotron radiation facilities are open to the public. Academic can use for free. Researchers need to follow the application procedure to obtain the user's status and beam times to use these facilities. As for the various workstation, each synchrotron radiation facility has and what kind of tests can be performed, check out their web page.

4.2 Nuclear Science Facility*

Nuclear science and technology is a multidisciplinary field of many subjects. We can not de-

家级的高端科技设施，包括了以下几个重要园区：①OPAL 研究用核反应炉 (OPAL Research Reactor)；②加速器科学中　心 (Center for Accelerator Science)；③澳大利亚中子散射中心 (Australian Centre for Neutron Scattering)；④回旋加速器 (Cyclotron Facility)；⑤澳大利亚同步辐射 (Australian Synchrotron)。本节仅介绍第一项，其他各项可上网查询。

发现放射性带来很多好处，但也须小心处理。一张纸或我们身体的皮肤，就可以阻止α粒子，薄的Perspex有机玻璃或玻璃，可以阻挡β辐射。但是，如果通过呼吸吞咽或伤口进入到人体，这两种的能量会损坏细胞。为了阻挡穿透力更强的γ射线，就必须用厚铅块、水和水泥（图4.2.1）。

核科技是从原子层次去了解宇宙世界和我们自己的关键，如果我们可以了解原子如何聚合、互动或以最好方式和其他原子合并，就可研发出新的更有效的材料和药品。

因为中子有很多特性，非常适合用来研究从一纳米到几百纳米的原子和分

scribe it in details here. Instead, only the OPAL reactor of Australian Nuclear Science and Technology Organization (ANSTO) will be described. The information here is from their web pages. For further information, please search the internet.

ANSTO is an Australian national level research facility. It includes the following key units: ① OPAL Research Reactor, ② Center for Accelerator Science, ③ Australian Centre for Neutron Scattering, ④ Cyclotron Facility, and ⑤ Australian Synchrotron. This section will introduce item #1. For the rest, please check the internet.

The discovery of radioactivity has delivered many benefits, but it must be handled with care. A sheet of paper, or even the skin of our bodies, will stop alpha particles, while a thin sheet of perspex or glass will stop beta radiation. However, the energy of both can cause damage to cells if they enter the body through inhalation, swallowing, or wounds. Thick barriers of lead, water, and concrete are necessary to stop much more penetrating and damaging gamma radiation (Figure 4.2.1).

Nuclear science is crucial to understanding our universe, our world and ourselves at the atomic level. If we can understand how atoms

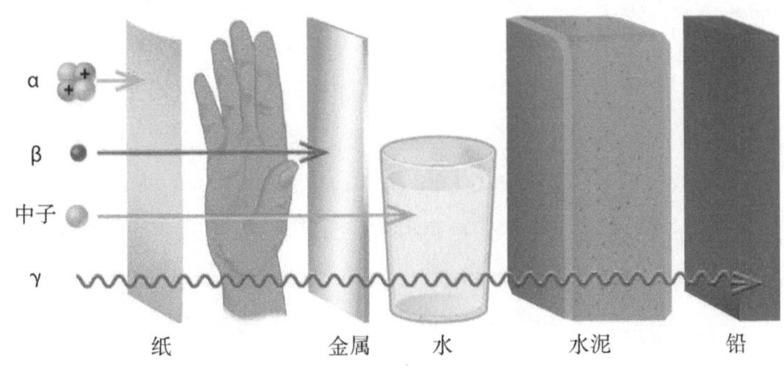

▲ 图 4.2.1　各种辐射线的穿透力

▲ Figure 4.2.1　The Penetration power of various radiations

子构造。它们可以看成粒子或波动，其波长相当于固体液体原子之间的距离，它们的能量等同于分子振动的量阶。

冷（慢）中子能量低波长长，热中子有中阶能量和波长，高热（快）中子能量高波长短。

热中子和冷中子*

热中子是从反应炉和散裂加速器产生的，它们的平均能量热中子只有 25 meV，冷中子只有 4 meV。因为中子不带电，可以穿透大部分的物质，只会和样本的原子核相互作用，可因散射或吸收减弱。吸收则依照标靶原子内部结构而定，亦即原子核内的质子和中子数量，同样元素的同位素，可能产生很大的反差，在周期表内并没有规则性，不像X射线，对于很多轻的元素，会有很好的反差，而可穿透很多金属（图4.2.2）。

氢会散射中子，因此透过等质量散射中子减弱特别好，产生化石内保存有机物的巨大反差。若要检查被围岩包裹的化石，中子可穿透含铁岩石，其内的化石会有很大的反差，在化石牙齿的珐琅质和牙本质都可看到好的反差。在X射线这边，依照石化过程的矿化程度，往往没有反差。

中子技术可以使用单一波长或一个范围的波长。中子相较于X射线作为工具，在分析分子结构及材料中的原子排列方面有些优势。比方说，中子可用于

get together, interact, or can be best combined with other atoms, new, more efficient materials and drugs can be developed.

Neutrons have many properties that make them useful for studying atomic and molecular structures ranging in size from one nanometer to several hundred nanometers. They can be considered as particles or waves, with wavelengths comparable to the interatomic distances found in solids and liquids. Their energies are of similar magnitude to those associated with molecular vibrations.

Cold (slow) neutrons have low energy and long wavelengths. Thermal neutrons have intermediate energy and wavelengths. Hot (fast) neutrons have high energy and short wavelengths.

Thermal and cold neutrons*

Thermal neutrons are generated in a reactor and accelerator (spallation) sources. Their average energy is only 25 meV for thermal, and about 4 meV for cold neutrons. As uncharged particles, neutrons can penetrate most materials easily, and interact only with the nuclei of the sample atoms. Attenuation can be caused by scattering or by absorption. The latter depends on the internal structure of the target atoms, i.e., the number of protons and neutrons. A significant difference, in contrast, may exist even between isotopes of the same element. There is no obvious regularity across the periodic system, but opposed to X-rays, there is often good contrast for many light elements, while most metals can be penetrated as

* 参考文献见：*Neutron Tomography and X-ray Tomography as Tools for the Morphological Investigation of Non-mammalian Synapsids, Laaß Michael, Burkhard Schillinger, Ingmar Werneburg, https://doi.org/10.1016/j.phpro.2017.06.013.*

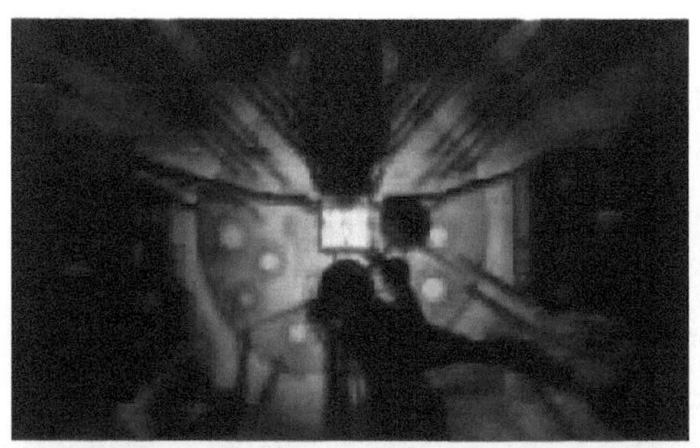

▲ 图 4.2.2　世界最有效多用途研究反应炉，也是澳大利亚核科技组织研究设施的主角

▲ Figure 4.2.2　One of the world's most effective multi-purpose research reactors and the centerpiece of ANSTO's research facilities

探讨提供计算机用半导体和磁性材料的独特信息。

不像电子和质子，中子没有电荷，它能把原子里面很小的原子核位置显示出来；相对地，X射线会被电子散射，只能显示电子云，而非原子核的位置，重的原子比轻的原子更有效地散射X射线，而中子的散射，则依原子核不同而有差异。

X射线无法用于鉴定靠近重原子附近的氢原子或较轻原子，而中子可以；同位素如氢和氘 (deuterium) 可以在中子散射中很容易区分，而X射线则不能。因此，科学家可以透过中子散射，把聚合物和生物重要分子内的氢换成氘，以凸显某特定特征。

因为中子散射是发生于原子核，而非电子云，因此它们很有穿透力，能够深入大物件样本的内部，如飞机引擎和有不同压力温度环境条件的管子。

到某个程度而言，中子行为像小块

well (Figure 4.2.2).

Hydrogen scatters and thus attenuates neutrons especially well by equal-mass scattering and delivers huge contrast depending on the content of organic matter in fossils. For the examination of embedded fossils, ferrous rock can be penetrated well while high contrast exists for the fossilized bones. Contrast is visible even for distinguishing enamel and dentine in fossilized teeth, where X-rays, depending on the mineralization process during fossilization, often show no contrast at all.

Neutron techniques can use a single wavelength or a range of wavelengths. Neutrons have some advantages over X-rays as tools for determining the structure of both molecules and the arrangement of molecules within materials. For example, neutrons can be used to investigate and provide unique information about semiconductors and magnetic materials used in computers.

Unlike electrons and protons, neutrons have no electric charge. This means that they can reveal the position of the nucleus itself, which makes up a tiny fraction of the volume of an atom. In contrast, X-rays are scattered by electrons and reveal the position of the electron clouds. Heavy atoms scatter X-rays more effectively than light atoms. Neutron scattering varies from the nucleus to nucleus.

X-rays cannot be used to determine the positions of hydrogen atoms or light atoms close to heavy atoms but neutrons can. Isotopes (atoms of the same element with different numbers of neutrons) such as hydrogen and deuterium can readily be distinguished by neutron scattering but not by X-rays. Scientists can, therefore, sub-

的磁铁条，因此可以用来研究材料的磁性，如超导体和计算机内存。当中子束打到某个样本，80%~90%的中子从样本穿过去，有一些“散射”，很小一部分被吸收，中子打到样本的角度，影响“散射”，因此，可以收集这种信息。大部分的中子散射是基于弹性散射，散射中子的能量没有改变；非弹性散射用于研究分子振动和磁性，散射中子的能量有改变。

OPAL 反应炉是全球最有效率多目标研究设备之一，也是澳大利亚核科技组织研究设施的主轴（图4.2.3）。研究用反应炉主要的目的在于提供中子源，中子是铀原子分裂产生的次原子粒子，有很多应用，包括检验材料、工业上照射硅晶和生产核医疗用的同位素。

澳大利亚核科技组织科学家们探索很广泛的领域，如材料、生命科学、气候变迁、采矿工程等等。其研究能力主要是 OPAL 核子研究用反应炉和相关设施，包括加速器和X射线仪器与辐射药物实验室。澳大利亚轻水开放反应炉 (Open Pool Australian Lightwater)，是个最尖端20 MW的反应炉，使用低浓缩铀

▲ 图 4.2.3 研究用反应炉

▲ Figure 4.2.3 Research Reactor

stitute deuterium for hydrogen in polymers and biologically relevant molecules to highlight particular features by neutron scattering.

As neutrons scatter from nuclei and not electrons, they are highly penetrating. This makes it possible to study samples deep inside large pieces of equipment (such as aircraft engines), and inside vessels that have different conditions of pressure, temperature and environment.

Neutrons behave to some extent, like tiny bar magnets and can, therefore, be used to investigate the magnetic properties of materials such as superconductors and computer memories. When a neutron beam hits a sample, 80 to 90 percent of the neutrons pass through the sample, some 'scatter', and a very small number are absorbed. The angle at which the neutron beam hits the sample affects the 'scattering' and, hence, the type of information that can be gained. Most neutron scattering techniques are based on elastic scattering, in which the energy of the scattered neutrons does not change. Inelastic scattering, in which the energy of the scattered neutrons changes as a result of interaction with the sample, is used to investigate molecular vibrations and magnetic properties.

One of the world's most effective multi-purpose research reactors and the centerpiece of ANSTO's research facilities (Figure 4.2.3). Research reactors have the primary purpose of providing a source of neutrons - subatomic particles produced when uranium atoms split - for a wide range of applications, including the investigation of materials, the irradiation of silicon for industrial uses, and for the production of radioisotopes used in nuclear medicine.

ANSTO scientists investigate areas as diverse

(low enriched uranium, LEU) 燃料，达成诸多领域核医药、研究、科学、工业和生产的目标。

它在 2007 年由澳大利亚总理启动，是少数可以生产商业量辐射同位素的反应炉，这种能力，配合开放式水池设计、使用低浓缩铀燃料和广泛的应用，使得OPAL跻身于世界最好的研究用反应炉。OPAL是澳大利亚核科技组织研究设备的主角，放在反应炉四周建筑物的各种中子束仪器，由澳大利亚核科技组织中子散射中心操作，展现出澳大利亚核科技组织更多重大的研究能力。

研究用反应炉的角色

OPAL是全球少数这类生产设施之一，其他还包括南非的 Safari-1、荷兰贝登的HFR反应炉和加拿大查克河的NRU 反应炉。这些反应炉在社会上扮演关键的“中子工厂”功能角色，生产检测治疗癌症的辐射同位素以及基础材料研究的中子束。OPAL操作人员与国际同僚合作，透过直接正式合作协议，或各种国际组织平台，分享信息和知识。

虽然OPAL非常多样性，中子科学的应用也几乎无限制，但OPAL主要用于：照射靶物产生医学和工业应用的辐射同位素；使用中子束和相关仪器的材料科学研究；使用中子活化技术和延迟中子活化技术分析矿物和样本；照射硅晶圆，称之为中子变化掺杂 (neutron

as materials, life sciences, climate variability, mining, and engineering. ANSTO's research capabilities, led by the OPAL nuclear research reactor and associated instruments, include accelerator-based and X-ray instruments and radiopharmaceutical laboratories. Australia's Open Pool Australian Lightwater (OPAL) reactor is a state-of-the-art 20 Megawatt reactor that uses low enriched uranium (LEU) fuel to achieve a range of nuclear medicine, research, scientific, industrial and production goals.

Opened by the Prime Minister in 2007, OPAL is one of a small number of reactors with the capacity to produce commercial quantities of radioisotopes. This capacity, combined with the open pool design, the use of LEU fuel and the wide range of applications, places OPAL amongst the best research reactors in the world. While OPAL is the centerpiece of ANSTO's research facilities, the suite of neutron beam instruments housed next to the reactor building and operated by ANSTO's Centre for Neutron Scattering represents a significant addition to ANSTO's research capabilities.

The role of research reactors

OPAL is one of several similar production facilities around the world, including the Safari-1 reactor in South Africa, the HFR reactor at Petten in the Netherlands and the NRU reactor at Chalk River in Canada. These reactors play a vital role in society by functioning as "neutron factories," producing radioisotopes for cancer detection and treatment, and neutron beams for fundamental materials research. OPAL's operation staff cooperate with their international

transmutation doping, NTD)，制造电子半导体。

OPAL内部

反应炉的核心是一个 4 × 4 阵列 16 束紧密燃料组合，五根控制棒控制反应炉功率，也方便关机，OPAL 使用低浓缩铀，含稍低于20%的铀-235。OPAL的燃料组合（核心）用净水冷却，并以锆合金“反射器”容器包围，其内含有重水。反射器容器放置于 13 米深的清水池底部，这种开放式水池设计，使得容易观察并操作放在反应炉池内的物件。水的深度作为很有效的辐射屏障，重水维持核心的核子反应，将中子“反射”回核心。

设备

现有 14 条中子束仪器，正在建构另一条，也有各种样本的环境器材，包括极低温、熔炉、磁铁、压力舱、张力设备、剪力设备和电化学器材。为了了解中子散射数据，也可提供一些制模协助。给中子散射用样本的分子氘化反应，也可通过国家氘化设施进行。除了世界级的中子设施之外，也有很多X射线仪器（反射仪和小角度X射线散射, SAXS）以及提供相关计算机设备。

仪器

OPAL反应炉（图4.2.4）有15条使用中和建构中的中子束仪器，在未来五

colleagues in sharing information and knowledge both directly through formal collaboration agreements and via various international organizations and forums.

While OPAL is exceptionally versatile, and the uses of neutron science are virtually unlimited, OPAL's main uses are: Irradiation of target materials to produce radioisotopes for medical and industrial applications; Research in the field of materials science using neutron beams and associated instruments; Analysis of minerals and samples using neutron activation techniques and delayed neutron activation techniques; Irradiation of silicon ingots (termed Neutron Transmutation Doping or NTD) for use in the manufacture of electronic semiconductor devices.

Inside OPAL

The heart of the reactor is a compact core of 16 fuel assemblies arranged in a 4×4 array, with five control rods controlling the reactor power and facilitating shutdown. OPAL uses low enriched uranium fuel containing just under 20 percent uranium-235. OPAL's fuel assemblies (core) are cooled by purified water and are surrounded by a zirconium alloy “reflector” vessel that contains a special type of water called heavy water. The reflector vessel is positioned at the bottom of a 13-meter-deep pool of light water. The open pool design makes it easy to see and manipulate items inside the reactor pool. The depth of the water acts as a very effective radiation shield. The heavy water maintains the nuclear reaction in the core by “reflecting” neutrons back towards the core.

年内，澳大利亚核科技组织计划添加新的仪器，整个设施可以继续扩充，包括第二个中子导引厅。这14个运转中的仪器如下，都是以澳大利亚和海外动物群命名：

ECHIDNA: 高分辨率粉末衍射仪（澳大利亚针鼹）；

WOMBAT: 高强度粉末扰射仪（塔斯马尼亚袋熊）；

KOALA: 劳厄衍射仪（无尾熊）；

KOWARI: 拉力扫描器（脊尾袋鼠）；

PLATYPUS: 中子反射仪（水平样本）（鸭嘴兽）；

QUOKKA: 小角度中子散射（短尾矮袋鼠）；

TAIPAN: 热中子三轴光谱仪（太攀蛇属），可选用铍滤片；

KOOKABURRA: 极小角度中子散射（笑翠鸟）；

PELICAN: 飞行时间质量分析仪（澳大利亚鹈鹕）；

DINGO: 中子辐射/影像/断层扫描（澳大利亚野犬）；

SIKA: 冷三轴光谱仪（梅花鹿）；

BILBY: 第二台小角度中子散射仪器（兔耳袋狸）；

EMU: 高分辨率背散射光谱仪（鸸鹋）；

JOEY: 中子劳厄相机,用于单晶对准；

SPATZ: 第二台中子反射仪（垂直样本）（家麻雀）；

X射线仪三台，计算机基础设施和其他设施；

Facilities

A suite of 14 neutron beam instruments is available, with one instrument under construction. A broad selection of sample-environment equipment is also available, including cryostats, furnaces, magnets, pressure cells, stress rigs, shear cells, and electrochemical apparatus. We can also offer some help with modeling in order to understand neutron-scattering data. Molecular deuteration of samples for neutron-scattering studies is also available through the National Deuteration Facility. In addition to our world-class neutron instruments, we have a range of in-house X-ray (reflectometer and SAXS) and computing facilities available for use.

Instruments

Fifteen neutron beam instruments are either operational or commissioning at the new OPAL reactor (Figure 4.2.4). ANSTO expects to add more instruments within five years. The facility has the capacity for further expansion, including the potential for a second neutron guide hall. 14 Operational instruments (named after Australian and overseas fauna)

High-Resolution Powder Diffractometer (*Tachyglossus aculeatus*);

High-Intensity Powder Diffractometer (*Vombatus ursinus*);

Laue Diffractometer (*Phascolarctos cinereus*);

Strain Scanner (*Dasyuroides byrnei*);

Neutron Reflectometer (with horizontal sample) (*Ornitho-rhynchus anatinus*);

Small-Angle Neutron Scattering (*Setonix*

SAXS：小角度X射线散射仪 (Bruker and Hecus)；

X射线反射仪；

物理特性测量系统；

使用模型方法分析中子散射数据；

样本-环境载具；

氦-3 极化系统。

在过去几年内，我们多次使用澳大利亚野犬 (DINGO) 工作站进行化石的断层扫描，得到很多的惊喜，特别是看到化石和围岩内保存的有机残留物，非常振奋人心，如第1章图1.4.2等。

▲ 图 4.2.4　站在 OPAL 反应炉（身后下方），两人中间为同样大小的 OPAL 反应炉核心的模型

▲ Figure 4.2.4　Standing in front of the OPAL reactors (behind lower) and the replica of OPAL Reactor Core

Brachyurus);

Thermal Neutron 3-Axis Spectrometer (*Oxyuranus scutellatus*), Beryllium-filter option on TAIPAN;

Ultra Small-Angle Neutron Scattering (*Dacelo nova-eguineae*);

Time-of-Flight Spectrometer (*Pelecanus conspicillatus*);

Neutron Radiography/Imaging/Tomography (*Canis lupus dingo*);

Cold Neutron 3-Axis Spectrometer (*Cervus nippon*);

2nd Small-Angle Neutron Scattering Instrument (*Macrotis lagotis*);

High-Resolution Backscattering Spectrometer (*Dromaius novaehollandiae*);

Neutron Laue Camera for single-crystal alignment;

2nd neutron reflectometer (with vertical sample) (*Passer domesticus*);

3 X-ray Instruments, Computing Infrastructure and other Facilities;

SAXS (Bruker and Hecus Small-Angle X-ray Scattering) Instruments;

X-ray Reflectometer;

Physical Properties Measurement System;

Analysis of neutron scattering data using modeling methods;

Sample-Environment Apparatus;

Helium-3 Polarization System.

In the past years, we used the DINGO workstation to conduct many tomographic scans of fossils and got many surprises, particularly seeing organic remains preserved inside fossil and matrix. Very encouraging as Figure 1.4.2 of Chapter 1.

4.3 微型计算机断层扫描 / 计算机断层扫描*

计算机断层扫描（computed tomography，CT），是一种常见的医学影像诊断设备和方法，它利用 X 射线很强的穿透力，通过单一轴面的X射线旋转照射样本，再经由计算机利用数位几何处理，重建样本的三维放射线医学影像。由于不同组织成分对X射线的吸收能力（或称阻射率）有所不同，可以用计算机的三维技术重建出断层面影像，再经由调整窗宽（window width, WW）、窗位（window location, WL）等处理，可以得到相应组织的断层影像，将断层影像层层堆叠即可形成立体影像。X 射线断层面的数据，是由 X 射线源环绕物体一圈得来，感应器放置于射源的对面位置，随着物体慢慢地被推入内侧端，数据也不断地处理，经由一系列的数字运算，也就是所谓的断层面重建来得到影像，如图 4.3.1。

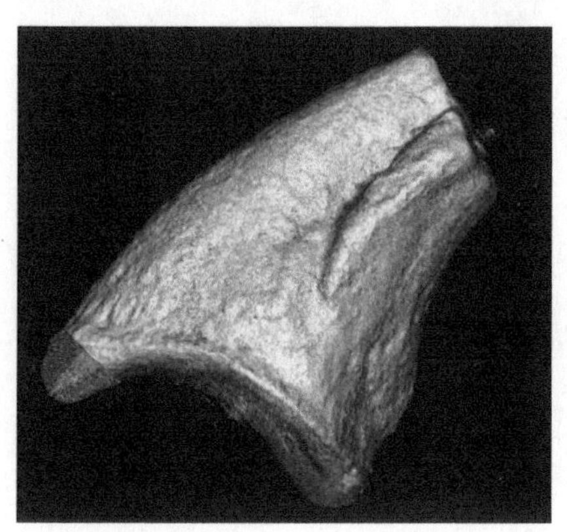

图 4.3.1 计算机断层扫描断掉的禄丰龙爪子

Figure 4.3.1 CT scan of a broken *Lufengosaurus* claw

* 取自维基。
From Wiki.

4.3 Micro Computer Tomography/Computer Tomography*

The Computed Tomography (CT) is a commonly used medical diagnosis facility and method. It uses the high penetration power of X-ray circling around a single axial to radiate the sample and uses a computer to process the digital data for reconstruction of three-dimensional radiology medical image. Due to the different X-ray absorption of different matter (or blockage), computerized 3D reconstruction can provide images of thin slices. Then after adjusting the Window Width (WW) and Window Location (WL), corresponding tomographic images of a given tissue. By piling up the tomographic images, a 3D model can be obtained. The data of X-ray slice image is obtained from the X-ray circles the sample. The detector is on the opposite side. When the object is gradually pushed into the inside, data was processed. A serious of numerical operations were carried out. Also called as the reconstruction of topography, as Figure 4.3.1.

X-ray computerized tomographic scan provides complete 3D information of organ to doctor. Due to the higher resolution of CT, the radiodensity of different tissues, even with the differences of less than 1% can be distinguished by software. CT provides 3D images and can

X射线计算机断层扫描，给今日医生提供器官的完整三维信息，由于计算机断层的分辨率较高，不同组织阻射过所得的放射强度 (radiodensity)，即便小于1%的差异，也可以通过软件操作区分出来。断层成像技术提供三维图像，依诊断需要不同，可以看到轴切面 (axial)、冠状面 (coronal)、矢切面 (sagittal) 的影像，称之为“多平面数位重建 (multi-plane reformat imaging, MPR)”，如图 4.3.2。除此之外，任意切面的图像，均可通过插值技术产生，给诊断和科研带来了极大的便利。

放射强度 (radiodensity) 的阈值是可以调整的（例如对应于骨头的值），当阈值一定，便可使用“边缘检测 (edge detection)”影像处理法，如此一来，一个三维的物体就可以成像了，不同的物

be viewed with Axial, Coronal, and Sagittal cut images depend on the need. This is called Multi-Plane Reformat imaging (MPR), as shown in Figure 4.3.2. Besides, any cut plane image can be interpolationally generated. This brings a great deal of convenience for diagnosis and scientific research.

The threshold of Radiodensity can be adjusted, such as corresponding to the bone. When the threshold is fixed, the edge detection image processing method can be used. By doing so, a 3D image model can be presented with a different threshold represented by a different color to indicate different anatomic structure. Volume Render is limited to a range of threshold to show the surface image volume and as a close resemblance to the idea surface volume. However, with pixel rendering by using transparency and color can show the unique feature in a single image. That is to show more. For example,

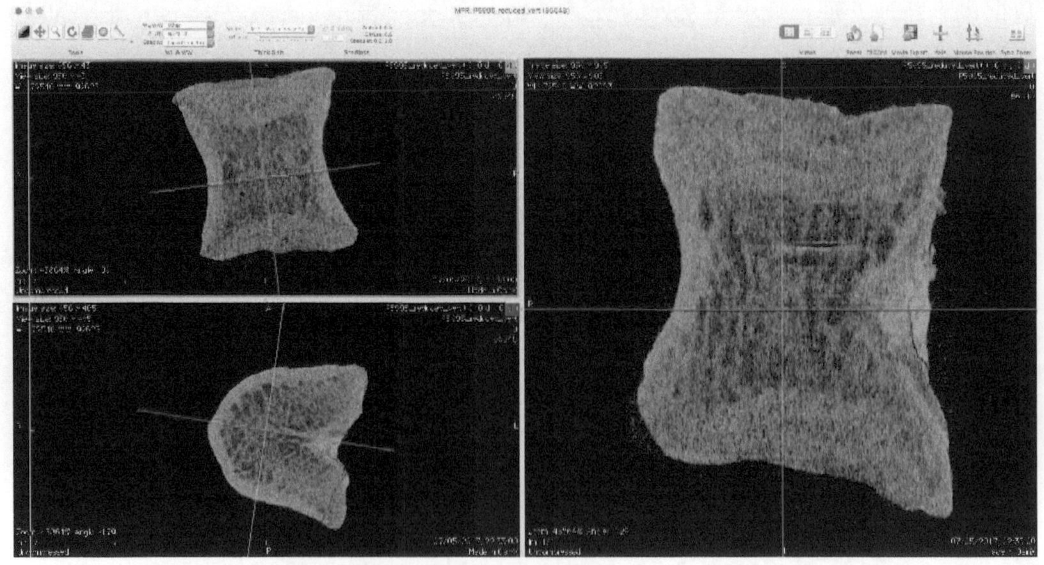

▲ 图 4.3.2　计算机断层扫描多平面数位重建禄丰龙脊椎体

▲ Figure 4.3.2　CT Multi-Plane Reformat (MPR) reconstruction of a *Lufengosaurus* vertebra

体可以用不同的阈值成像，使用不同的颜色来代表不同的解剖构造。体积成像 (volume render) 只限于在一定的阈值下，表现物体的表面体积像，也止于呈现接近想象的表面体积，而在体素成像 (pixel render) 中，利用透明度和颜色可以在单一影像中的特色，就可以呈现更多的东西，例如，骨盆就可以用半透明的方式显现，那么即使是斜位角，小部分其他的解剖成像并不会挡住其他重要的部分。

所谓的窗宽 (window width)，就是指用Hounsfield单位(Hounsfield unit，HU)所得的数据，用来计算出影像的过程，不同的放射强度对应从 0 到 255 (8 bit)不同数位程度的灰阶值，有些先进的设备，甚至可提供到16 bit的灰阶，这些不同的灰阶值，可以依计算机断层扫描值的不同范围，来重新定义衰减值；假设计算机断层扫描范围的中心值不变，定义的范围变窄后，我们称为窄窗位 (narrow window)，比较细部的小变化就可以分辨出来了，在影像处理的观念上，称为对比压缩。

三维重建是用数学的方法，从断层成像器测量到的信号 (X 射线通过样本后的衰减)，重建出样本的三维影像。目前常用的重建方法主要有两种：滤波反投影法 (filtered backprojection) 和卷积反投影法 (convolution backprojection)。

由于目前的X射线计算机断层扫描，都是等方性，亦即*x*，*y*，*z*轴的分辨率都一样，或是接近等方性的分辨率，所以

the pelvis can be shown as semi-transparent, so even if it is tilted, the image of other small anatomic images will not block the view of other important parts.

The so-called Windowing is the process of using the value of Hounsfield Unit (HU) to calculate the image. Different radiodensity corresponds from 0 to 255 (8 bit) gray levels. In some advanced instrument, 16 bit of grey scale is provided. These different grey scales can redefine the attenuation value according to CT scan range. If the value in the CT scan center is kept constant, once the defined range narrowed, what we call Narrow Window, the smaller fine changes can be distinguished. This is called contrast compression.

The three-dimensional reconstruction means to use a mathematic method on the measured signal from the image sensor (X-ray decay after passing thru the sample) to reconstruct the 3D model of the sample. There are two main types: Filtered Backprojection and Convolution Backprojection.

Due to the nature of current X-ray CT is equality, i.e., the resolution of *x*-, *y*-, and *z*-axes are the same or very close to equality; thus the display form does not limit to cross cut section. The software can be used to pile up all Volume Pixels (Voxel, Vox), viewing the image from a different angle can be achieved.

A volumetric pixel (Volume Pixel or Voxel) is the three-dimensional (3D) equivalent of a pixel and the tiniest distinguishable element of a 3D object. It is a volume element that represents a specific grid value in 3D space. However, like pixels, voxels do not contain information about their position in 3D space. Instead, coordinates

显示的方式，不一定只限于横切面，还可借着软件的帮忙，只要把所有的像素体 (volume pixel, voxel, vox) 堆叠起来，就可以用不同的视点来看影像。

所谓的像素体 (volume pixel or voxel)，是一个像素的三维等值，也是三维物件可区分的最小单位，在三维空间的某个位置的单位元素。但是就如像素，像素体并不含它在三维空间位置的数值，而坐标只是相对于周围像素体，可以把像素体想象成砖块，用来堆砌，构成更大的结构，每一块砖块紧邻着旁边的砖块，但是没有定义砖块。像素体能包含多种标尺数值（向量值），如密度、透明度、颜色和体积流速，因此广泛地被应用在科学和医学数据的分析，如计算机断层扫描等。

多层面重建 (multi-plane reformat, MPR) 是最简单的重建方式，它把所有的横切面数据堆叠起来，软件可以用不同的平面来切割物体（大部分是垂直面），或是特别的一些影像，例如最大强度投射成像 (maximum-intensity projection, MIP) 或是最低强度投射成像 (minimum-intensity projection, mIP)；现代的软件可以重建斜位的影像，所以经由自由的选择平面，可以看到想看的解剖构造，比如样本内有的管是斜的，就可以借由这个技术达到所要的目的，即便弯曲的平面，也可透过软件操作重建，使得弯曲的管子或物件可以被“拉直”，如把牙齿排成直线。

体积成像 (volume rendering, VR) 是

are inferred based on their designated positions relative to other surrounding voxels. One may compare volume pixels to bricks, which are stacked and used to build bigger structures. In this scenario, each brick is placed next to each other, but the bricks are not defined. Voxels can contain multiple scalar values (Vector Data), such as density, opacity, color, and volumetric flow rate. Thus, they are used extensively for visualization and analysis of scientific and medical data from devices like CT scanners and X-ray/ ultrasound machines.

Multi-Plane Reformat (MPR) is the easiest reconstruction method. It stacks up all the cross cut planes. The software can use a different plane to cut the object, usually vertical plane, or some special images, such as the Maximum-Intensity Projection (MIP), or Minimum-Intensity Projection (mIP). Modern software can rebuild incline image. Thus desirable anatomic structure can be seen by a freely selected plane. For example, if a tube inside the sample is slanted, it can be done in the desired goal by this technique, even with the curved plane, it can be manipulated by software to show the bent tube to be “straighten”, such as to line up teeth to a straight line.

Volume Rendering (VR): Volume rendering is a technique used to display a 2D projection of a 3D discretely sampled data set, as produced by a microtomography scanner. Usually, these are acquired in a regular pattern (e.g., one slice every millimeter) and usually have a regular number of image pixels in a regular pattern. This is an example of a regular volumetric grid, with each volume element, or voxel represented by a single value that is obtained by sampling

一种用来将二维投射成三维个别采样数据集合的技术，如显微断层扫描。通常这些是以常规格式取得，如每毫米一个切片，而且有固定格式的影像像素，这是定规体积格栅，每个体积元素，以单一数值表示，从该成像体紧邻区域取得。它只展示被扫描物件的最外表面，相当于常见的三维扫描，不过若用微型计算机断层扫描所得的体积成像，分辨率提高很多。

影像分割 (segmentation)：有一些部位虽然结构不同，但是有相似的阻射性，只是单纯地改变像素体成像的参数，可能不是这么简单就可以区分它们，解决的方式称为影像分割，就是用手动或是自动的方式，去除不想要的部分，或者把想要的部分用显著的颜色标示出来，如图 4.3.3。当不同的构造有近似的阈值密度 (threshold density)，就无法简单地调整体积展现参数来区分它们，解决的方法称为影像分割，这是一个手工或自动的过程，把不要的构造除去，如图 4.3.4。

the immediate area surrounding the voxel. It only displays the most outside surface of the scanned object, equivalent to conventional 3D scan, except μ-CT can provide much better resolution.

Image Segmentation: Where different structures have similar threshold density, it can become impossible to separate them simply by adjusting volume rendering parameters. The solution is called segmentation, a manual or automatic procedure that can remove the unwanted structures from the image, as Figure 4.3.3. If the different structures have similar Threshold Density, it can not be distinguished by simple adjustment of whole volume rendering parameters. The solution is called image segmentation. This is done automatically or by hand to remove the unwanted structure, as Figure 4.3.4.

The principle of 3D image reconstruction: Because microtomography scanners offer isotropic, or near isotropic, resolution, display of images does not need to be restricted to the conventional axial images. Instead, it is possible for a software program to build a volume by "stacking" the individual slices one on top of the other. The program may then display the vol-

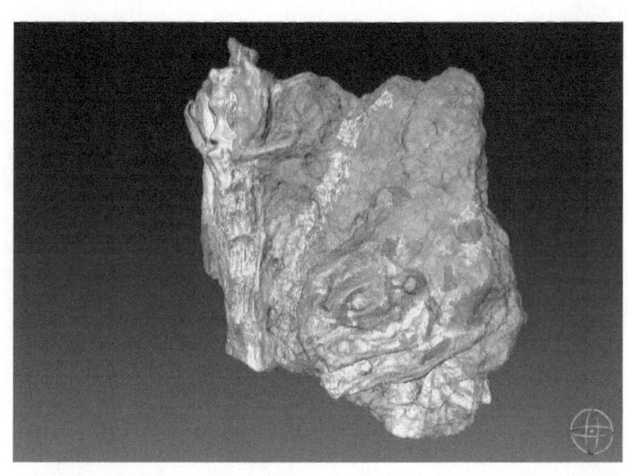

◀ 图 4.3.3 云南禄丰大洼出土的世界最古老、1.95 亿年前禄丰龙胚胎的微型计算机断层扫描，体积成像

◀ Figure 4.3.3 The CT volume rendering of the oldest dinosaur *Lufengosaurus* embryo of 195 million years old from DaWa, Lufeng, YunNan, China

◀ 图 4.3.4 微型计算机断层扫描后的影像数据，经过“影像分割”之后，可以把围岩“去除”，只显示禄丰龙胚胎不同部位，并以不同颜色区分，还可做成视频，仔细观看—这就是“数位清理”可做到的成果

◀ Figure 4.3.4 After segmentation of μ-CT scan. The matrix can be removed and show only the different parts of *Lufengosaurus* embryo with different colors. Video can also be made for careful observation. This is the result that segmentation can do

三维重建的原理：因为显微断层扫描仪提供各向同性 (isotropic) 或几乎各向同性和分辨率，因此展现影像可不必受限于传统的轴向影像，反而可以透过软件以片片层叠来建立某体积，而让该体积的显示以不同的方法展现。

相对于一般医学或工业用计算机断层扫描，微型计算机断层扫描，提供高分辨率或非常高分辨率的X射线扫描，如今它的像素体分辨率，可高达200 nm，而其X射线源，可使用亮度更大的同步辐射光源，对于更微细的形态探讨，这是一种很棒的设备。因为它提供很高分辨率的影像，扫描所得的影像文件大小，也比医学断层扫描者更大很多，因此需要更有力的计算机来处理，也需要更大的硬盘空间来储存所产生的结果影像。

X射线微断层扫描 (X-ray microtomography)，就如断层扫描和X射线计算机断层扫描，使用X射线产生样本横切面影像，用来重建三维虚拟模型，而不会破坏原来的样本，使用前置“微，μ”表

ume in an alternative manner.

μ-CT is a tool using X-ray to scan the object with high or very high resolution in comparison to medical or industrial CT. Nowadays, the VOX (Volume Pixel) resolution can be as high as 200 nm. The X-ray could be provided by synchrotron radiation, which can provide a much much stronger light source. This is an excellent tool for finer morphological details. Since it provides very high-resolution scan images, the image file size is much much more significant than that of conventional CT and requires a much more powerful computer and much bigger storage to post-process the resulting image set.

X-ray Microtomography, like tomography and X-ray computed tomography, uses X-rays, to create cross-sections of a physical object that can be used to recreate a virtual model (3D model) without destroying the original object. The prefix micro- (symbol: μ) is used to indicate that the pixel sizes of the cross-sections are in the micrometer range. These pixel sizes have also resulted in the terms high-resolution X-ray tomography, micro-computed tomography (mi-

示横切面的像素尺寸为微米级，因此也导致“高分辨率X射线断层扫描”、“微型计算机断层扫描”等名称，有时候这两个明称有所区别，但是在其他情况下，使用“高分辨率微型计算机断层扫描”，今日几乎所有的断层扫描，都是计算机断层扫描。

微型计算机断层扫描广泛应用于医学影像和工业方面，基本上有两种设置，其中一种是X射线和检测器在扫描过程中固定，而样本旋转，第二种方式，比较像一般的断层扫描，有个桥架将样本固定，而X射线和检测器旋转，这种通常用于扫描小动物（活体扫描器，*in vivo* scanners）、生物医学样本、食品、小型化石（图4.3.3至图4.3.5）和其他需要很高细节的样本。

因为同步辐射设施可以提供比一般商业独立显微计算机断层扫描更强一两个量级的X射线源，因此利用这种强烈光

cro-CT or μCT), and similar terms. Sometimes the terms high-resolution CT (HRCT) and micro-CT are differentiated, but in other cases, the term high-resolution micro-CT is used. Virtually all tomography today is computed tomography.

Micro-CT has applications both in medical imaging and in industrial computed tomography. In general, there are two types of scanner setups. In one setup, the X-ray source and detector are typically stationary during the scan while the sample/animal rotates. The second setup, much more like a clinical CT scanner, is gantry based where the animal/specimen is stationary in space while the X-ray tube and detector rotate around. These scanners are typically used for small animals (*in vivo* scanners), biomedical samples, foods, microfossils (Figure 4.3.3–4.3.5), and other studies for which minute detail is desired.

Since synchrotron radiation facility can provide one or two orders of magnitude stronger X-ray than the commercial μ-CT, so by using the

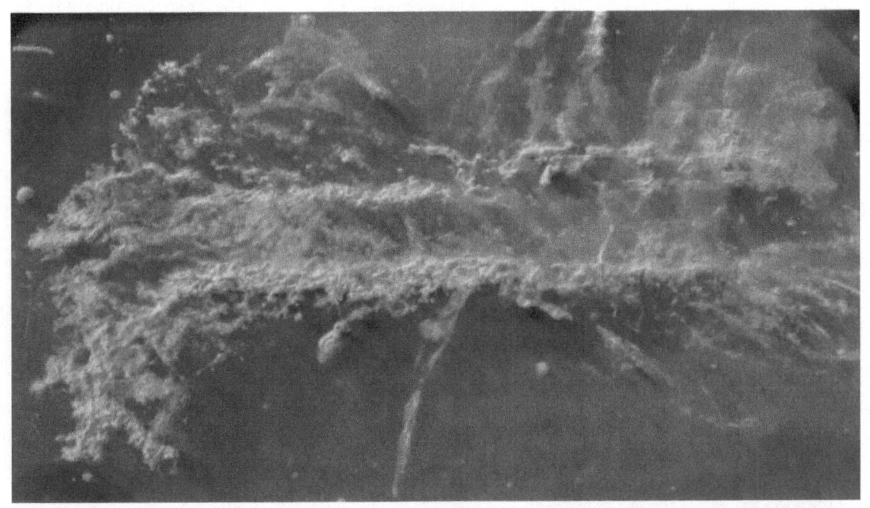

▲ 图 4.3.5 寒武纪纳罗虫微型计算机断层扫描之影像分割，显示其身体内构造

▲ Figure 4.3.5 Segmentation of μ-CT of *Naraoia* of Cambrian explosion showing the internal morphology

源的特性，诸多同步辐射微型计算机断层扫描 (sr-μ-CT)，可提供更大厚样本的显微计算机断层扫描，如图4.3.6，是一般商业仪器X射线无法打得过去的。最早的X射线显微计算机断层扫描系统是 Jim Elliott 在八十年代早期设计建构的，最早发表的X射线显微计算机断层扫描论文，以大约 50 μm 像素重建一个热带蜗牛。

影像重建软件：重建X射线显微断层扫描，有很多商业软件，它们的价格非常昂贵，也只限于授权者可以使用。但是也有很多很好用的公共版权 (public domain) 软件，如美国 (National Institutes of Health, NIH) 所提供的 ImageJ/Fiji 和相关的衍生软件，就是一个开放源码 (open source) 影像处理程序，用于处理科学多领域影像，它很能延伸，也有几千个外挂 (plugins) 和文稿 (scripts) 来处理各种工作，使用者的人数也众多。另外一个也是公共版权的微型计算机断层扫描软件，由国立澳大利亚大学推出的“凝聚视 Drishti”，它从使用者的角度设计：将断层扫描数据形象化、电子显微镜等等。了解数据集合很重要，而将该了解传递给研界者或一般使用者，也是同等重要，凝聚视就是针对这两群使用者，它的基本信念是，科学家应该可以用它来探讨测量体积的数据集合 (volumetric datasets)，也可用于展示（图4.3.6）。

在本节最后，提出一个重要的注意要点：虽然微型计算机断层扫描在古生物形态研究方面，非常具有威力与重要性，但是万万要记得：它并非可治百病

feature of such strong light source, many sr-μ-CT can provide a much thicker μ-CT scan. Such as Figure 4.3.6 cannot be done by the commercial X-ray. The first X-ray microtomography system was conceived and built by Jim Elliott in the early 1980s. The first published X-ray microtomographic images were reconstructed slices of a small tropical snail, with pixel size about 50 micrometers.

Image reconstruction software: There are many commercially available software. They are very expensive and limited for licensed user. However, there are also many excellent public domain software packages. For example, ImageJ / Fiji is an open source image processing program designed for multidimensional scientific images by the National Institutes of Health (NIH). ImageJ is highly extensible, with thousands of plugins and scripts for performing a wide variety of tasks, and a large user community. Drishti has been developed keeping in mind the end-user: visualizing tomography data, electron-microscopy data, etc. Understanding the data set is important and conveying that understanding to the research community or a layperson is equally important. Drishti is aiming for both. The central idea about Drishti is that the scientists should be able to use it for exploring Volumetric Datasets as well as use it in presentations (Figure 4.3.6).

At the end of this section, please keep in mind an important key point: μ-CT is very useful and essential on the morphological paleontology; however, it is NOT a cure-all solution. There are plenty of fossils do not work well, such as paleobotany specimens. This is what I repeatedly emphasized that what was NOT seen could

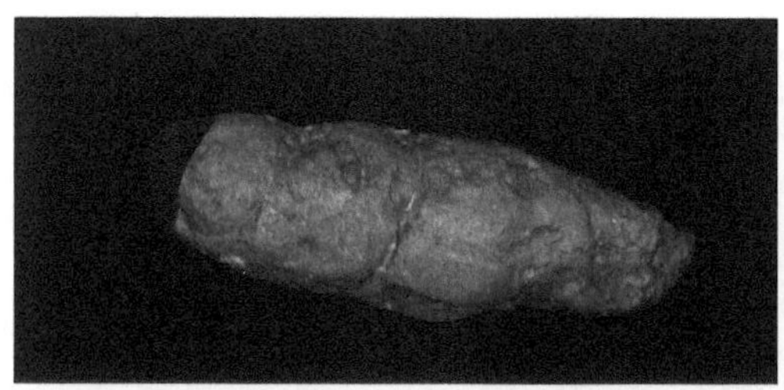

▲ 图 4.3.6 同步辐射显微计算机断层扫描黑龙江粪便化石：左边以体积展现，右边以最大强度投射成像。注意到，体积展现只能看到化石的表面，而最大强度投影成像，则可看到化石里面去。以这个例子来说，此粪便化石里面，包藏着一些东西，透过探讨粪便的内含物或可判断这是什么动物的粪便，它吃什么

▲ Figure 4.3.6 sr-μ-CT Coprolite from HeiLongJiang. The left image is volume rendering, while the right shows the MIP reconstruction. Note that volume rendering just shows the surface of the fossil, while MIP provides the Look into the Fossil. There is something inside the poop! By studying of this coprolite, we can figure out who's poop and what it ate

的仙丹，有很大一堆化石，如古植物，它就显出黔驴技穷的窘状。这也是笔者一再强调：所没有看到的，可能比所看到的更重要。本书第1章就说：用X射线为照射光源的微型计算机断层扫描影像所看到的和用中子为照射源的结果不同？为什么？说穿了，原因很简单：因为样本内不同成分的原子，对于不同照射源有不同的反应啊！举个实际的例子来说，用微型计算机断层扫描古植物化石（和压得扁扁的昆虫化石），所得的结果惨不忍睹，完全看不到啥！有图为证，图 4.3.7，至于为何会如此，请看本章下一节，就会明白。

4.4 中子断层扫描

相对于使用X射线，中子断层扫描用中子束为照射源，不像以X射线为基础的计算机断层扫描和微型计算机断层

be more important than what can be seen. In Chapter one, we described that what the X-ray based μ-CT sees is different from that of using neutron as the radiation source. Why? It's very straightforward: Different atoms inside the sample react differently to different radiation sources. As a real example, using μ-CT to scan plant (and highly flatten insect) fossils, the results are horrible. See Figure 4.3.7 as proof. As for the reason why, please read the next section of this chapter.

4.4 Neutron Tomography

Instead of using X-ray as the radiation source, a neutron beam is used. Unlike the CT and μ-CT, which is based on X-ray density, i.e., heavier atoms will have bigger cross sections. On the other hand, neutron scattering is different for each element and more sensitive to lighter elements, such as C, H, O, P, which are the main constitutional elements for organic remains. So, for searching for organic

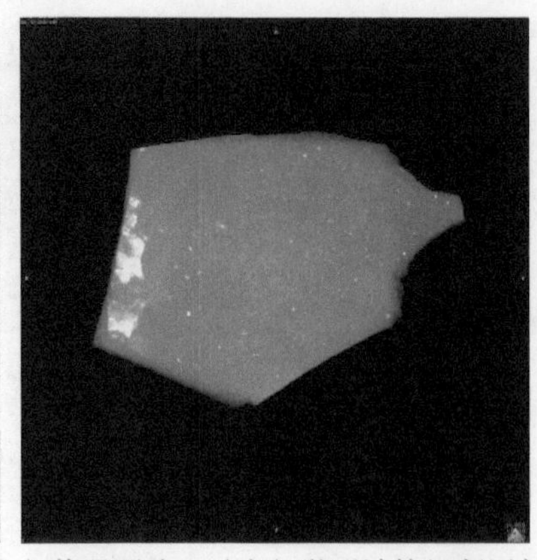

▲ 图 4.3.7 辽西古植物化石（左）普通照片，（右）微型计算机断层扫描 MIP 结果。普通照片中很明显的针状叶片，在微型计算机断层扫描中完全破功，完全看不到

▲ Figure 4.3.7 (Left) Regular optical image of a plant fossil from LiaoXi. Many needle shape leaves can be seen clearly. (Right) MIP reconstruction of μ-CT of the same specimen. Nothing can be seen

扫描，元素随着原子量增加而有更大的截面积，中子对于各元素有不同截面积，较轻的元素如构成有机残留物主要元素的碳、氢、氧、磷等，更为灵敏有效。因此，若要找寻化石或围岩内的有机残留物，中子计算机断层扫描是个必要的尖端威力工具，在微型计算机断层扫描看不到的东西，在中子扫描看得很清楚，这两者相辅相成得很好。

中子断层扫描是一种计算机断层扫描，检测各元素的不同中子截面积，合并已知距离的多个平面影像，产生三维影像。它的最高分辨率已经可达 16 μm，虽然低于高端X射线断层扫描，但可用于要检测低反差的物件和基质的样本，如含有高碳含量的化石，如植物或脊椎动物遗骸，在分辨率方面，根据最近的进展，已经进入 6 μm等级了。

中子扫描不幸的副作用是，如果被

remains inside fossil or matrix, neutron tomography is a potent and necessary cutting edge tool. What cannot be seen in μ-CT can be seen clearly in the neutron scan. These two are excellent complementary to each other.

Neutron tomography is a form of computed tomography involving the production of three-dimensional images by the detection of the cross-section of various elements by neutron radiation. It created a three-dimensional image of an object by combining multiple planar images with a known separation. It has the highest resolution of down to 16 μm as now. While its resolution is lower than that of high X-ray tomography, it can be useful for specimens containing low contrast between the matrix and object of interest; for instance, fossils with high carbon content, such as plants or vertebrate remains. As far as the resolution is concerned, the latest advancement reaches 6μm.

Neutron tomography can have an unfortunate side-effect of leaving imaged samples

检测的样本含有足够量的某些元素，中子照射后的样本，会有辐射性，因此被中子照射过的样本，可能要几个月才能"冷下来"。计算机断层扫描 (tomography) 可用任何对样本有不同穿透吸收的辐射源，中子断层扫描，就是如此之一。中子的吸收情况，和X射线有很大的不同，比方说，中子被有机物吸收强于被岩石吸收，因此中子扫描更适合于检测保存着有机残留物的化石，如植物，也可用于较大的脊椎材料，与X射线计算机断层扫描相辅相成。然而中子扫描的分辨率，若以最小像素体积来说（大约100 mm），比微型计算机断层扫描 (XMT, µ-CT) 和同步辐射X射线断层扫描显微术 (sr-XMT) 来得较低（可达纳米级）。不过随着科技进步，此差距也日渐减少，再者对于某些地质材料，被中子轰炸可能引发危险的放射辐射，特别是含有钴 (Co) 和铕 (Eu) (M. Dawson 2008, personal communication)，中子照射过后，样品可能要放着好几个月或好几年（图4.4.1至图4.4.3）。(*Tomographic techniques for the study of exceptionally preserved fossils, Mark D. Sutton*)

radioactive if they contain significant levels of certain elements. So the neutron radiated specimen may have to let it "cold down" for several months. Computed axial tomography can use any penetrative radiation differentially absorbed by a sample; neutron tomography (NT) represents one such alternative. The absorption profile of neutrons is very different from that of X-rays; they are, for instance, more strongly attenuated by organic material than by most rock. NT is hence appropriate for the imaging of organically preserved fossils (e.g., plants); it may also be useful for larger vertebrate material as a complement to X-ray CT. NT resolution is, however, lower than that of XMT (X-ray micro-tomography (XMT)) and SRXTM (synchrotron radiation X-ray tomographic microscopy) in terms of minimum voxel size (approx. 100 mm). However, it is catching up quickly along with the advancement of technology. The high-intensity neutron bombardment can induce hazardous levels of radioactivity in some geological materials, for example, those containing cobalt or europium; samples may thus need to be interred for months or years after NT study (Figure 4.4.1, Figure 4.4.2 and Figure 4.4.3). (*Tomographic techniques for the study of exceptionally preserved fossils, Mark D. Sutton*)

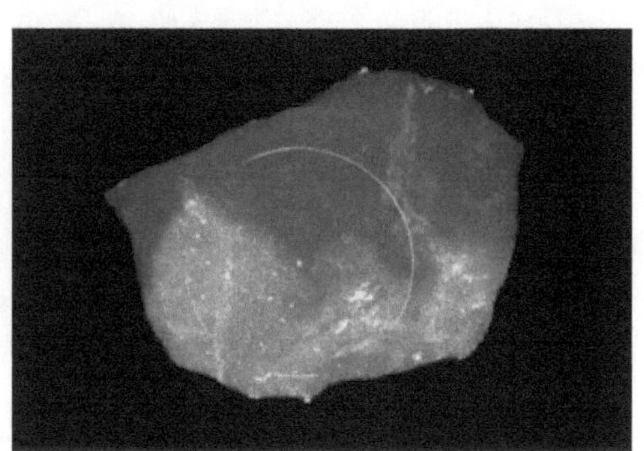

◀ 图 4.4.1 在澳大利亚核科技组织做的中子扫描，一块大椎龙胚胎围岩，显示可能是在蛋窝内残留的植物或蛋壳，这些在 µ-CT 是看不到的

◀ Figure 4.4.1 Neutron scan (done at ANSTO) of a Massospondylus embryo matrix. It shows likely the remains of plant material or eggshell fragments inside the nest. These could not be seen by µ-CT

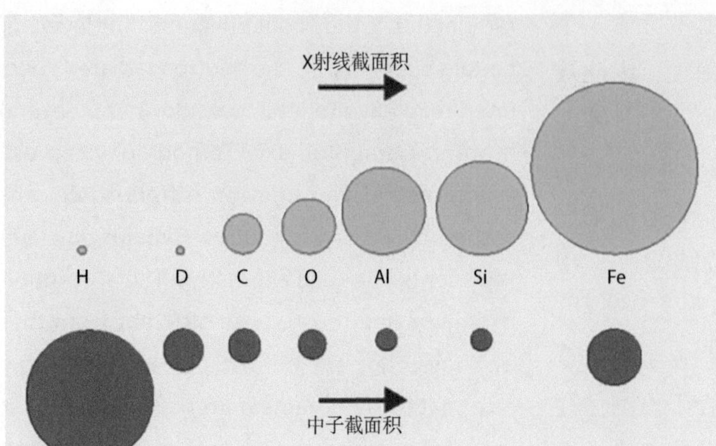

◀ 图 4.4.2　几种元素的中子与 X 射线截面积示意图

◀ Figure 4.4.2　Some elemental cross areas of neutron and X-ray

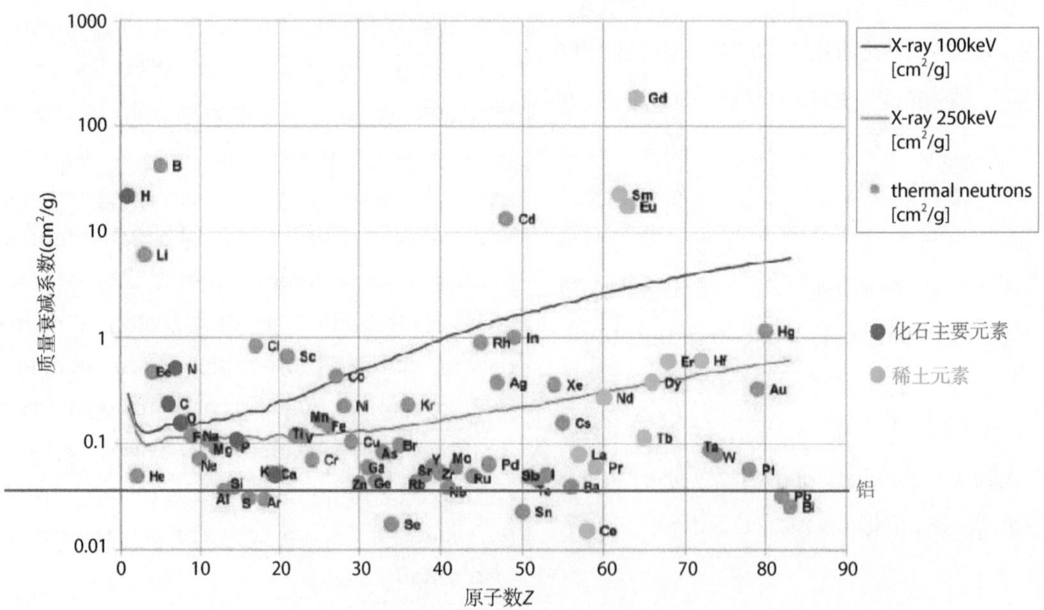

▲ 图 4.4.3　诸元素的中子截面图。请注意图下端的红线，标示为铝（罐），亦即对中子来说，铝是"透明"的，在此线以下的元素，都会是中子扫描看不到的元素。而在此线以上的元素，以绿色点标示的为镧系稀土元素，以红色点标示的为化石内常见的元素，这些都是在骨头化石内常见的成分，特别是有机残留物，可在中子扫描中清楚地显示出来

▲ Figure 4.4.3　Neutron cross area of various elements. Note: the red line around the bottom represents aluminum (can). That is as far as the neutron is concerned, Aluminum is "transparent". The neutron won't be seen any element below this line. Among the elements above this line, Rare Earth Elements are colored in green while the major common elements inside fossils are in red, particularly the elements of organic remains can be seen by neutron

4.5 小角度中子 /X 射线散射*

小角度散射 (small-angle scattering, SAS) 是一种基于使用准直、远比辐射波长更长的辐照到样本构造而从正直照射轨迹所产生的折射，折射角度很小（0.1°~10°），因此称为小角度。小角度散射可提供样本内结构的大小、形状和排列。

探讨从10 Å到甚至好几万Å大尺寸结构，小角度散射是个很有力的技术，最重要的特性是，它能分析混乱系统的内部构造，这种分析方法的应用，经常是一种最直接取得不均匀密度随意排列系统构造信息独特的方法。

目前，小角度技术，有很好的实验和理论基础及广泛的研究物件，是物质结构分析自给自足的领域。小角度散射很有用，因为大幅增加相转换 (phase transitions)（又称临界乳光，critical opalescence）前向散射，又因为很多材料、物质和生物系统，在结构上含着有趣复杂特征，刚好对应于这项术有用的长度尺寸范围，这种技术提供广泛科学和技术方面很有用的信息，包括化学聚合、材料内缺陷、表面剂、胶体、磁铁铁磁关系、合金分离、聚合物、蛋白质、生物膜、病毒、核糖体和巨大分子；虽然数据分析可提供大小、形状等信息，而不必做任何模型假设，初步的数据分析只能使用Guinier公式提供某粒子旋转

* 取自维基。
From Wiki.

4.5 Small Angle Neutron/X-ray Scattering*

Small-Angle Scattering (SAS) is a scattering technique based on the deflection of collimated radiation away from the straight trajectory after it interacts with structures that are much larger than the wavelength of the radiation. The deflection is small (0.1°-10°) hence the name small-angle. SAS techniques can give information about the size, shape, and orientation of structures in a sample.

Small-angle scattering (SAS) is a powerful technique for investigating large-scale structures from 10 Å up to thousands and even several tens of thousands of angstroms. The most important feature of the SAS method is its potential for analyzing the inner structure of disordered systems, and frequently the application of this method is a unique way to obtain direct structural information on systems with a random arrangement of density inhomogeneities in such large-scales.

Currently, the SAS technique, with its well-developed experimental and theoretical procedures and a wide range of studied objects, is a self-contained branch of the structural analysis of the matter. Small-angle scattering is particularly useful because of the dramatic increase in forwarding scattering that occurs at phase transitions, known as critical opalescence, and because many materials, substances, and biological systems possess exciting and sophisticated features in their structure, which match the useful length scale ranges that these tech-

半径的信息。

小角度中子散射 (SANS) 是一种实验性的技术，使用小角度弹性中子散射，来探讨诸多物质约在1~100 nm的介观尺寸 (mesoscopic scale)。小角度中子散射，在很多方面类似小角度X射线散射 (SAXS)，两者合称小角度散射 (SAS)；小角度中子散射比小角度X射线散射好的地方，是在它对于轻元素的灵敏度、同位素标示和对磁矩有很强的散射。

小角度X射线散射，对于纳米级不均匀样本产生对波长 0.1~0.2 nm X射线的弹性散射，以很低角度，通常在0.1°~10°做记录，在这些角度范围，包含了巨大分子的形状和大小、部分顺序物质的特征距离、孔隙大小和其他数据。SAXS能提供部分最大为150 nm的顺序系统巨大分子5~25 nm重复性距离。特小角度X射线散射 (ultra-small angle X-ray scattering, USAXS) 可以提供更大尺寸的解析。

小角度X射线散射 (SAXS) 和特小角度X射线散射 (USAXS) 都属于X射线散射技术家族，用来描述材料，对于生物大分子，如蛋白质，由于不必有结晶样本，小角度X射线散射胜于结晶学。核磁共振方法，当高分子量巨大分子 (> 30~40 kDa) 时，就会出现问题，然而因为溶解或部分顺序的分子随意方向，方向上的平均导致小角度X射线散射，这导致小角度X射线散射相对应于结晶学，损失了一些信息。

小角度X射线散射用于检测显微级

niques probe. The technique provides valuable information over a wide variety of scientific and technological applications including chemical aggregation, defects in materials, surfactants, colloids, ferromagnetic correlations in magnetism, alloy segregation, polymers, proteins, biological membranes, viruses, ribosome, and macromolecules. While analysis of the data can give information on size, shape, etc., without making any model assumptions a preliminary analysis of the data can only give information on the radius of gyration for a particle using Guinier's equation.

Small-angle neutron scattering (SANS) is an experimental technique that uses elastic neutron scattering at small scattering angles to investigate the structure of various substances at a mesoscopic scale of about 1–100 nm. Small angle neutron scattering is in many respects very similar to small-angle X-ray scattering (SAXS); both techniques are jointly referred to as small-angle scattering (SAS). Advantages of SANS over SAXS are its sensitivity to light elements, the possibility of isotope labeling, and the strong scattering by magnetic moments.

Small-angle X-ray scattering (SAXS) is a technique where the elastic scattering of X-rays (wavelength 0.1–0.2 nm) by a sample which has inhomogeneities in the nm-range, is recorded at shallow angles (typically 0.1° - 10°). This angular range contains information about the shape and size of macromolecules, characteristic distances of partially ordered materials, pore sizes, and other data. SAXS is capable of delivering structural information of macromolecules between 5 and 25 nm, of repeat distances in partially ordered systems of up to 150 nm. USAXS

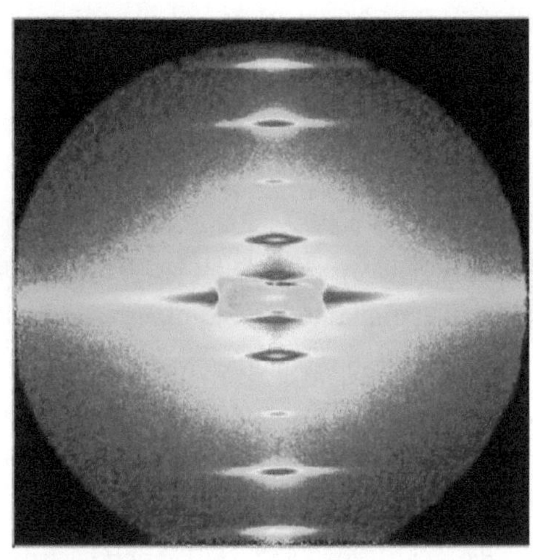

Rat tail tendon
Camera length: 9.35 m
Wavelength: 0.154 nm

图 4.5.1 胶原蛋白的小角度中子/X 射线散射

Figure 4.5.1 SANS/SAXS of collagen

或纳米级颗粒体系的一些参数，如粒子大小、形状、分布和表面与体积比例，材料可以是固体或液体，能含有同样物质或与其他物质任何混合的固态、液态或气态粒子。不止粒子，如薄板和碎形材料，都可以它来研究。这个方法精确、非破坏性，只需小量样本制备，应用范围很广泛，包括各种胶质、金属、水泥、油、聚合物、塑胶、蛋白质、食物和药品（图4.5.1），很多研究单位也将其用做品质控制。X射线源可以是实验室的或能提供更高X射线流明的同步辐射光。

生物小角度散射用来做生物材料的结构分析。小角度散射用于研究诸多物体的结构，如生物大分子溶液、纳米组合、合金和合成聚合物。小角度X射线散射和小角度中子散射，是两种相辅相成的技术，合称为小角度散射。小角度

(ultra-small angle X-ray scattering) can resolve even larger dimensions.

SAXS and USAXS belong to a family of X-ray scattering techniques that are used in the characterization of materials. In the case of biological macromolecules such as proteins, the advantage of SAXS over crystallography is that a crystalline sample is not needed. Nuclear magnetic resonance spectroscopy methods encounter problems with macromolecules of higher molecular mass (> 30–40 kDa). However, owing to the random orientation of dissolved or partially ordered molecules, the spatial averaging leads to a loss of information in SAXS compared to crystallography.

SAXS is used for the determination of the microscale or nanoscale structure of particle systems in terms of such parameters as averaged particle sizes, shapes, distribution, and surface-to-volume ratio. The materials can be

散射等同于X射线和中子折射，广角X射线散射和静态光（static light）散射。但是该方法和其他X射线和中子散射方法的区分，在于它产生结晶和非结晶体粒子的大小和形状信息，当用来研究生物材料，通常为水溶液，散射格式是平均方向。

小角度散射格式在几度小角范围收集，能提供 1~25 nm 分辨率的结构资料，在部分顺序系统，重复距离可到 150 nm 大小。超小角度散射 (ultra small-angle scattering, USAS) 可以解析更大尺寸，掠射小角度散射 (grazing-incidence small-angle scattering, GISAS) 是研究生物表面分子层的有力技术。

在生物学应用方面，小角度散射用来决定粒子的结构，平均粒子大小和形状，也可用来取得表面积体积比例。通常生物巨大分子分散于液体中，这种方法很精准，大部分非破坏性，只需要很少量的样本制备，然而生物分子都要考虑辐射破坏。

从观念上来说，小角度散射实验很简单，样本被X射线或中子照射，检测器记录散射辐射，因为小角度测量，非常接近于主要光束（小角度），这种技术需要一个很准直或聚焦X射线或中子束，生物小角度X射线散射通常在同步辐射设施进行，因为生物分子通常微弱散射，所测量的溶液也很稀薄。生物小角度X射线散射，可由同步辐射所提供的高强度X射线得益，X射线或中子散射曲线（强度对散射角度）用来产生蛋白

solid or liquid, and they can contain solid, liquid, or gaseous domains (so-called particles) of the same or another material in any combination. Not only particles, but also the structure of ordered systems like lamellae and fractal-like materials can be studied. The method is accurate, non-destructive, and usually requires only a minimum of sample preparation. Applications are vast and include colloids of all types, metals, cement, oil, polymers, plastics, proteins, foods and pharmaceuticals and can be found in research as well as in quality control (Figure 4.5.1). The X-ray source can be a laboratory source or synchrotron light which provides a higher X-ray flux.

Biological small-angle scattering is a small-angle scattering method for structure analysis of biological materials. Small-angle scattering is used to study the structure of a variety of objects such as solutions of biological macromolecules, nanocomposites, alloys, and synthetic polymers. Small-angle X-ray scattering (SAXS) and small-angle neutron scattering (SANS) are the two complementary techniques known jointly as small-angle scattering (SAS). SAS is an analogous method to X-ray and neutron diffraction, wide-angle X-ray scattering, as well as to static light scattering. In separation to the other X-ray and neutron scattering methods, SAS yields information on the sizes and shapes of both crystalline and non-crystalline particles. When used to study biological materials, which are very often in aqueous solution, the scattering pattern is orientation averaged.

SAS patterns are collected at small angles of a few degrees. SAS is capable of delivering structural information in the resolution range between 1 and 25 nm, and of repeat distances in partially ordered systems of up to 150 nm in

质低分辨率模型。

相对于其他结构测定方法，如液态核磁共振 (NMR) 或X射线晶体学，小角度散射可以免去一些限制，比方说，液体核磁共振，受限于蛋白质大小，而小角度散射可用于小分子，也可用于大多分子组合，固态核磁共振在测定大于 40 kDa 巨大分子或非结晶样本（如类淀粉蛋白小纤维）提供原子层面信息，还是不可或缺的。使用X射线晶体学决定构造，可能需要好几个礼拜或好几年，而小角度只要几天，然而，小角度散射无法测量在分子内的原子位置。

4.6　傅里叶变换红外光谱 / 同步辐射傅里叶变换红外光谱

红外光谱是个成熟而且经常用于化学分析的工具。不同的官能团 (functional group) 会在不同波长位置出现不同的波峰，同步辐射设施提供更强的红外线光源，因而可提高分辨率，是一般商业产品无法比较的。

（来自维基）在有机化学，官能团是分子内代表某特定化学反应的原子群或键群（部分），不管它所属的分子大小，同样的官能团会有相同或类似的化学反应，然而它的相对反应性，会受到邻近官能团的影响。官能团的原子连接在一起，再以共价键连接到分子的其他部位。某化合物的原子次群，也可叫做游离基 (radical)，如果某共价键被均裂，所产生

size. Ultra-small-angle-scattering (USAS) can resolve even larger dimensions. The grazing-incidence small-angle scattering (GISAS) is a powerful technique for studying biological molecule layers on surfaces.

In biological applications, SAS is used to determine the structure of a particle in terms of average particle size and shape. One can also get information on the surface-to-volume ratio. Typically, the biological macromolecules are dispersed in a liquid. The method is accurate, mostly non-destructive and usually requires only a minimum of sample preparation. However, biological molecules are always susceptible to radiation damage.

Conceptually, small-angle scattering experiments are simple: the sample is exposed to X-rays or neutrons, and a detector registers the scattered radiation. As the SAS measurements are performed very close to the primary beam ("small angles"), the technique needs a highly collimated or focused X-ray or neutron beam. The biological small-angle X-ray scattering is often performed at synchrotron radiation sources because biological molecules normally scatter weakly and the measured solutions are dilute. The biological SAXS method profits from the high intensity of X-ray photon beams provided by the synchrotron storage rings. The X-ray or neutron scattering curve (intensity versus scattering angle) is used to create a low-resolution model of a protein.

In comparison to other structure determination methods, such as solution NMR or X-ray crystallography, SAS allows one to overcome some restraints. For example, solution NMR is limited to protein size, whereas SAS can be used

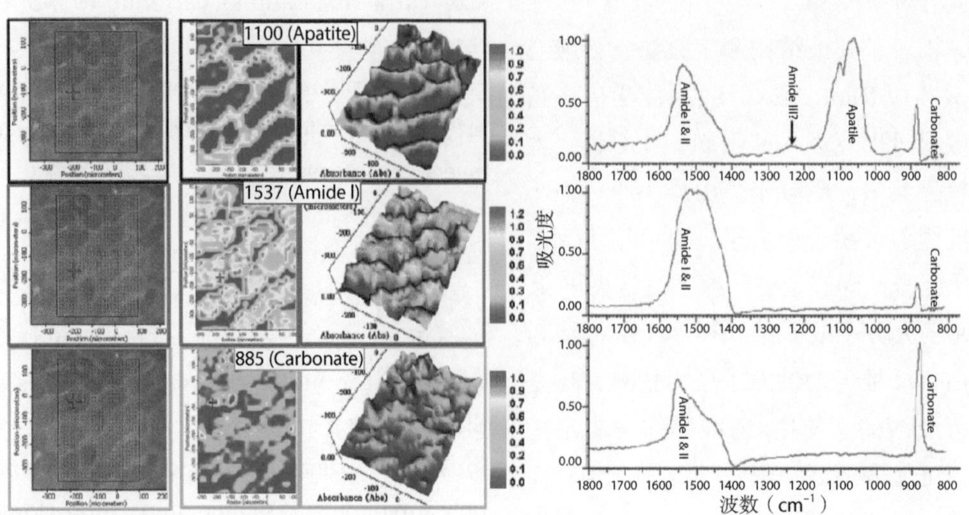

▲ 图 4.6.1 1.95 亿年前禄丰龙胚胎的同步辐射傅里叶变换红外光谱，明显展示保存在原位的第一类胶原蛋白

▲ Figure 4.6.1 sr-FTIR spectra of 195 million years old *Lufengosaurus* embryonic bone. It clearly shows the preservation of organic remains, collagen type I *in situ*

的自由基片段称之为“自由游离基”；官能团可能带电，如碳酸根 ($—COO^-$)，它把分子转变成多原子离子或一个复合离子。

傅里叶变换红外光谱 (FTIR) 是一种取得固体、液体或气体的吸收或发射光谱，傅里叶变换红外光谱仪同时收集广泛光谱范围的高分辨率数据，这就比一次只能测量某狭窄波长色散光谱仪 (dispersive spectroscopy) 有更大的优点。

傅里叶变换红外光谱这个名词，来自于它需要经过数学运算傅里叶变换 (fourier transform)，将原始数据转换成光谱。任何吸收光谱，傅里叶变换红外光谱、紫外–可见光谱 (UV-Vis) 等等，

for small molecules as well as for large multi-molecular assemblies. Solid-State NMR is still an indispensable tool for determining atomic level information of macromolecules greater than 40 kDa or non-crystalline samples such as amyloid fibrils. Structure determination by X-ray crystallography may take several weeks or even years, whereas SAS measurements take days. However, with SAS, it is not possible to measure the positions of the atoms within the molecule.

4.6 FTIR/sr-FTIR

Infrared spectroscopy is a well developed and commonly used chemical analysis tool. The different functional group will yield a different peak(s) at a different wavelength. Synchrotron Radiation source provides much stronger IR

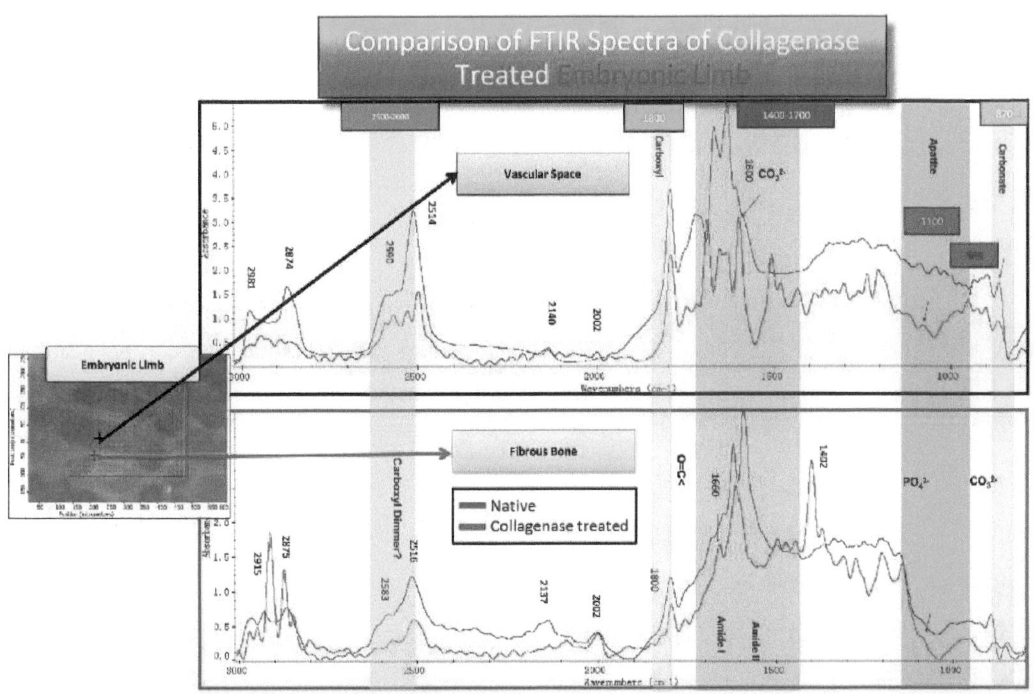

▲ 图 4.6.2　胶原蛋白经过胶原蛋白分解酶处理前（蓝）后（红）FTIR 光谱比较

▲ Figure 4.6.2　Comparison of FTIR spectra of collagen before (blue) and after (red) collagenase digestion

需要测量样本在每个波长吸收有多好，最直接的方法是“色散光谱 (dispersive spectroscopy)”，以单色光照射样本，测量多少光被吸收，然后重复每个波长。

傅里叶变换红外线光谱，用比较不直接的方法取得同样的信息，它照射到样本的不是单色光，而是同时照射很多频率的光，测量该光束被样本吸收多少，接着修正光束为不同频率组合，取得第二个据点，这个程序重复很多次，之后借由计算机反向从所有数据计算出每个波长的吸收（图4.6.1和图4.6.2）。

上述描述的光束是用宽带光源产

light source, hence increases the resolution, which the commercial unit can't match.

(From Wiki) In organic chemistry, functional groups are specific groups (moieties) of atoms or bonds within molecules that are responsible for the characteristic chemical reactions of those molecules. The same functional group will undergo the same or similar chemical reaction(s) regardless of the size of the molecule it is a part of. However, its relative reactivity can be modified by other functional groups nearby. The atoms of functional groups are linked to each other and to the rest of the molecule by covalent bonds. Any subgroup of atoms of a compound also may be called a radical, and if a covalent bond is broken homolytically, the resulting fragment radicals are referred to as free

生，包含整个要测量所有光谱波长，这个光照射到一系列镜子其中一个以马达移动的“迈克耳孙干涉仪 (Michelson interferometer)”，当这个镜子移动，光束里面的每个光波长，由于波干涉周期性地被干涉仪遮掩、传送、遮掩、传送。不同的波长以不同速率被调整，因此在某个瞬间，从干涉仪出来的光束，有一个不同的光谱。

如上述,需要计算机将原始数据（每个镜子位置的光吸收）转换成所要的结果，亦即每个波长的光吸收，这个所需要的过程是个普通演绎式，叫做傅里叶变换，因此称为傅里叶变换光谱，而原始数据则称为“干涉图”。

北达科他州地狱溪地层鸭嘴龙木乃伊之软组织矿构造与化学

http://rspb.royalsocietypublishing.org/content/276/1672/3429

4.7 拉曼计算机断层扫描

随着科技进步，拉曼光谱不再受限于检测单一个点或二维面积的光谱，而是能深入 200 μm 的三维空间，提供检测化石表面的非破坏性光谱分析。

（来自维基）拉曼光谱是一种用来观测系统振动、滚动和低频模式的光谱技术。经常使用于化学，提供指纹来鉴定分子。它依据单色光的非弹性散射或称为拉曼散射 (Raman scattering)，通常用可见、近红外或近紫外光激光，激光与系统内分子振动、声子(phonons)或其他激发 (excitation)，导致激光光子能量

radicals. Functional groups can also be charged, e.g., in carboxylate salts (—COO^-), which turns the molecule into a polyatomic ion or a complex ion.

Fourier transform infrared spectroscopy (FTIR) is a technique which is used to obtain an infrared spectrum of absorption or emission of a solid, liquid or gas. An FTIR spectrometer simultaneously collects high spectral resolution data over a wide spectral range. This confers a significant advantage over a dispersive spectrometer which measures intensity over a narrow range of wavelengths at a time.

The term *Fourier transform infrared spectroscopy* originates from the fact that a Fourier transform (a mathematical process) is required to convert the raw data into the actual spectrum. The goal of any absorption spectroscopy (FTIR, ultraviolet-visible ("UV-Vis") spectroscopy, etc.) is to measure how well a sample absorbs light at each wavelength. The most straightforward way to do this, the "dispersive spectroscopy" technique, is to shine a monochromatic light beam at a sample, measure how much of the light is absorbed, and repeat for each different wavelength.

Fourier transform spectroscopy is a less intuitive way to obtain the same information. Rather than shining a monochromatic beam of light at the sample, this technique shines a beam containing many frequencies of light at once, and measures how much of that beam is absorbed by the sample. Next, the beam is modified to contain a different combination of frequencies, giving a second data point. This process is repeated many times. Afterward, a computer takes all this data and works backward to infer what the absorption is at each wavelength (Fig-

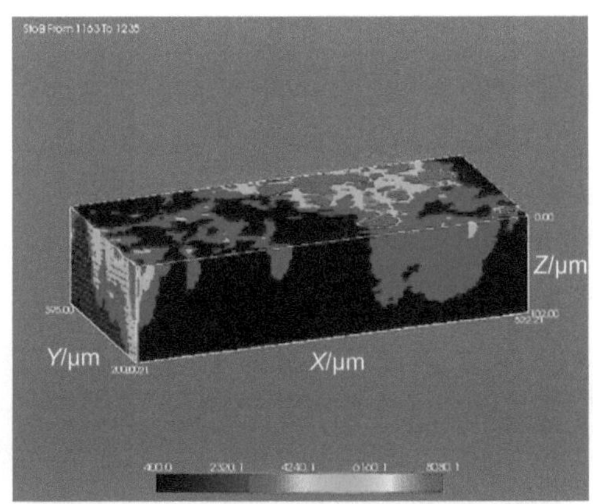

▲ 图 4.7.1 1.95 亿年前禄丰龙胚胎之酰胺 3 (Amide III) 三维拉曼光谱，显示此官能团的三维分布；（胶原蛋白的）酰胺 3 官能团三维分布，与高分辨率 μ-CT 结果之原始管状空间（第 1 章图 1.1.7），对应得很好；黑色表示无，红色为很强

▲ Figure 4.7.1 3D Raman spectra of 195 million years old *Lufengosaurus* embryonic bone of the Amide III peak. It shows the 3D distribution of this particular functional group. The 3D distribution of Amide III functional group (of collagen) matched very well with the results from high-resolution μ-CT of the Primary Tubular Cavity (as Figure 1.1.7). Peak strength: Black is none, Red is very strong

往上或往下移动，这个能量的移动，提供了系统振动模式的信息；红外线也提供类似但互补的信息（图4.7.1）。

通常样本被激光照射，被照射点的电磁辐射被透镜记录下来，传送到单光器。在此波长弹性散射的辐射对应到激光线（瑞利散射，Rayleigh scattering）被槽滤光片 (Notch filter)、高低通滤光片 (edge pass filter) 或带通滤光片 (band pass filter) 过滤掉，收集剩下来的光分散于检测器。

ure 4.6.1 and Figure 4.6.2).

The beam described above is generated by starting with a broadband light source—one containing the full spectrum of wavelengths to be measured. The light shines into a Michelson interferometer—a specific configuration of mirrors, one of which is moved by a motor. As this mirror moves, each wavelength of light in the beam is periodically blocked, transmitted, blocked, transmitted, by the interferometer, due to wave interference. Different wavelengths are modulated at different rates so that at each moment, the beam coming out of the interferometer has a different spectrum.

As mentioned, computer processing is required to turn the raw data (light absorption for each mirror position) into the desired result (light absorption for each wavelength). The processing required turns out to be a standard algorithm called the Fourier Transform (hence the name, "Fourier Transform Spectroscopy"). The raw data is sometimes called an "interferogram".

http://rspb.royalsocietypublishing.org/content/276/1672/3429

Mineralized soft-tissue structure and chemistry in a mummified hadrosaur from the Hell Creek Formation, North Dakota (USA), Phillip L. Manning, Peter M. Morris, Adam McMahon, Emrys Jones, Andy Gize, Joe H. S. Macquaker, George Wolff, Anu Thompson, Jim Marshall, Kevin G. Taylor, Tyler Lyson, Simon Gaskell, Onrapak Reamtong, William I. Sellers, Bart E. van Dongen, Mike Buckley, Roy A. Wogelius

4.7 Raman-CT

With the advancement of technology, Raman spectroscopy is no longer limited to measure the

同时发生的拉曼散射通常很弱，因此拉曼光谱的主要困难在于将弱的非弹性散射光，从强烈瑞利散射激光分开。历史上的拉曼光谱仪，使用全息栅 (holographic grating) 和多重分散阶来达到高度排除激光，过去也用光电倍增器 (photomultiplier) 作为拉曼设备的分散检测器，因此取得数据的时间很长。然而，现代的仪器几乎都用槽滤光片或带通滤光片来排除激光，光谱仪或是轴发射 (axial transmissive, AT)、CT单 光 器 (Czerny–Turner (CT) monochromator)、傅里叶变换或电荷耦合器件 (CCD) 检测器。

请看图 4.7.2，这是从 Raman-CT 取得的禄丰龙胚胎骨头化学成分三维分布。蓝色的是磷灰石群（以磷酸根波峰表示）、绿色的是碳酸盐（以碳酸根波峰表示）、红色的是第一类胶原蛋白（以酰胺3波峰表示）。从此图可看出至少两个有趣的要点：①磷灰石群矿物形成于第一类胶原蛋白的表面，这是已知的，本图很清晰明显地在原位在同点显示在胚胎阶段磷灰石群矿物的形成。此外，有好多红色胶原蛋白区域，附着着蓝色的磷灰石群，显示高百分比的原始管状空间 (primary tubular cavity, PTC)；②绿色的碳酸盐并没有从表面进入太深，大约只有一半的深度而已，这和第7章7.2节所说的碳酸盐胶合 (carbonate cementing) 吻合，碳酸盐形成包裹层于化石表面，并未深入。在恐龙牙齿化石，也看到类似的情况，因此有机残留物的保存是否和碳酸盐胶合有没有关系？还

spectrum at a single spot or two-dimensional area, but in 3D with penetration of 200μm is available. This provides a non-destructive spectroscopic analysis on the surface of fossils.

(From Wiki) Raman spectroscopy is a spectroscopic technique used to observe vibrational, rotational, and other low-frequency modes in a system. Raman spectroscopy is commonly used in chemistry to provide a fingerprint by which molecules can be identified. It relies on inelastic scattering, or Raman scattering, of monochromatic light, usually from a laser in the visible, near infrared, or near ultraviolet range. The laser light interacts with molecular vibrations, phonons or other excitations in the system, resulting in the energy of the laser photons being shifted up or down. The shift in energy gives information about the vibrational modes in the system. Infrared spectroscopy yields similar, but complementary information (Figure 4.7.1).

Typically, a sample is illuminated with a laser beam. Electromagnetic radiation from the illuminated spot is collected with a lens and sent through a mono-chromator. Elastically scattered radiation at the wavelength corresponding to the laser line (Rayleigh scattering) is filtered out by either a notch filter, edge pass filter, or a bandpass filter, while the rest of the collected light is dispersed onto a detector.

Spontaneous Raman scattering is typically very weak, and as a result, the main difficulty of Raman spectroscopy is separating the weak inelastically scattered light from the intense Rayleigh scattered laser light. Historically, Raman spectrometers used holographic gratings and multiple dispersion stages to achieve a high degree of laser rejection. In the past, photomulti-

Lufengosaurus Embryo Raman-CT
Apatile, 1010~1040 cm^{-1}
Amide_III, 1239~1320cm^{-1}
Calcite, 1078~1083 cm^{-1}

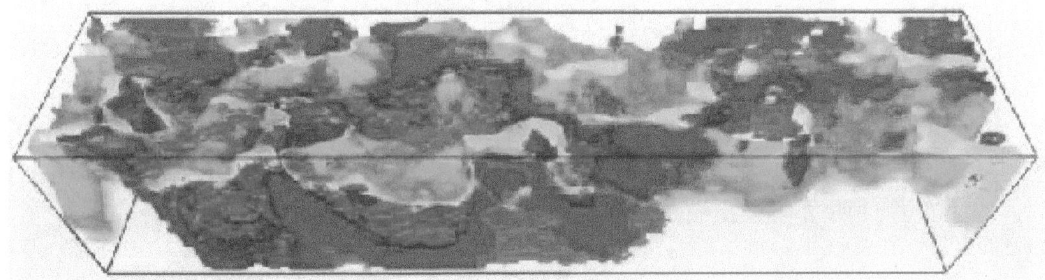

▲ 图 4.7.2 1.95 亿年前禄丰龙胚胎三种化学成分（磷灰石、第一类胶原蛋白、方解石）三维拉曼光谱分布。这应该是首次能看到化石内不同化学成分三维分布的纪录，此图与上述图 4.7.1 不同之处在于，前者只是单一化学成分（酰胺 3）的分布，而本图是三种不同化合物的分布，可看出三种化合物彼此的相对位置与关系

▲ Figure 4.7.2 3D Raman spectra of apatite, collagen type I, and calcite 3D distribution of a 195 million years old *Lufengosaurus* embryonic bone. The difference of this from Figure 4.7.1 is that the previous figure just shows the 3D distribution of only one compound (Amide III), but this figure shows three. From this 3D of three compounds, their relative distribution and relationship can be easily seen

需深入探讨。

4.8 多频显微

多频显微是台湾大学孙启光教授很杰出的医学用发明，在强力 1230 nm 红外线激光照射下，组合了二光子 (two photon, TPH)、二倍频 (second harmonic generation, SHG) 和三倍频 (third harmonic generation, THG) 非线

pliers were the detectors of choice for dispersive Raman setups, which resulted in long acquisition times. However, modern instrumentation almost universally employs notch or edge filters for laser rejection and spectrographs either axial transmissive (AT), Czerny–Turner (CT) monochromator, or FT (Fourier transform spectroscopy based), and CCD detectors.

Please look at the above Figure 4.7.2. This is a 3D chemical composition distribution image obtained from the Raman-CT scan. The blue is the mineral apatites (represented by phosphate peaks), green is the carbonates (represented by carbonate peaks), and red is the collagen type I (represented by Amide III peaks). From this image at least two important and interesting things can be seen: ① As known, apatites minerals were formed on the surface of collagen type I. This image clearly show *in situ in loco* formation of apatites during the embryonic stage. Furthermore, there are plenty of red collagen areas with blue apatites attached showing high percentage primary tubular cavity. ② The green carbonates did not penetrate too deeply, just about half way or less in this image. This

性光学反应，它是个亚微米 (0.47 μm)、非破坏性、快速、无光漂白、大面积三维的影像系统，对于化石也很好用。孙教授的设备，原本是为活体组织设计的，我们恐龙胚胎团队在高档期刊不只发表了一篇而是两篇论文，其中一篇还是封面论文。

虽然已经证明了它对我们是个很有

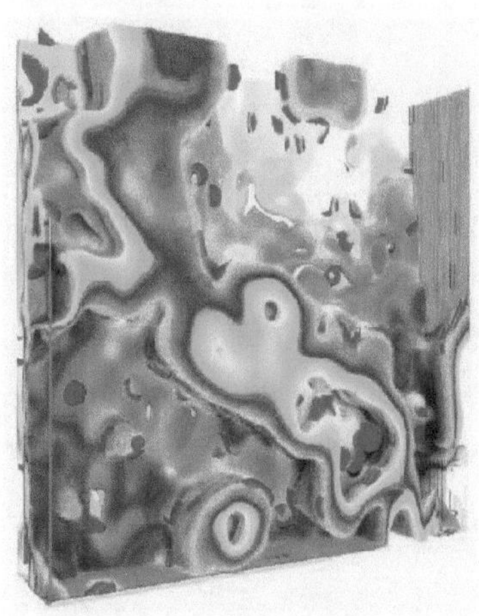

▲ 图 4.7.3 二十多亿年前古元古代多细胞生物的拉曼三维扫描，以不同假色，显示不同化学成分的立体分布

▲ Figure 4.7.3 3D-Raman-CT using false color to show the chemical distribution of a 2+ Ga multicellular organism

agrees with the Carbonate Cementing Chapter 7 Section 2. Carbonates are the main ingredient of cement. They formed an encasing surrounding layer outside the fossil without going too deeply. We saw the same phenomenon in the dinosaur tooth fossil. Thus, is any relationship between this carbonate cementing phenomenon and the preservation of organic remains? Further investigations shall be conducted.

4.8 Multiple Harmonic Generation Microscopy

Multiple Harmonic Generation Microscopy is a very genius invention and development by Professor Sun, ChiKuang of National Taiwan University for medical usages. It combines Two-Photon (TPH), Second Harmonic Generation (SHG), and Third Harmonic Generation (THG) nonlinear optical responses from the matter under strong and powerful Infrared Laser (1230 nm). It is a sub-micron (0.47μm), non-destructive, fast, non-photo bleaching, big area 3D imaging system. Dr. Sun's setup was designed for living tissues. It works very well for fossil study too. Our dinosaur embryology teams published not one

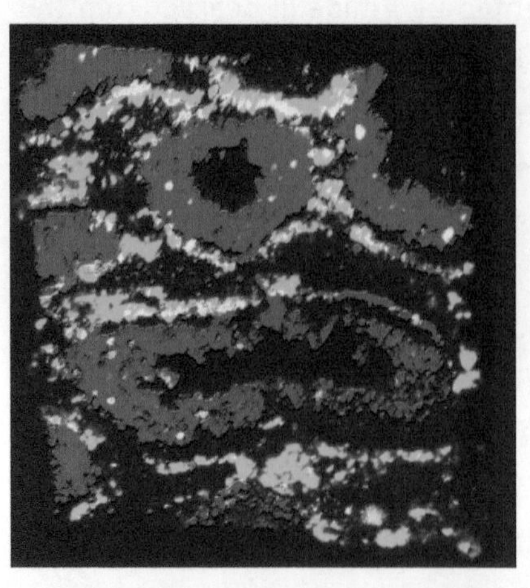

◀ 图 4.8.1 1.95 亿年前禄丰龙胚胎股骨的多频显微，黄色二光子，绿色二倍频，洋红三倍频，注意到白色区域，表示二倍频和三倍频共存

◀ Figure 4.8.1 MHGM of 195 million years old *Lufengosaurus* embryonic femur bone. Yellow is the TPH, Green SHG, and Magenta THG. Notice the white areas meaning coexistence of both SHG and THG

力很有用的工具，但是这个新仪器的诸多课题，还等待我们进一步了解，简单来说，我们看到了什么？这是一个我们可以发挥的未曾探索领域，比方说：依照光学理论，羟磷灰石 (HO-apatite) 不应该有二倍频反应，因为它是个中央对称 (central-symmetric) 的物质，然而我们却很明确观察到恐龙骨头和牙齿化石的二倍频反应，如图4.8.1。

除了用于古生物研究，它也可期望用于其他领域，如古人类学和考古学等等。

4.8.1　多频显微的幽怨

图4.8.2所示论文的摘要。

牙齿是了解现生和已灭绝脊椎动物暖食生态的关键。新近研究点出以前不知道的恐龙牙齿复杂性，不同种类、吃不同食物恐龙的牙齿组织与形状不同。然而并未了解到

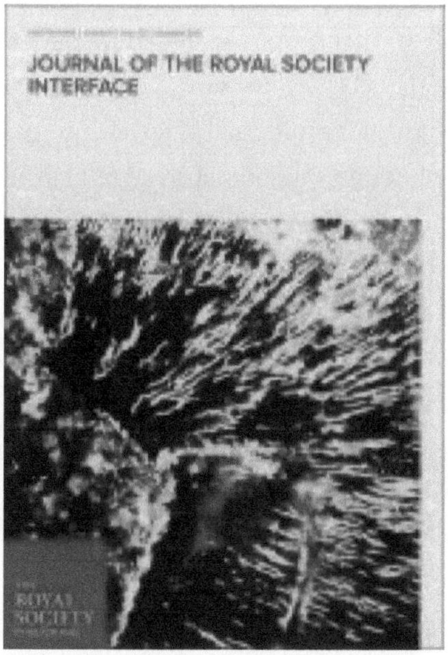

but two papers in high caliber journals, one of which is a cover paper.

Even though it was proved that it is a very powerful and useful tool for us, many issues of this new instrument are still pending for us to understand in more details. In short, what are we seeing? This is an unexplored area that we can shine on. For example, according to the optical theory that HO-apatite will not have SHG response, due to it is a central-symmetric compound. However, undeniable SHG images were observed unambiguously (Figure 4.8.1) from dinosaur teeth and bones.

Besides for paleontological studies, usages in other fields, such as paleoanthropology and archeology, can be expected.

4.8.1　Far Cry of MHGM

Abstract of the cover paper of Figure 4.8.2.

Teeth are vital to understanding the feeding ecology of both extant and extinct vertebrates. Recent studies have highlighted the previously unrecognized complexity of dinosaur dentitions and how specific tooth tissues and tooth shapes differ between taxa with different diets. However, it is unknown how the ultrastructure of these tooth tissues contributes to the differences in feeding style

◀ 图 4.8.2　（英国）皇家学会界面期刊的封面论文

◀ Figure 4.8.2　Royal Society - Interface Journal cover paper

这些牙齿组织的细微构造，如何影响不同恐龙分类之间食性的差异。

本研究，我们使用三倍频显微和扫描电子显微镜，来检查植食性与肉食性恐龙牙齿的细微结构，以便了解这些组织结构如何影响它们使用牙齿。牙小管直径、密度和分支率的形态分析，强烈显示食物偏好的选择。植食性的蜥脚类和鸟脚类恐龙，一致性地比肉食性的恐龙有更高的牙小管密度。我们提出假说，认为这和牙本质的硬度有关系，在植食性类恐龙的牙本质珐琅质界面 (Dentin-Enamal-Junction, DEJ)，比肉食性恐龙更能抗拒损坏和磨损。本研究倡导利用先进的显微术详细研究牙本质，以便了解已灭绝脊椎动物的牙本质的演化和啜食生态。

终于，经过了四五年的挣扎斗争，关于多频显微 (multiple harmonic generation microscopy, MHGM) 的第二篇文章发表了，而且被选为封面论文（图 4.8.2）。这是我们研究团队的第二篇封面论文，上一篇是在 2013 年 4 月 13 日 *Nature*，这次是在英国皇家学会界面期刊*Journal of Royal Society-Interface*。最近这篇在学术期刊界的分量，虽然与前者等高不一，但也是佼佼者，它的分量，也是很让人羡慕的。

有关多频显微应用于古生物研究，在2015年4月1日 于*Optics Letters*我们已经发表了第一篇论文 Third-harmonic generation microscopy reveals dental anatomy in ancient fossils（透过三倍频显微显示古化石牙齿解剖构造，如图 4.8.3）。从表面上来看，一个课题可以发表两篇稍有重量级的论文，应该算是不

between taxa.

In this study, we use third harmonic generation microscopy and scanning electron microscopy to examine the ultrastructure of the dentine in herbivorous and carnivorous dinosaurs to understand how the structure of this tissue contributes to the overall utility of the tooth. Morphometric analyses of dentinal tubule diameter, density, and branching rates reveal a strong signal for dietary preferences, with herbivorous saurischian and ornithischian dinosaurs consistently having higher dentinal tubule density than their carnivorous relatives. We hypothesize that this relates to the hardness of the dentine, where herbivorous taxa have dentine that is more resistant to breakage and wear at the dentine-enamel junction than carnivorous taxa. This study advocates the detailed study of dentine and the use of advanced microscopy techniques to understand the evolution of dentition and feeding ecology in extinct vertebrates.

After 4 to 5-year struggles, our second MHGM paper finally published as cover paper (Figure 4.8.2). This is our second cover paper. The previous one was on Nature on April 13, 2013. This time it is on Journal of Royal Society - Interface. Although this journal is not as high as the previous one, it still is an enviable one.

About MHGM for fossil study, we published a paper on Optics Letters 40(7): 40(7):1354–7 April 1, 2015, entitled Third-harmonic generation microscopy reveals dental anatomy in ancient fossils, Figure 4.8.3. So, one topic can yield two higher caliber papers; it is not bad at all. However, who knows the hot and bitter tears in the process?

Back in the second half of 2011 or early 2012,

错的，可是，我这个第一个吃螃蟹的人的辛酸，有谁知道？

话说回头，应该是在 2011 下半年或 2012 年初期间，一个偶然的机会，我得知台湾大学孙启光教授发明了一台医疗用多频显微的仪器，我觉得这是孙教授非常聪明伟大的发明，至今举世无双。这套设备伟大之处，从个人所了解认知的角度来说，在光学、光电、物理和显微镜学等几方面：①透过激光强光产生“二光子*(two-photon, TPH)”的荧光，已经有很多这方面的研究和好成果，这些日子，我也在搞化石的激光荧光检测，“没啥了不起”，人家已经搞过，不能说是最先进科技；②透过激光和光电科技产生“二倍频 (second harmonic generation, SHG)”方面，也可上网查询到看不完的文献，甚至现在已经有商业荧光二倍频显微镜可买得到，所以也“没啥了不起”；③同样使用激光和光电原理，单独产生“三倍频”(third harmonic generation, THG)，也有很多学者很多发表的论文，真的“没啥了不起”，亦即各别来说都“没啥了不起”。但是，却没有

I learned by accident that Professor Sun, Chi-Kuang constructed a medical use MHGM. I feel this is his genius and great invention and is the only one in the world as now. The greatness of this facility, from my knowledge on optics, photonics, physic, and microscopy lay on the following sides: ① to produce strong Two-Photon by a powerful laser is nothing new. I am working on Laser Stimulated Fluorescence (LSF) recently. It is not the cutting edge technology. ② Producing Second Harmonic Generation thru laser and photonics material is also not new at all. Tons of literature can be found thru a internet search. ③ Same for the Third Harmonic Generation is similar to SHG. Not a big deal. Each of these individually is not a big deal by itself at all. However, nobody combined all these three in one useful instrument. Moreover, that is the uniqueness and genius of Professor Sun. He merged all these three photonics into one great seamless equipment providing TPH, SHG, and THG in one shot. Combining with a Confocal Laser Scanning Microscope (CLSM), it can scan an area of 240 μm×240 μm× ~100 μm and produce a 3D model in about 10 minutes. It is THE very first for material study.

Although Professor Sun designed this for

* “二光子”是指被照射物质吸收了两个光子的（红外线激光波长）能量，再以可见光波辐射出来的荧光。“二倍频”是一种虚拟的光学反应，如用 1230 nm 的红外线激光，则有“非中心对称 (non-centrosymetric)”成分的被照射物质，就会发出 615 nm 波长的反应。这种二倍频的应用在我们生活中，就是现在大家普遍使用的绿色激光笔，它就是用 1040 nm 的红外线激光二级发光体 (LED)，透过二倍频显像，产生 520 nm 的绿色激光。“三倍频”的原理也是同样，只是将照射激光的波长除以三，就可得到三倍频反应的光。因为二倍频和三倍频都是虚拟的光学反应，因此对于样本来说，没有所谓的光漂白 (photo bleaching) 破坏性。也就是说，孙老师这个合并三种光学于一体的伟大发明，真是天才！
Two-Photon means the visible response (fluorescence) from the object absorbed the energy of two photos from an infrared laser. Second Harmonic Generation is a pseudo optical response from the irradiated object. Such as when a non-centrosymmetric material is radiated by 1230 nm infrared laser, it will produce 615 nm light. The green laser pointer commonly used in the classroom is produced by using 1040 nm laser LED with a non-centrosymmetric material to produce 520 nm light. The same principle applies to the Third Harmonic Generation. Just divide the wavelength of the incoming laser by 3. Since both SHG and THG are pseudo optical responses, they will not cause any destructive photobleaching on the specimen under examination. So, it is a very genius invention of Professor Sun.

任何人聪明到结合这三种光学于一体。孙老师独创伟大之处，在于他把这三种物质的光学反应结合起来，整合于一台仪器设备之内，在同样的激光扫描之际，同时提供“二光子”、“二倍频”和“三倍频”等光学资料，利用配合有扫描平台的共轭焦激光扫描显微镜（confocal

nondestructive examination medical living tissues, it never tried on the sample already dead for many million years (fossil). However, I think using this facility to play on fossils is the first in the world. On one side, there is just only one of such at Taiwan University. On the other side, these instruments are very busy doing various medical studies. Perhaps, Professor Sun never

1354 OPTICS LETTERS / Vol. 40, No. 7 / April 1, 2015

Third-harmonic generation microscopy reveals dental anatomy in ancient fossils

Yu-Cheng Chen,[1] Szu-Yu Lee,[2] Yana Wu,[3] Kirstin Brink,[4] Dar-Bin Shieh,[3] Timothy D. Huang,[5] Robert R. Reisz,[4,5] and Chi-Kuang Sun[1,2,6,*]

[1]Molecular Imaging Center, National Taiwan University, Taipei 10617, Taiwan
[2]Department of Electrical Engineering and Graduate Institute of Photonics and Optoelectronics, National Taiwan University, Taipei City 10617, Taiwan
[3]Institute of Oral Medicine, National Cheng Kung University College of Medicine and Hospital, Tainan City 701, Taiwan
[4]Department of Biology, University of Toronto Mississauga, Mississauga, Ontario L5L 1C6, Canada
[5]National Chung Hsing University, Taichung City 402, Taiwan
[6]Department of Optics and Photonics, National Central University, Jhongli, Taoyuan City 320, Taiwan
*Corresponding author: sun@ntu.edu.tw

Received October 24, 2014; revised February 12, 2015; accepted February 15, 2015; posted February 17, 2015 (Doc. ID 225362); published March 23, 2015

Fossil teeth are primary tools in the study of vertebrate evolution, but standard imaging modalities have not been capable of providing high-quality images in dentin, the main component of teeth, owing to small refractive index differences in the fossilized dentin. Our first attempt to use third-harmonic generation (THG) microscopy in fossil teeth has yielded significant submicrometer level anatomy, with an unexpectedly strong signal contrasting fossilized tubules from the surrounding dentin. Comparison between fossilized and extant teeth of crocodilians reveals a consistent evolutionary signature through time, indicating the great significance of THG microscopy in the evolutionary studies of dental anatomy in fossil teeth. © 2015 Optical Society of America

OCIS codes: (180.4315) Nonlinear microscopy; (190.4160) Multiharmonic generation; (320.7110) Ultrafast nonlinear optics; (170.1850) Dentistry; (170.3880) Medical and biological imaging; (170.1420) Biology.
http://dx.doi.org/10.1364/OL.40.001354

Modern paleontology investigates the evolution of organisms and communities through time, and their interactions in the biosphere. A particularly rich fossil record has allowed paleontologists to study in great detail more than 300 million years (Ma) of terrestrial vertebrate evolution. Of particular significance is their recent focus on reconstructing the feeding behavior of ancient vertebrates in an attempt to understand the evolution of communities of organisms, their trophic structure, and the overall structure of their ecosystem. The most common inferences for the study of feeding behavior are based on preserved hard tissues of the vertebrates, with fossil teeth providing the best indicators of diet. Most studies to date have focused on enamel wear or overall tooth morphology [1], and more recently tooth microstructures [2]. However, few studies have focused on the ultrastructure of dentin, despite their critical function in shaping and maintaining the overall integrity of the tooth [3–5]. A major obstacle in the study of dentin is in image modality, especially the image contrast.

Recent advances in image modality [6–10] have been widely applied to both living and fossil teeth. Among all, optical microscopy is considered an ideal tool in imaging extant teeth (extracted teeth from living animals) for its noninvasiveness, convenience, and resolution. Nonetheless, no studies have been able to demonstrate clearly the anatomy of dentinal tubules in fossil teeth, while most modalities rely strongly on refractive index differences, staining, or sample-dependent auto-fluorescence [9]. A method not yet used is third-harmonic generation (THG) microscopy. THG is a virtual-transition-based nonlinear process that deposits no energy [11] to the interacted specimens. Due to its third-order nonlinear nature, THG microscopy provides submicron resolution under reasonable depth and is extremely sensitive to fine structural variations [12–14]. It has been widely applied for noninvasive visualizations of 3D sub-cellular structures inside human [14–17] and living animals [18–20]. THG contrast was found to be enhanced by real electronic transitions in elastin [20], lipid [11,21,22], melanin [14], and hemoglobin [23].

In this Letter, we first apply THG microscopy to the study of fossil teeth of crocodilians and reveal that fossilization process in teeth provides remarkable difference in third-order nonlinear susceptibility $\chi^{(3)}$ between the fossilized tubules and adjacent surrounding tissues. Thus, the enhanced THG contrast enables us to study fossilized tubules regardless of similar refractive indices [24,25] throughout the fossilized dentine. Taking advantage of the unique contrast, we investigate the dentinal tissues of an extant *Alligator mississippiensis*, and compare its anatomy to the 1.5 million-year-old fossilized tissues of the same species, and with those of a 98 million-year-old crocodilian from the Kem Kem Beds of Morocco. Despite the difference in linear and nonlinear optical properties, correlation studies between extant and fossil teeth show a consistent morphology. This novel approach opens new research opportunities for paleontologists to resolve major issues in the evolutionary history of vertebrates by providing analyzable patterns and morphometric quantification of tubules, and allows for insights into the complex interactions between teeth and food within an evolutionary context [25].

THG, second-harmonic generation (SHG), and two-photon fluorescence (TPF) microscopies were simultaneously performed with a Cr:forsterite laser centered at 1230 nm with a 140-fs pulsewidth [23]. The laser output was focused 30 μm beneath the tooth surfaces by a

0146-9592/15/071354-04$15.00/0

▲ 图 4.8.3 发表在 *Optics Letters* 上的第一篇多频显微论文

▲ Figure 4.8.3 First MHGM paper published on *Optics Letters*

laser scanning microscope，CLSM），在短短的十分钟左右，提供 240 μm × 240 μm × ~100 μm 的三套立体影像数据，这些数据可很快地后制作成样本内部的三维构造模型，对于研究物质的结构来说，绝对是创举。

虽然孙老师原本设计是为了非破坏性的医学活体检验，从来未曾尝试过已经死了几千万几亿年前的样本（化石）。不过，从理论上来说，我认为以此来玩玩化石，应该属于全球首创，一方面是全世界仅有这么一台设备在台湾大学，另一方面台大那边忙着进行医学方面的研究，孙老师等人，也没想过他伟大的发明，除了医学应用之外，还可应用于化石和其他材料的研究和检测。所以，经过安排，拿了十几个样本去当小白鼠，玩玩看是否多频显微可用于化石研究，因为多频显微检测是完全非破坏性，不必担心样本需要被切成试片等等。

初期的尝试性扫描，由孙老师的学生刘卫民协助，扫了一些我的化石样本，主要有三类：埃迪卡拉实体化石（图 4.8.4）、中国龙（以前称为双脊龙）牙齿（图 4.8.5）和禄丰龙胚胎骨头化石（图 4.8.6），当我收到扫描结果，花了很短的后制作时间，看到了这些化石从表面深入大约 100 μm 深度、240 μm × 240 μm 面积的内部形态构造，分辨率高达 0.47 μm，让我跌破了好几副眼镜，难以相信的好结果，我认为这是一个很值得探索的领域，甚至建议孙老师可找个博士班的学生来研究，博士论文题目为：

thought to use his great invention for fossil or other material besides the medical application. Thus, after some arrangements, more than ten specimens were tried to see if MHGM can be used for fossil study. Since MHGM is a complete nondestructive examination, I do not have to worry about making thin slices.

Sun's student, WeiMing Liu, did the initial trail scans. He scanned some of my fossils. Mainly three types: Ediacaran body fossils (Figure 4.8.4), *Sinosaurus* (old name *Dilophosaurus*) (Figure 4.8.5), and *Lufengosaurus* embryonic bones (Figure 4.8.6). When I got the results and spent just a little bit of time to post-processing the images, I was very shocked to see the internal morphological structure of 100 μm ×240 μm ×240 μm with a resolution of 0.47 μm. Incredible excellent results. I think this is an excellent field for exploration. I even suggest to Professor Sun to have a Ph.D. student to do the Ph.D. research with the title of "The application of MHGM on Paleontology Study", because this is a significant break-thru for paleontology, and is the secret weapon nobody else has. Professor Sun's setup is leading the world.

When I saw the dentin tubules inside the dentin of *Sinosaurus*, I saw some tubules are: at one end of the same tubule is green SHG while the other end magenta THG (Figure 4.8.5). On top of this, from what I saw on DaWa *Lufengosaurus* embryonic bone: SHG corresponds to the hardy bones while THG for the Primary Tubular Cavity (PTC) (organic remains) as shown in Figure 4.8.6. So from these observations, for the time being, I proposed: "SHG can be considered as corresponding to apatiteS (notice the big S: meaning it's not just a single mineral, but a

"多频显微之应用于古生物研究"，因为这是一个古生物学的重大突破利器，而且是个独门秘密武器，别人没有这么先进的仪器设备，世界上就只有孙老师这套独步全球。

当看到中国龙牙齿牙本质内的牙小管，有好些同样一根牙小管一端是绿色的二倍频，而另一端是洋红的三倍频（图 4.8.5），加上从大洼禄丰龙胚胎骨头所看到，二倍频对应于硬骨头，而三倍频对应于"原始管状空间（有机残留物）"（图 4.8.6）。从这些观察，暂且提出："二倍频可以被认为对应于磷灰石群矿物，而三倍频可能对应于有机残留物"，因为在此时段，我们在同步辐射进行的傅里叶变换红外光谱分析，从形态和位置来说，两种完全不同理论基础领域的实验影像，非常吻合。但是谁能肯定无机成分一定只能是羟磷灰石 (HO-apatite)，羟磷灰石是六方晶系有中央对称性的物质，绝对不可能有二倍频出现。因此，

group) minerals, while THG may corresponding to organic remains" because we were conducting sr-FTIR at the same time. From morphology and position, the result from these two different theoretical background experimental matched very well. However, I did not expect that some people insist that the inorganic component inside the bone is just hydroxyapatite (HO-apatite), which is a hexagonal system centrosymmetric mineral. Thus, it should NOT have SHG. Therefore, the statement I got from my observation cannot be possible from physical optics theory.

Finally, Professor Robert Reisz made a compromise by cutting down from many types of fossils to just dinosaur teeth and compare with extant and extinct crocodile teeth. I supply the *Sinosaurus* teeth from Lufeng. Also, it will split one paper into two papers. The first paper would talk on the THG optics for fossil application (Figure 4.8.3). The second paper would discuss fossil teeth and explore the diet of the dinosaur. This is the cover paper. (Figure

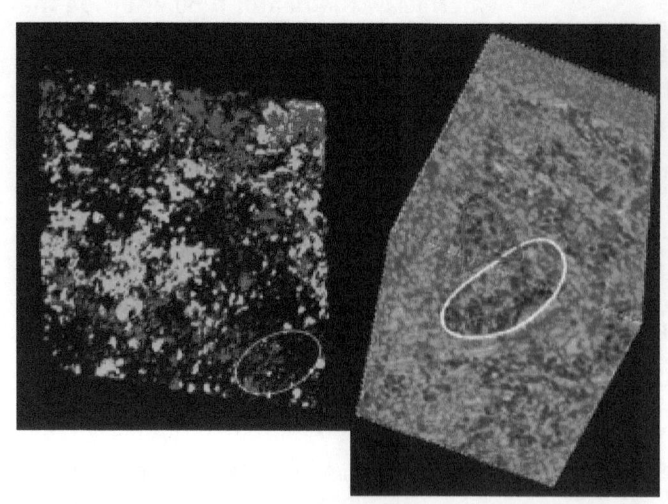

◀ 图 4.8.4 埃迪卡拉纪实体化石的多频显微，请注意到三维影像中的椭圆体，有可能是这生物的细胞。左边是整体结果，右边是经过影像处理，取出看起来像细胞构造重做的放大三维模型

◀ Figure 4.8.4 MHGM of Ediacaran body fossil. Note that the oval in the 3D image could be the cell of this creature. Left: the whole scan image, Right: enlarged portion showing cell-like structure

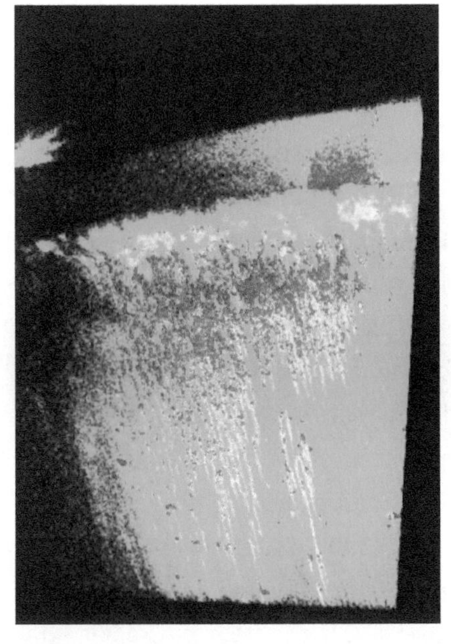

◀ 图 4.8.5 中国龙牙齿的多频显微，请注意到在同样的一根牙小管中，同时有二倍频（绿色）和三倍频（洋红）

◀ Figure 4.8.5 MHGM of *Sinosaurus*. Note that the same dentin tubule has SHG (green) and THG (magenta) simultaneously

▶ 图 4.8.6 禄丰龙胚胎化石的多频显微，请注意到二倍频（绿色）出现于硬骨头，三倍频（洋红）对应于胚胎骨头内保存的有机残留物（胶原蛋白）

▶ Figure 4.8.6 MHGM of *Lufengosaurus* embryonic bone. Note that SHG (green) corresponds to the hardy bone while THG (magenta) to the organic remains (collagen) preserved inside the embryonic bone

我的观点是，从最基础的物理光学理论角度来说，完全不可能成立。

最后Reisz教授折中，把我原来设计探讨多种化石，浓缩到只针对恐龙牙齿，并用现生鳄鱼牙齿和鳄鱼牙齿化石当比对，我提供禄丰出土中国龙 (*Sinosaurus*) 的牙齿，把论文一拆为二，前面一篇主要介绍多频显微的三倍频等光学原理在

4.8.2)

One of the critical points is the misunderstanding of the main inorganic component of bone. This is a very common mistake among medical, biology, and paleontology fields. From the past till now, people would say the main inorganic compound inside extant and extinct bones is hydroxylapatite first proposed by WF DeJong. He expressed apatite as $Ca_{10}(OH)_2(PO_4)_6$.

化石方面的应用（图4.8.3），而第二篇则以化石牙齿为主体，透过多频显微来研究探讨恐龙的食性，这就是这篇封面论文（图4.8.2）。

这其中的一个关键，在于对于骨头内主要无机成分的错误认知，其实，犯同样错误的人，在医学界和生物、古生物界相当普遍：从过去到如今，说到骨头内现生和化石的无机矿物成分，通常都会说是羟磷灰石（或称“氢氧磷灰石”）而已，最早做出如此荒谬认知最早的是DeJong，他把磷灰石写成 $Ca_{10}(OH)_2(PO_4)_6$，后面的学者也没详究，就跟着人云亦云，把骨头内无机矿物确定为（只是）羟磷灰石。然而，事实上生物性的磷灰石并非“剂量化学(stoichiometric)”的单纯羟磷灰石而已，从先进化学分析和透过傅里叶变换红外光谱分析所看到的，骨头内的磷灰石是一大群矿物，其中以甲种碳酸磷灰石(Carbonatoapaptite-A, CO_3^{2-} 取代 OH^-) 和乙种碳酸磷灰石 (Carbonatoapaptite-B, CO_3^{2-} 取代 PO_4^{3-}) 为大宗成分，羟磷灰石的含量，并没有比这两种碳酸磷灰石来得多！

如果从矿物学磷灰石群的化学式来探讨，它的化学式为：$Ca_{10}(PO_4)_6(OH,F,Cl)_2$，以通式来说，可看成三个部分，$A_5(XO_4)_3Zq$：最前面的阳离子 A，常见是二价的钙 ($Ca^{2+}$)，中间 $(XO_4)_3$ 是磷酸根 $(PO_4^{3-})_3$，最右边是 Z 阴离子，往往用 (OH) 代表。在活体骨头内和化石内，这个磷灰石通式里面的三

Scholars after him did not carefully look into this matter and parrot-talk the same mistake-inorganic component inside the bone is (just) hydroxylapatite. However, in fact, the biogenic apatite is not simple stoichiometric hydroxylapatite. What we saw from recent advance studies and sr-FTIR observations, the apatite inside the bone is a big group of phosphate minerals, mainly Carbonatoapaptite-A, CO_3^{2-} replacing OH^-, and Carbonatoapaptite-B, CO_3^{2-} replacing PO_4^{3-}. The amount of hydroxylapatite is far less than these two carbonatoapatites.

Let us look at this issue from mineralogy of apatite group point of view. Chemical formula $Ca_{10}(PO_4)_6(OH, F, Cl)_2$, can be expressed as $A_5(XO_4)_3Zq$. The cation A is commonly bivalent Ca^{2+}, the middle $(XO_4)_3$ is $(PO_4^{3-})_3$, and the anion at the right commonly expressed as (OH). In the bone of living animal, all these three can be replaced, such as Cation A been replaced by other bivalent metal elements, such as Sr, Mg, Mn, etc., or even trivalent elements such as the element of two Rare Earth Element series in the fossil. There are at least 24 possible replacements. The phosphate in the middle has 3 possible replacements, and the anion at the end could have 4 replacements. Thus, the possibility of the combinations of apatites is 24×3×4 = 288! So up to this point, we are still talking stoichiometric, in which the ratio(s) of elements inside a formula are integers, such as water H_2O, there are two hydrogen atoms and one oxygen. Everybody learned this in middle school. However, in nature, scientists found that stoichiometric chemistry does not match with the actual situation too well. When looking up a chemical formula of a mineral, it is common to see decimal

个部分，都可被取代，比方说，最前面的阳离子 A，可被其他两价的金属元素取代，如锶 (Sr)、镁 (Mg)、锰 (Mn) 等等常见的二价金属元素，甚至可被三价的元素，如化石内两个系列的稀土元素 (rare earth elements)，总共有 24 种可能的取代。中间的磷酸根 $(PO_4^{3-})_3$有 3 种可能的置换，后边的阴离子 Z 则有 4 种可能，所以用排列组合算一下，可能的磷灰石矿物群就有 $24 \times 3 \times 4 = 288$ 种之多！说到这里，唐三藏再把孙悟空的头箍拴紧一格，加入另外一个化学化合物的考量：一般所谓的计量化学 (stoichiometric)，化学式内原子的比例是整数，如水的H_2O，两个氢和一个氧，大家初中都学过了；可是，当碰到大自然的时候，科学家们发现剂量化学和实际的情况对不起来，如果翻翻矿物学辞典去查某个矿物的化学式，经常会看到小数点的比例，这称为“非剂量化学 (non-stoichiometric)”，也就是说诸原子或根并非整数的比例；好了，剂量化学说有 288 种可能的磷灰石（复数），如果再把不知道有多少可能的非计量化学乘上去，请问天然界实际上有多少种磷灰石？怎么可以用一种六方晶系（中央对称）羟磷灰石来以偏概全呢？因此，当谈到磷灰石矿物的时候，比较正确的名称应该为“磷灰石群 (ApatiteS)”，切记切记；这也就是说，多频显微二倍频可看到磷灰石群是无法否认的现象，在这几百甚至上千种的磷灰石家族内成员，并非都是乖乖牌中央对称六方晶系的矿

fractions. This is non-stoichiometric. From stoichiometric, there are 288 apatites already. Now, throw in the unknown number of possibility of non-stoichiometry into the already complex arena, how many kinds of apatites? Thus, how can we simplify such a big mess to just one hexagonal centrosymmetric hydroxylapatite to represent all of the apatites? The correct name shall be apatiteS. Make sure to keep this in mind. This also means that for SHG to see apatites is a normal phenomenon. Among these hundreds or even thousands of apatite groups member minerals, not each of them well behave centrosymmetric hexagonal mineral. MHGM Figure 4.8.4 thru 4.8.6 demonstrate this very well in front of our eyes. How can we deny them?

Back to Figure 4.8.4 to Figure 4.8.6. In these photos, the color used is false color, yellow for TPH, green for SHG, and magenta for THG. Figure 4.8.4 is an MHGM scan of an Ediacaran body fossil I found. Please note in the left photo there are many oval structures. One of them was marked with a white line. The size of these ovals is ~20μm, close to the average size of a commonly seen cell. When these ovals were cut and made to 3D models, viewing them from different angles, they look like cells -Right photo. However, up to now, the study of Ediacaran body fossils is not formally started. I just did some initial works as favors given to me to gain some necessary fundamental information. Discovery of the Ediacaran body fossils and the research following will yield significant impacts. It could likely rewrite Darwin's evolution hypothesis because these about 560 million years old life could be ancient multiple cellular or-

物！图4.8.4到图4.8.6活生生的多频显微扫描结果证据摆在眼前。

回头来说说图4.8.4到图4.8.6这几张照片，这些照片的颜色都是用“假色”，二光子用黄色表示，二倍频用绿色，三倍频用洋红表示。图4.8.4是我发现的埃迪卡拉纪实体化石的多频显微扫描，请注意到左边画面内，有很多如白色椭圆标示的椭圆球状构造（只标示一个），进一步测量这些椭圆球状的大小，大约在20 μm左右，和一般的细胞大小形状很接近，当把这些橄榄球“（数位）切下”做成三维的模型，转过来倒过去来看，真的很像是细胞，见右边。不过，埃迪卡拉纪实体化石的研究探讨，到现在为止，不算正式进行，只能这边能做一点算一点，人家给我机会，我就塞入一两个样本做测量，摸蛤兼洗裤，先取得一些基本的资料。埃迪卡拉纪实体化石的发现和接下来的研究，其影响的重大性，很有可能改写达尔文的进化学说，因为这些大约 5.6 亿年前的古生物，是地球上很古老的多细胞生物，而且到目前为止，世界上其他地方所发现的埃迪卡拉纪化石，都只是印模铸模 (imprint, cast) 而已，没有实体留下来，亦即原本的生物体腐烂掉，留下空间，后来泥沙跑进去填充的。因此这方面的研究，就受到很大的限制，就如医学院的学生学习解剖学，不能拿公仔来切割，一定要人尸体。同样的道理，想要进一步了解埃迪卡拉纪生物的奥秘，就必须有实体化石，才能深入探究，我所发现的就是实体化

ganisms. Up to now, the Ediacaran fossils found worldwide are just imprints/casts. The original body decayed and left a halo for sand and mud to fill in. No body fossils were ever found. Thus, the study of them is limited, just like the medical school students cannot use a mannequin for human body anatomy. They have to use real human corpse. For the same reason, in order to further understanding of the secret of Ediacaran creatures, body fossils shall be used to “Look into the Bone.” What I found are body fossils to do so. However, as stated, this is not formally started.

Figure 4.8.5 is an MHGM image of *Sinosaurus* tooth from DaWa, Lufeng of YunNan province. As mentioned previously, there are many dentin tubules with one end of green SHG and the other end magenta THG. The diameter of the dentin tubules is about 1~2 μm. They can be seen very clearly under MHGM, but not by regular optical microscope. This is discussed in details in our first published paper. This paper can be downloaded from *https://www.osapublishing.org/ol/abstract.cfm?uri=ol-40-7-1354*. Please note that some tubules are in white? What does this mean? Back to basic optics. What is the complementary color of green? Magenta! So, when green is mixed with magenta, what will be the resulting color? White! So, this means that at these white color areas have both SHG and THG responses. The main minerals of dentin are apatites. Having SHG responses is common. It is also known from dentistry that the dentin tubules are very long and thin tubes. While the tooth is still alive, nerve (the reason for causing toothache) and various nutrients are inside. During the fossilization process, both ends of

石，可以用来“看进骨头内”。不过，这个科研一直没有正式展开。

图 4.8.5 是云南禄丰大洼中国龙牙齿的多频显微影像，前面有说过，牙本质内的牙小管，有好多一端是绿色的二倍频，另一端是洋红的三倍频，牙小管的直径大约只有 1~2 μm，看得非常清楚，这是一般光学显微镜没法看到的，这一点在第一篇多频显微的论文中，已有详细解说，有兴趣的人可上网去把那篇论文抓下来看（详见网址：*https://www.osapublishing.org/ol/abstract.cfm?uri=ol-40-7-1354*）。不过，不知道有没有注意到，有些牙小管部分是白色的？哇！这是什么意思？回想一下基本光学，绿色的互补色是什么？不就是洋红吗？如果把绿色和洋红混在一起，会得到什么颜色？白色！也就是说，在这些白色部位，同时有二倍频和三倍频反应！牙本质的主要矿物成分是磷灰石群，出现二倍频很正常，而从牙医学来的知识，牙小管很细长，牙齿活着的时候，管内有神经（所以会牙酸）和各种有机的养分，而在形成化石的过程中，牙小管的两端很容易封闭起来，而保存了管内的有机物质，这就是搞古人类 DNA 最喜欢拔死人牙齿的原因。Bingo！在此是一个我前面提到“二倍频对应于磷灰石群、三倍频对应于有机残留物”的无法否认证据！

再看图 4.8.6，这是我发现并亲自捡拾几近两亿年前禄丰龙胚胎脊椎骨切面的多频显微扫描影像。这个图像，也有

a dentin tubules could be sealed easily and preserved the organic remains inside. This is one of the reasons that anthropologists like to yank dead people's tooth. Bingo! this is an undeniable proof of what I stated: "SHG can be considered as corresponding to apatites minerals, while THG may corresponding to organic remains."

Now, let us look at Figure 4.8.6. This is an MHGM scan image of 195 million years old *Lufengosaurus* embryonic vertebra transversal cut. Good story about this too. When we published the Nature cover paper, one of the team members dug out a paper denying what Mary Schweitzer's discovery of collagen inside a *T. rex* femur. That paper stated that during the fossilization process, bacteria play a significant role. Thus, the organic remains Schweitzer found could be produced by or the remains of bacteria. They run the experiment by putting bones into lake water for a certain time, and then analyzed those bacteria. They use the diagram and text to deny Schweitzer's discovery. However, when our team was doing sr-FTIR at NSRRC, we were very sure that in these dead bones, organic remains were preserved, and knew they are collagen type I. But, we did not have time to run enough experiments to have solid enough evidence to tag them as type I collagen. Thus, in our paper, the catch-all safety net phrase of "Organic Remains" was used. Then, in our January 31, 2017, paper on Nature Communications paper, we nailed down this to collagen type I preserved in *Lufengosaurus* of DaWa.

So this photo, Figure 4.8.6, is a piece of very solid evidence to rebuttal the bacteria guys. During the embryonic development stage,

好故事可说，看官姑且看下去：当年我们撰写发表那篇《自然》封面论文之际，有团员找出了以前有人发表过、用来否定Mary Schweitzer发现霸王龙大腿骨内保存着胶原蛋白的论文，他们的说法是，在形成化石的过程中，细菌扮演着关键的角色，因此玛丽所发现的有机物，很有可能是细菌的遗骸或产生的，并非原本恐龙骨头内的有机物，他们还做了实验，把骨头泡到湖内一段时间，再分析那些细菌，图文并茂地反驳玛丽。但是从我们团队透过同步辐射的傅里叶变换外光谱分析，我们非常确定这些死骨头内保存着有机残留物，那个时候我们事实上已经知道是第一种胶原蛋白，但是论证还不够严谨，因此我在论文内打个迷糊仗，说是“有机残留物 (organic remains)”，而没说死是第一种胶原蛋白 (collagen type I)，哈！如今我们已经在《自然通讯》(*Nature Communications*) 发表明确证明大洼禄丰龙骨头化石内保存着第一种胶原蛋白的论文。

话说回来，这张图像 (图4.8.6) 就是反驳那些细菌论说者最有力的证据。在胚胎阶段的骨头成长，先从中央部位开始形成骨质，逐渐往两端成长过去，亦即骨头中央部分会有较多的骨头构造，胚胎硬骨头里面会有好些我们称之为“原始管状空间 (primary tubular cavity, PTC)”夹杂在硬骨头之间，其直径在20~40 μm，而两端会有较多大孔隙的钙化软骨 (calcified cartilage)，钙化软骨区内的孔洞，长度可以达到600~700 μm，

the bone started to grow from the middle and gradually toward the two terminals. That is, in the center portion, there are more bone structures. Inside the hardy embryonic bones, there are many Primary Tubular Cavity within, with a diameter of 20–40μm. At both terminal ends, there is more Calcified Cartilage with holes of 600–700μm in length and diameter of more than 100 μm. OK, in the center portion of the PTC, there are plenty of THG (organics). However, at the calcified cartilage terminals, there is almost no magenta THG. This is strange. If the organic remains were the bacteria or produce by bacteria, how to explain this fact? Could it possible that bacteria are “smart” enough to pick on the little guys and avoid big bully? In the cartilage, there is a more organic matter to chew. The bacteria carried into the bones from the same groundwater and with such a small distance, how could they have such a “smart” choice? This cannot be so! This reversal evidence overturned the bacteria claim and established the endogenous of the preservation of organic remains (collagen type I) inside the embryonic bones.

From the technical angle to express my thought of “Look into the Bone” more, if possible due to the rareness of our study fossil, the quantity usually is limited to one and one only. There are very precious. Other possible reasons could also prevent us from conducting destructive analyses. We can only use nondestructive methods to gain as much information as possible for us to listen to whatever the dead bones trying to tell us. Therefore, if possible to conduct many nondestructive analyses on the very same spot of the same specimen (*in situ in loco*) to obtain the secret information, won’t that be the best

直径上百微米。好了，请注意，在图像中央部位原始管状空间内，看到好多洋红的三倍频（有机物质），而在两个终端钙化软骨的大孔洞里面，几乎没有洋红表示的有机物质，这就很奇怪了，如果这些有机残留物是细菌本身遗留或产生的，怎么解释这个事实呢？难道细菌已经聪明到会欺小怕大，只找小洞钻，而不敢跑到更宽敞的大洞里面舒服躺？软骨有更多细菌喜欢吃的有机质啊！同样的地下矿泉水带进的细菌，相隔就这么一点点距离，竟然会有如此“聪明”的选择？怎么说也说不过去！这个反证，一方面推翻了细菌说，另一方面也确认这些胚胎化石内所保存的有机残留物（第一种胶原蛋白）是原生的，而非外来物。

再从技术的角度来说些我对于“看进骨头内”的看法：如果可能的话，因为我们研究的化石样本，几乎都有罕见性的问题，世界上就这么一个样本，或是非常珍贵，或是其他种种原因，不容许对样本进行破坏性的研究，只能透过非破坏性的探索，能够取得越多信息，越能听到这些死骨头想对我们说什么。因此，如果能在同一块样本同一个地方，进行多种完全没有破坏性的研究，撷取里面的奥秘信息，那不是最好最理想吗？

这个“看进骨头内”思维、古生物新进的理想化研究方法，以今日现有的尖端科技来说，是完全可行的。图 4.8.7 显示我初步所做到的，在一块大洼恐龙胚胎股骨中央横切面，显示一般光学影像，看到胚胎腿骨快速发展中的磷灰石

and ideal?

This "look into the Bone" strategy is the advanced ideal and workable paleontology study approach with the state of art science and technology. Figure 4.8.7 shows what I initially already done. On the transversally cut of a DaWa *Lufengosaurus* embryonic femur, the conventional optical microscopic image shows the rapid evolving hardy bone (white pale brown) and PTC (dark brown). Then, the pinkish FTIR infrared image was superimposed on top. They align very well. Then, MHGM (greenish magenta) image was added. All these three matched so well. In other words, we can say that the hardy bone saw by optics correspond to infrared images as well as what MHGM shows. This is to prove that the apatites inside the hardy dinosaur bones will definitely have SHG response. Moreover, that explains the mineral components of dinosaur bones are not the misunderstood hydroxylapatite alone. They are apatiteS. Similarly, whatever the THG of MHGM show correspondence perfectly to the PTC seen from infrared and conventional optics (Figure 4.8.8).

Now, the next big question: what do we see from the SHG and the THG? What kind of chemical matter will produce these two harmonic generation responses?

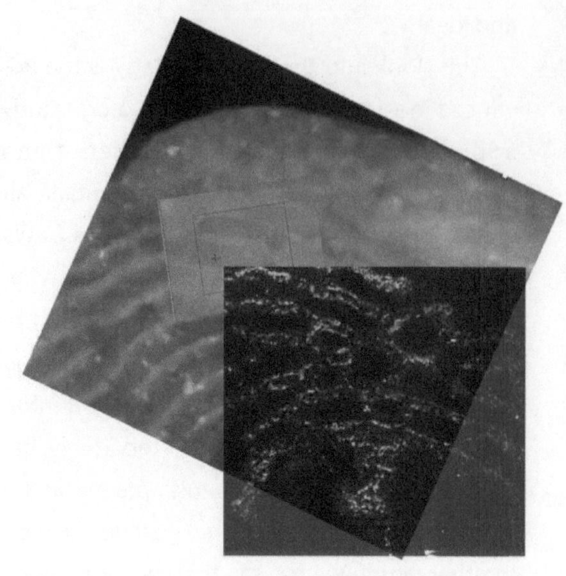

◀ 图 4.8.7 在同一个样本同样位置，重叠不同光学检测，最底下的是一般光学显微照片，可看到大洼恐龙胚胎股骨轮状硬骨和原始管状空间。比较粉红这张是红外线影像，绿色这张是多频显微的影像，这三者完全重叠

◀ Figure 4.8.7 On the same spot of the same specimen, *in situ in loco*, of DaWa *Lufengosaurus* embryo femur overlapping different optical images. The very bottom one is common optical microscope image. Ring shape hard bone and PTC can be seen. The reddish photo is the IR image, while the greenish one was from MHGM. These three images superimposed very well

群硬骨头（白淡棕色），以及原始管状空间（深棕色）。我们也取得了同步辐射傅里叶变换红外光谱红外线影像（粉红色），把它重叠到光学影样上，两者吻合得密密密，再把多频显微的影像（绿色洋红）也加进来，三者都非常吻合；换句话说，可以这么看，在光学影像所看到的硬骨头，对应于红外线影像，更也是多频显微所显示的，这就证明了恐龙的硬骨头里面的磷灰石群，的确会有二倍频反应，也就是说明白了，恐龙骨头的矿物成分，不是一般误解的羟磷灰石而已，而是磷灰石群；同样地，在多频显微三倍频所显示的，完美地对应于红外线和光学影像的原始管状空间（图4.8.8）。

现在，接下来的一个大问题：到底从二倍频和三倍频，我们到底看到了什么？化石内的什么样化学物质会产生这两种倍频反应？

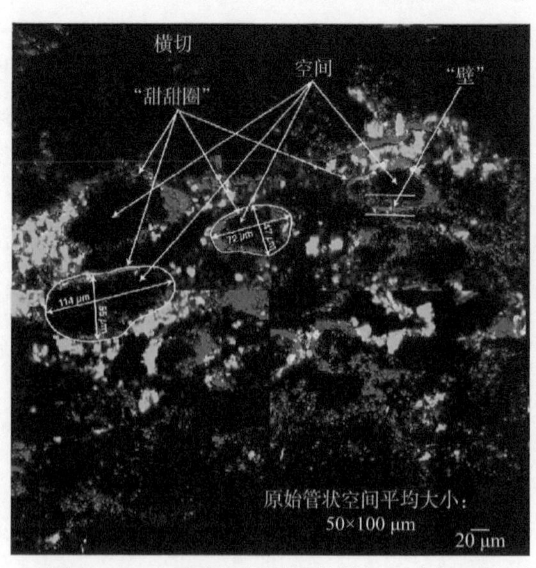

▲ 图 4.8.8 大洼恐龙胚胎股骨横切面多频显微所得的原始管状空间图示

▲ Figure 4.8.8 The MHGM of a transversal cut embryonic femur showing the Primary Tubular Cavity

4.9 直接质谱分析与解吸电喷雾电离

（取自维基并修正）质谱分析法(mass spectrometry) 是将化合物形成离子和碎片离子，按质荷比 (*m/z*) 的不同进行分离测定，来进行成分和结构分析的一种方法，简言之，质谱就是测定某样本内的诸质量，所得结果用质谱图（亦称质谱，mass spectrum）表示。它以质量对电荷 (*m/z*) 比例画出离子信号，这些质谱用以测定样本的元素或同位素，粒子和分子的质量，阐明分子的化学结构，如多肽和其他化合物。按照惯例，一般将质谱分析方法依照其测试对象的不同而主要划分为无机质谱、有机质谱和生物质谱三大类，这三大类的质谱分析，我们古生物化学都会用得到：无机质谱用于元素鉴定，如用电感耦合等离子体质谱进行骨头内稀土元素分析（下一节）；有机质谱用于鉴定化石和围岩内的有机残留物，生物质谱则会在以现生生物作为有机残留物的比对。

在典型的质谱测试，气态、液态或固态的样本，首先被离子化，比方说被电子轰炸，这有可能使得样本的分子被打断成为带电的碎片，这些离子依照质量和电荷比例不同而被分离，通常是把它们加速度，然后放入电场或磁场里面，同样质量和电荷比例的离子，会有同量的偏折，最终，这些离子被如电子倍增管等检测器检测到，所得的结果以质量和电荷比例为横轴，相对数量多寡为纵

4.9 Direct Mass Spectrometry and Desorption Electrospray Ionization (DESI)

(Modified from Wiki) Mass spectrometry (MS) is an analytical technique that ionizes chemical species and sorts the ions based on their mass-to-charge (*m/z*) ratio. In simpler terms, a mass spectrum measures the masses within a sample. Mass spectrometry is used in many different fields and is applied to pure samples as well as complex mixtures. A mass spectrum is a plot of the ion signal as a function of the mass-to-charge ratio. These spectra are used to determine the elemental or isotopic signature of a sample, the masses of particles and molecules, and to elucidate the chemical structures of molecules, such as peptides and other chemical compounds. Customary, they are classified as inorganic, organic, and biological mass spectrometry, according to the nature of the analyzed target. In PaleoChemistry studies, we will use all these three. For the inorganic MS, we use ICP-MS for elemental analyses, organic MS will be used for the fossils and matrix, and biological MS will be needed when we use extant organism as comparisons to organic remains.

In a typical MS procedure, a sample, which may be solid, liquid, or gas, is ionized, for example by bombarding it with electrons. This may cause some of the sample's molecules to break into charged fragments. These ions are then separated according to their mass-to-charge ratio, typically by accelerating them and subjecting them to an electric or magnetic field: ions of the same mass-to-charge ratio will undergo

轴之质谱显示。样本内的原子或分子可以通过比对已知质量（如某整个分子）来鉴定，或通过碎片特征来检定。（从维基修正到此）

以上的简述，可以归纳为离子化—分离—检测三个阶段，其中后两阶段虽然有好些各自特有的优劣，但是相对来说，是比较成熟的部分，本书就不细究。但是最先的这个“离子化”步骤，却是最最关键，物质没有被离子化，后端的质谱分离和检测，英雄无用武之地；幸运地说，这方面在近年来有非常关键性的进展，特别是从古生物化学非破坏性在原位在同点的角度来说，非常关键，我们希望罕见珍贵的样本，受到最小的伤害，最好肉眼看不到，就能把化石和围岩内的有机成分检测出来。因此，在此简介一些较新的离子化技术。

如果上网去查的话，很容易发现这些年来在离子化方面的新发展，多到令人眼花缭乱，光是缩写英文的专有名称，就有 84 种之多，光是要把这些名称搞清楚记下来，就够累人的，我们也无法细说，因此，本节只挑选几个关键名词和作者认为比较重要的做简介。

由于离子化所需要的能量随分子不同差异很大，因此，对于不同的分子应选择不同的离解方法。

硬电离：一般说来，内能大的分子能够碎裂，得到许多不同的碎片离子，适合对有机物进行精确的结构鉴定和同位素丰度测定。

软电离：如果需要测定物质的分子

the same amount of deflection. The ions are detected by a mechanism capable of detecting charged particles, such as an electron multiplier. Results are displayed as spectra of the relative abundance of detected ions as a function of the mass-to-charge ratio. The atoms or molecules in the sample can be identified by correlating known masses (e.g., an entire molecule) to the identified masses or through a characteristic fragmentation pattern. (Modified Wiki up to here.)

The brief description above can be expressed in three stages as Ionization - Separation - Detection. Despite various strong and limitation of each of the methods used in the last two stages, they are more mature technologies relatively and will not be discussed further here. However, the first stage of Ionization is the most crucial step. Without ionized the matter, the following separation and detection cannot do a thing. Fortunately, there are very significant advancements in recent years, particularly in the view of *in situ in loco* we are pursuing in the PaleoChemistry. This is very important. We want to conduct the identification of organic remains inside the fossil and surrounding matrix with the minimum damage to the specimens, not visible to the naked eyes will be very desirable. Therefore, some of the newer ionization methods will be briefly introduced here.

From surfing the internet, it can be easily found many many new advancements on the ionization techniques, to the point of mind-boggling. There are at least 84 acronyms! Trying to understand and remember all of these names is not an easy task. We cannot go into details in this book. But some of the key terminologies

量，则需要在离子化后得到尽可能多的分子离子，适用于易破裂或易电离的样品。

电子轰击电离 (electron ionization, EI)：是最古老而又最简单的一种离子化源，其最大优点是能够获得重现性非常好的质谱图，其最大的缺点是仅对能够不分裂而产生蒸气的样品有效；通常适用于低分子量（≤1000 Da）和低极性有机化合物。

化学电离 (chemical ionization, CI)：与电子轰击电离相比，化学电离显然是一种比较软的电离源，它用的电离试剂是气相离子而非电子。

喷雾电离与电喷雾电离：喷雾电离 (spray ionization, SI) 能够将液体样品，如生物试样的水溶液，直接引入离子化源中，形成气态离子，需要形成一种细小的雾状液粒，这些细小的液粒被干燥后则立即形成孤立的气相离子；电喷雾电离 (electrospray ionization, ESI)，其机理目前尚无统一的认识；过程中大致可以分为液滴的形成，去溶剂化，气相离子的形成三个阶段。

电喷雾解吸电离 (desorption electrospray ionization, DESI)：典型装置由电喷雾电离装置和承载样品表面两部分组成，是电喷雾电离和解吸附作用的综合体现，能够对非粉尘状固体表面进行直接分析。

常压化学电离：常压化学电离 (ambient pressure chemical ionization, APCI)，是一种与电喷雾电离过程类似的化学电离方法，一般需要在常压下采用

and essential methods will be briefly described here.

Due to the significant differences of molecules the energy needed to ionized them also has a big difference. Thus, it is crucial to select the appropriate ionization method for a different molecule.

Hard Ionization: In general, the big high energy molecules can be cracked into many fragmented ions. It is useful for precise structural determination of organic compounds and richness of isotopes.

Soft Ionization: If the molecular weight is needed, as many as molecular ions shall be obtained. This is suitable for fragile or easily ionizable specimens.

Electron Ionization, EI: This is the oldest and easiest ionization source. The most significant advantage is to have excellent mass spectra repeatedly. The major disadvantage is that it can be used only for no crackable to vapor specimens, usually for low molecular weight (≤1000 Da) and low polarity organic compounds.

Chemical Ionization, CI: In comparison to the Electron Ionization, this is a relatively softer ionization source. The reagent to ionization is gas ions, not electrons.

Spray Ionization, SI: It can be ionized the liquid sample, such as an aqueous solution of biomaterial directly into the ionization source to form gaseous ions. It needs to be formed in very small misty droplets, and then the droplets are dried to gaseous ions. There is no nonsense on the ElectroSpray Ionization, ESI yet. There are three steps: forming the droplets, desolventization, and gaseous ion formation.

Desorption ElectroSpray Ionization, DESI: It

直流等离子体（direct current plasma）作为初级的离子源，使得一般在负压下进行的离子-分子反应或电子-分子反应能够在常压下进行，从而保证了进行这种反应的充足的时间，通常为数秒，能够发生足够多的有效碰撞将样品分子充分离子化，因而具有很高的灵敏度，特别适宜在环境监测和生物表征中发挥作用。

表面解吸常压化学电离 (desorption atmospheric pressure chemical ionization, DAPCI)：以常压电晕放电产生初级能荷载体，通过气-固-气或液-固-气三相进行能荷传递，实现固体表面分子的高效电离，它充分结合了常压化学电离和电喷雾解吸电离的优点，在无须样品预处理的前提下，可以对各种不同表面吸附的痕量挥发性物质和非挥发物质进行常压解吸化学电离；主要用来分析中等极性的化合物。

电喷雾解吸电离 (DESI)，相对来说，这是一个较新有机物分析的仪器方法。

（取自维基）解吸电喷雾电离 (desorption electrospray ionization, DESI) 是一种常温离子化技术，可用于质谱仪来做化学分析。它用普通气压离子源在普通环境下离子化气体、液体、和固体。它是 2004 年 Zoltan Takáts 等在普渡大学 Graham Cooks 教授团队开发出来的，现在由 Prosolia 公司商业化。解吸电喷雾电离类似于即时直接分析 (direct analysis in real time, DART) 离子化技术，在多方面、应用和分析时间，都差不多。

解吸电喷雾电离合并电喷雾

composes of ElectroSpray Ionization and carrier. It is a combination of ElectroSpray Ionization and desorption. It can work directly on the surface of the non-powdery solid material.

Ambient Pressure Chemical Ionization, APCI: Its a chemical ionization method similar to the Desorption ElectroSpray Ionization. It usually needs to use Direct Current Plasma as the primary ionization source so that the typical negative pressure ion-molecular or electron-molecular reaction can be achieved under ambient pressure. To ensure enough time for these reactions, usually in seconds to have enough effective collisions to fully ionized the sample. Thus, it has very high sensitivity and very suitable for the environmental monitoring and biological survey.

Desorption Atmospheric Pressure Chemical Ionization, DAPCI: Under the ambient pressure corona discharge produces primary energy carriers and thru gas-solidi or liquid-solid-gas phases for energy transfer to achieve high ionization effect on the surface of the solid matter. It combines the advantages of atmospheric pressure chemical ionization and Desorption ElectroSpray Ionization. No sample preparation is needed. It can analyze the trace amount of volatile and non-volatile material under atmospheric desorption chemical ionization. It mainly used in the medium polarity compounds.

DESI, This is a relative new analytical instrumentation for chemist to analyze organic matters.

(From Wiki) Desorption electrospray ionization (DESI) is an ambient ionization technique that can be used in mass spectrometry for chemical analysis. It is an atmospheric pressure ion source that ionizes gases, liquids, and solids in the open air under ambient conditions. It

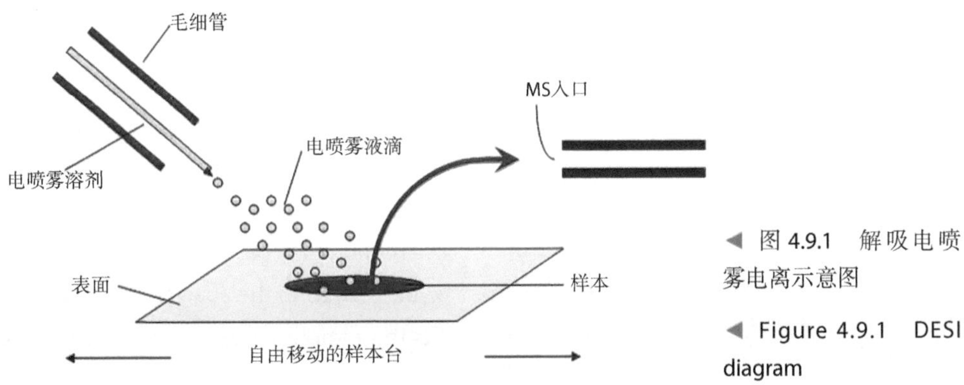

◀ 图 4.9.1 解吸电喷雾电离示意图

◀ Figure 4.9.1 DESI diagram

(electrospray, ESI) 和解吸 (desorption) 离子化技术，离子化透过导引一股带电的喷雾喷到几厘米近的样本，通过气压将电喷雾导引到样本，接下来的喷洒小滴携带解吸、离子化被分析物。在离子化之后，离子在空气中进入大气压界面，导入质谱仪；解吸电喷雾电离是一种技术，可在室温大气压下离子化微量样本，样本的准备也很少，它可用于在原位检验、二级代谢产物、特别检测空间和时间的分布。

离子化机制

解吸电喷雾电离有两种离子化机制：其一用于低分子量的分子，其二用于高分子量分子。高分子量分子，如蛋白质和多肽，显示电喷雾像光谱，可看到多电荷离子。这就是说，被分析物的解吸，喷雾里面的多电荷，可以转移到被分析物，带电的小滴打击到样本，散发开成比原本小滴更大的直径，溶解蛋白质并回弹，接着小滴进入质谱仪入口，

was developed in 2004 by Zoltan Takáts et al., in Professor Graham Cooks' group from Purdue University and is now available commercially by Prosolia Inc. DESI is a similar ionization technique to Direct Analysis in Real Time (DART) in its versatility, applications, and analysis time.

DESI is a combination of electrospray (ESI) and desorption (DI) ionization methods. Ionization takes place by directing an electrically charged mist to the sample surface that is a few millimeters away. The electrospray mist is pneumatically directed at the sample where subsequently splashed droplets carry desorbed, ionized analytes. After ionization, the ions travel through the air into the atmospheric pressure interface, which is connected to the mass spectrometer. DESI is a technique that allows for ambient ionization of a trace sample at atmospheric pressure, with little sample preparation. DESI can be used to investigate *in situ*, secondary metabolites specifically looking at both spatial and temporal distributions.

Ionization mechanism

In DESI, there are two kinds of ionization mech-

更进一步去溶液。通常电喷雾使用的溶剂为甲醇和水的混合。

对于低分子量的分子，离子化发生于电荷转换：一个电子或一个质子。电荷转换有三种可能：第一，电荷转换于溶剂离子和被分析物表面；第二，电荷转换于气相离子和被分析物的表面，在此状况下，溶剂离子在到达样本表面之前，已经挥发了，为达到如此，喷雾到样本表面距离大；第三，电荷转换于气相离子和气相被分析物分子，这是当分子有高的蒸气压力才发生。低分子量分子的解吸电喷雾电离机制，类似于即时直接分析的离子化机制，在气相中发生电荷转换。

离子化效率

解吸电喷雾电离的离子化效率很复杂，依赖几个参数，如表面作用、电喷雾参数、化学参数和几何参数；表面作用包含化学组成、温度和所用的电压；电喷雾参数包括电喷雾电压、气体和液体流速；化学参数指喷雾溶剂组成，例如加入氯化钠；几何参数为：α, β, d_1 和 d_2（图 4.9.2)。

再者，α 和 d_1 影响离子化效率，β 和 d_2 影响收集效率；为了取得 α 和 d_1 最佳数值，透过测试多种分子的结果显示，有两群分子：高分子量（蛋白质、多肽、寡糖等等）和低分子量（双偶氮色料、类固醇、咖啡因、硝基芳烃等)；高分子量群的最佳条件为高入射角，70°~90°，

anism, one that applies to low molecular weight molecules and another to high molecular weight molecules. High molecular weight molecules, such as proteins and peptides, show electrospray like spectra where multiply charged ions are observed. This suggests desorption of the analyte, where multiple charges in the droplet can easily be transferred to the analyte. The charged droplet hits the sample, spreads over a diameter greater than its original diameter, dissolves the protein and rebounds. The droplets travel to the mass spectrometer inlet and are further desolvated. The solvent typically used for the electrospray is a combination of methanol and water.

For the low molecular weight molecules, ionization occurs by charge transfer: an electron or a proton. There are three possibilities for the charge transfer. First, charge transfer between a solvent ion and an analyte on the surface. Second, charge transfer between a gas phase ion and analyte on the surface; in this case, the solvent ion is evaporated before reaching the sample surface. This is achieved when the spray to surface distance is significant. Third, charge transfer between a gas phase ion and a gas phase analyte molecule. This occurs when a sample has a high vapor pressure. The ionization mechanism of low molecular weight molecules in DESI is similar to DART's ionization mechanism, in that there is a charge transfer that occurs in the gas phase.

Ionization efficiency

The ionization efficiency of DESI is complicated and depends on several parameters, such as surface effects, electrospray parameters, chemical parameters, and geometric parameters.

短 d_1 距离，1~3 mm；低分子量群最佳条件相反，低入射角，35°~50°，长d_1距离，7~10 mm；这些测试结果指出，每一群的分子，有不同的离子化机制。

喷嘴尖端和表面支托器，都连接到一个三维移动云台，可以选择某特定数值，设定四个几何参数：α, β, d_1 和 d_2。

4.10　电感耦合等离子体质谱及激光剥蚀电感耦合等离子体质谱

这是一种很棒的新元素分析仪器，取代了原子吸收仪 (atomic absorption)，我们用过它分析镧系稀土元素、测量云南楚雄州恐龙地层的相对定年，也用它做了铀/铅同位素直接定年。

（取自维基）电感耦合等离子体质谱 (ICP-MS) 兼具多元素分析及高灵敏度（包括金属元素）之优点，适于样品中

Surface effects include chemical composition, temperature, and the electric potential applied. Electrospray parameters include electrospray voltage, gas, and liquid flow rates. Chemical parameters refer to the sprayed solvent composition, e.g., the addition of NaCl. Geometric parameters are α, β, d_1, and d_2 (see Figure 4.9.2).

Furthermore, α and d_1 affect ionization efficiency, while β and d_2 affect the collection efficiency. Results of a test performed on a variety of molecules to determine optimal α and d_1 values show that there are two sets of molecules: high molecular weight (proteins, peptides, oligosaccharide, etc.) and low molecular weight (diazo dye, steroids, caffeine, nitroaromatics, etc.). The optimal conditions for the high molecular weight group are high incident angles (70°-90°) and short d_1 distances (1–3 mm). The optimal conditions for the low molecular weight group are the opposite, low incident angles (35°-50°) and long d_1 distances (7–10 mm). These test results indicate that each group of

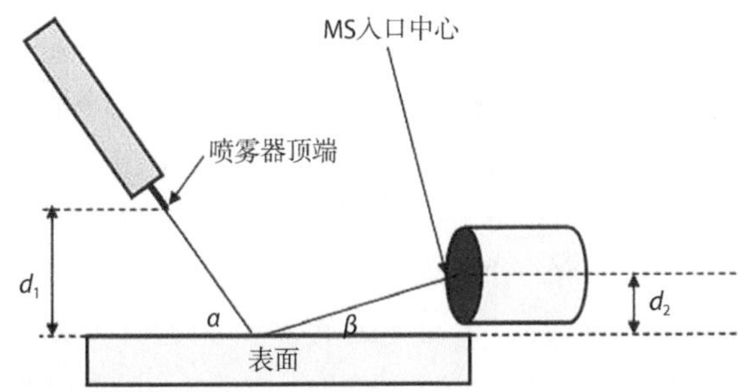

α	入射角	0~90°
β	收集角	5°~10°
d_1	顶端至表面的距离	1~10 mm
d_2	MS入口至顶端的距离	0~2 mm

◀ 图 4.9.2　DESI 离子源侧视图和包含典型数值的几何参数表格

◀ Figure 4.9.2　Side view of DESI ion source along with a table containing typical values for the geometric parameters

金属与非金属含量之精确浓度定量，在无干扰低背景同位素情况下可达10^{-15}。它透过电感耦合电浆离子化，然后用质谱仪分开并定量离子。

相对于原子吸收光谱仪(atomic absorption spectroscopy, AA)，电感耦合等离子体质谱测试速度较快、精准、灵敏，然而，比较于其他质谱仪，如热离子质谱仪(thermal ionization mass spectrometry, TIMS)和光释放质谱仪(glow discharge mass spectrometry, GD-MS)，它引入了好多干扰的物种：电浆的氩、从锥形孔漏进来的空气成分和玻璃器皿与锥子的污染物。

电感耦合等离子体质谱广泛地应用于地球化学领域，用来做同位素定年，测定不同同位素的相对丰量，特别是铀和铅，电感耦合等离子体质谱比以前使用的热离子化质谱更适合用于此分析，因为高离子化能量的核种，如锇和钨，可以容易地被离子化。为取得精准比例，经常会使用多接收器的设备，来减少计算比例的噪声。

元素分析

电感耦合等离子体质谱能测定原子量从7到250（锂到铀）的元素，有些质量，如40则被拒绝，因为在样本中高浓度的氩，其他被排除的质量为80（氩的二聚体）和质量56（因为ArO)，除非仪器加装一个反应腔，后者会妨碍铁的分析，这种干扰可透过高分辨率电感耦合

molecules has a different ionization mechanism.

The sprayer tip and the surface holder are both attached to a 3D moving stage which allows selecting specific values for the four geometric parameters: α, β, d_1, and d_2.

4.10 Inductively Coupled Plasma Mass Spectrometry (ICP-MS) Laser Ablation Inductively Coupled Plasma Mass Spectrometry (LA-ICP-MS)

This is an excellent new elemental analysis instrument, replacing Atomic Absorption. Some works were carried out to provide relative dating (by La-Series Rare Earth Elements) of dinosaur formations in ChuXiong Prefecture of YunNan Province, and absolute dating for DaWa Layer by U/Pb isotopes.

(From Wiki) Inductively coupled plasma mass spectrometry (ICP-MS) is a type of mass spectrometry which is capable of detecting metals and several non-metals at concentrations as low as one part in 10^{15} on non-interfered low-background isotopes. This is achieved by ionizing the sample with inductively coupled plasma and then using a mass spectrometer to separate and quantify those ions.

Compared to atomic absorption spectroscopy, ICP-MS has higher speed, precision, and sensitivity. However, compared with other types of mass spectrometry, such as thermal ionization mass spectrometry (TIMS) and glow discharge mass spectrometry (GD-MS), ICP-MS introduces many interfering species: argon from the plasma, compo-

等离子体质谱 (HR-ICP-MS) 来减少，这是使用两个或更多狭缝来紧缩喷束，区分相连的波峰，但是灵敏度会被打折。比方说，要分辨铁和氩，需要大约 10000 的区分能力，而导致铁灵敏降低 99%。

激光剥蚀是另外一种样本导入的方法，在过去比较不常见，但是已经开始热门起来，用来导入样本，这是受益于电感耦合等离子体质谱扫描速度的提高。一束脉冲紫外线激光照射于样本，产生一股剥蚀，被带入电浆中。如此一来，让地球科学家能测量岩石剖面的同位素组成分布，因为如果将样本溶解成液体导入的话，就无法达成；这种激光有精确的控制能力和均匀的相近功率分布，产生照射底部平整、大小和深度可选择的小洞。

样本制备

在大部分使用电感耦合等离子体质

nent gases of air that leak through the cone orifices, and contamination from glassware and the cones.

ICP-MS is also used widely in the geochemistry field for radiometric dating, in which it is used to analyze the relative abundance of different isotopes, in particular, uranium and lead. ICP-MS is more suitable for this application than the previously used thermal ionization mass spectrometry, as species with high ionization energy such as osmium and tungsten can be easily ionized. For high precision ratio work, multiple collector instruments are generally used to reduce the effect of noise on the calculated ratios.

Elemental analysis

The ICP-MS allows determination of elements with atomic mass ranges 7 to 250 (Li to U), and sometimes higher. Some masses are prohibited, such as 40 due to the abundance of argon in the sample. Other blocked regions may include mass 80 (due to the argon dimer), and mass 56 (due to ArO), the latter of which dramatically hinders Fe

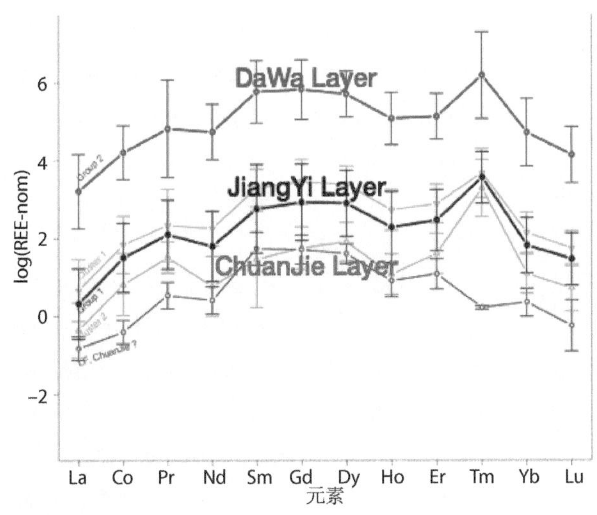

◀ 图 4.10.1 镧系稀土元素的蜘蛛线，很清晰地显示云南省楚雄州恐龙地层有“三”层

◀ Figure 4.10.1 Spidergram of La series Rare Earth Element analysis clearly shows three "layers" of dinosaur formations in ChuXiong Prefecture of YunNan Province

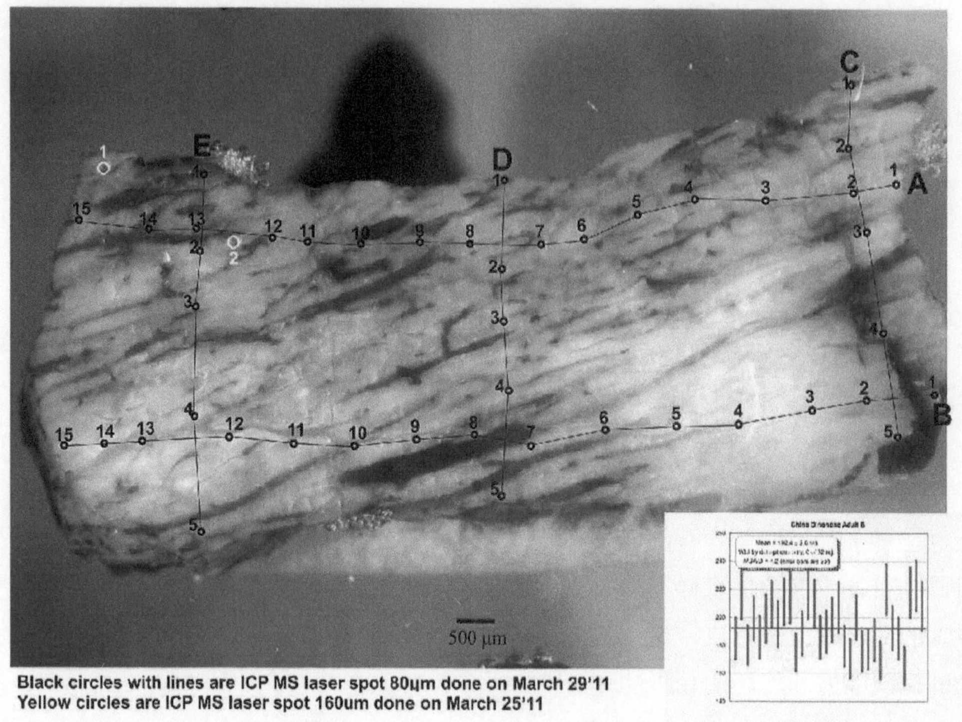
C
E
D
A
B
500 μm
Black circles with lines are ICP MS laser spot 80μm done on March 29'11
Yellow circles are ICP MS laser spot 160um done on March 25'11

▲ 图 4.10.2 禄丰龙成龙的激光剥蚀电感耦合等离子体质谱测试

▲ Figure 4.10.2 LA-ICP-MS of adult *Lufengosaurus* bone

谱的检测方法中，有一个相对来说简单快速的样本制备过程，样本中主要的成分为内在的标准，同时也当稀释剂，内在标准主要是去离子水，混合硝酸或盐酸，以及铟或镓标准液，依照样本种类，通常内在标准加入到含着 10~500 ml 样本的试管内，然后这个混合溶液摇动几秒钟，最后放入自动样本盘；对于其他的运用，可能包含黏稠样本，获有某种特殊物质的样本，可能就需要先执行样本消化，然后用吸管放入去分析。这就多增加了一道手续，使得样本制备需要更长的时间。

analysis unless the instrumentation is fitted with a reaction chamber. Such interferences can be reduced by using a high-resolution ICP-MS (HR-ICP-MS) which uses two or more slits to constrict the beam and distinguish between nearby peaks. This comes at the cost of sensitivity. For example, distinguishing iron from argon requires a resolving power of about 10,000, which may reduce the iron sensitivity by around 99%.

Laser ablation is another sample introduction method. While being less common in the past, is rapidly becoming popular has been used as a means of sample introduction, thanks to increased ICP-MS scanning speeds. In this method, a pulsed UV laser is focused on the sample and creates a

4.11 X射线荧光/同步辐射X射线荧光

我最早期的同步辐射X射线荧光是2013年在美国斯坦福大学(Stanford Linear Accelerator Center, SLAC)/(Stanford Synchrotron Radiation Lightsource, SSRL)，以大约10 μm的分辨率进行测试含有胚胎牙齿的禄丰龙骼骨，取得了二维的元素分布，从这些很有用的信息，不只限于埋藏学资料，而且还用于如石化动物的食性——肉食性动物在它们化石骨头内会有锌，而植食性动物则无。

（取自维基）X射线荧光是某物质经过高能X射线或γ射线照射所产生的典型特征“二级（荧光）”X射线，这种现象广泛使用于元素和化学分析，特别是在分析金属、玻璃、陶瓷和建筑材料，也用于地球化学 (Geochemistry)、鉴识科学、考古学的字画和壁画等方面的研究。

当物质在短波X射线或γ射线照射下，其成分的原子可能被离子化。离子化是将原子的一个或两个电子抛弃出，可发生在照射能量大于物质离子化能量的情况。X射线和γ射线可能有足够的能量，把紧紧藏在原子内层轨道的电子驱离产生“洞”，如此一来，原子的电子结构就不稳定，在较高层轨道的电子会“掉入”较低轨道,填充“洞”,在此“掉入”过程中，将等于两层电子轨道的能差以光子释放出来。故此物质所释放的辐射，就是存在物质内原子的能量特性。荧光

plume of ablated material which can be swept into the plasma. This allows geochemists to spatially map the isotope composition in cross-sections of rock samples, a tool which is lost if the rock is digested and introduced as a liquid sample. Lasers for this task are built to have highly controllable power outputs and uniform radial power distributions, to produce craters which are flat bottomed and of a chosen diameter and depth.

Sample preparation

For most clinical methods using ICP-MS, there is a relatively quick and straightforward sample prep process. The main component of the sample is an internal standard, which also serves as the diluent. This internal standard consists primarily of deionized water, with nitric or hydrochloric acid, and Indium and or Gallium. Depending on the sample type, usually, 5 ml of the internal standard is added to a test tube along with 10–500 microliters of sample. This mixture is then vortexed for several seconds or until mixed well and then loaded onto the autosampler tray. For other applications that may involve very viscous samples or samples that have a particulate matter, a process known as sample digestion may have to be carried out before it can be pipetted and analyzed. This adds an extra first step to the above process and therefore makes the sample prep more lengthy.

4.11 X-ray Fluorescence/sr-XRF

An initial trail sr-XRF was carried out in 2013 at SLAC on a *Lufengosaurus* embryonic jaw bone containing an embryonic tooth with a resolution of ~10μm. Two-dimensional elemental distribu-

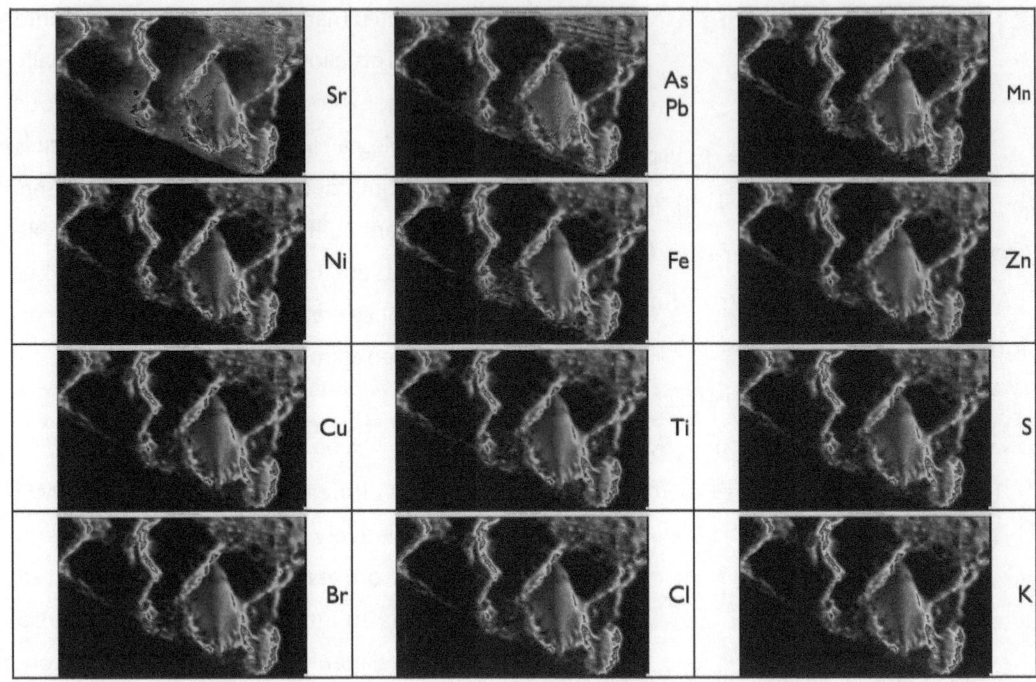

▲ 图 4.11.1 同步辐射 X 射线荧光提供化石内常见十四种元素的二维分布分析。这个样本是禄丰龙胚胎上骼骨，带着第三代的胚胎牙齿，左上角是磷和钙的重叠影像，代表磷灰石

▲ Figure 4.11.1 sr-XRF provides 2D elemental distribution of 14 commonly seen elements in the fossil. This particular specimen is a *Lufengosaurus* embryonic jaw bone with a third generation embryonic tooth. The image at the upper left is a composite of P and Ca representing apatite

这个词，是指某物质吸收某较高能量辐射照射，通常以较低能量再辐射的现象。

4.12 激光荧光光谱

这是一个很便宜每个古生物实验室都可自己架设的设备，所需要的只是高功率（大于半瓦）发光二极体激光 (LED Laser)，如 405 nm、450 nm 和 520 nm，现在这些都已经很便宜了；另外需要适当的滤光片来把激光本身的光过滤掉，只让它所引发的荧光通过到数位相机，滤光片该为 420 nm、470 nm或 480 nm

tion maps were obtained. From such many useful information can be derived not only taphonomically, but also for such as the diet of the fossilized animal. Meat-eater animal would have Zn in their fossilized bone, while plant-eater will not.

(From Wiki) X-Ray fluorescence (XRF) is the emission of characteristic "secondary" (or fluorescent) X-rays from a material that has been excited by bombarding with high-energy X-rays or gamma rays. The phenomenon is widely used for elemental analysis and chemical analysis, particularly in the investigation of metals, glass, ceramics and building materials, and for research in geochemistry, forensic science, archaeology and art objects

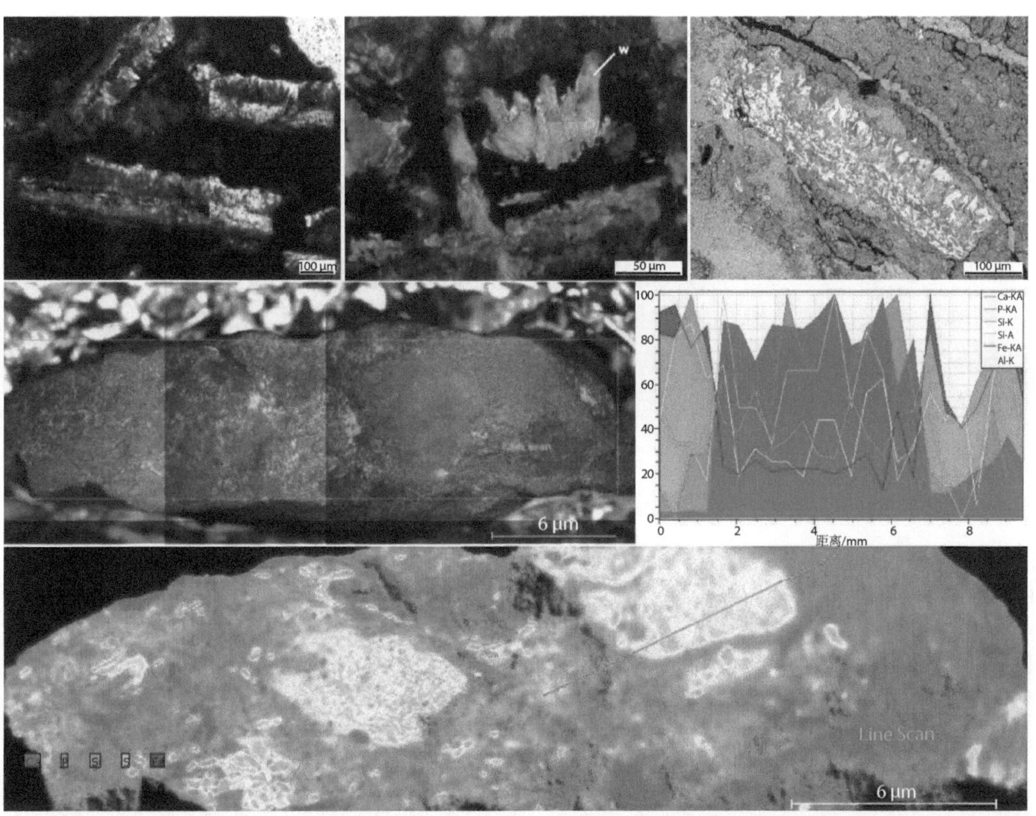

▲ 图 4.11.2　禄丰龙蛋壳之 X 射线荧光

▲ Figure 4.11.2　XRF of *Lufengosaurus* eggshell

（商业上叫 Y2 或 K2）和 560 nm（O2），可在老照相器材行买到以前黑白摄影用的，此外随着自己的需要添加配备。

如图 4.12.1 所示的显微观测，需要一个立体显微镜。若要扫描大型化时，则需要有个能投射一字形激光的镜头、暗房和一个可自动平移的云台，以便把整个化石扫描记录成数位相片。再者，因为自己做的手机光谱仪已经是现成的，可以容易地结合显微观测型，提供便宜的荧光元素光谱分析，比方说，若观察到化石上某个点有强烈的荧光反应，我们可以更近聚焦于这个点，进行

such as paintings and murals.

When materials are exposed to short-wavelength X-rays or gamma rays, ionization of their component atoms may take place. Ionization consists of the ejection of one or more electrons from the atom and may occur if the atom is exposed to radiation with an energy greater than its ionization energy. X-rays and gamma rays can be energetic enough to expel tightly held electrons from the inner orbitals of the atom. The removal of an electron in this way makes the electronic structure of the atom unstable, and electrons in higher orbitals "fall" into the lower orbital to fill the hole left behind. In fall-

元素光谱分析。元素光谱分析是太空物理常用来判断遥远星球的方法。借由这个简单的并用，可以取得该点构成元素的一些粗浅资料，再进行后续的分析。

图 4.12.2 为一对同样禄丰龙胚胎的微型计算机断层扫描和中子扫描比对，注意到下方微型计算机断层扫描影像中，有三个“漏洞”（红色圈），而在中子扫描影像里面却看不到，这一点显示了在不同照射源之下，会看到不同的景象，亦即前面提过的，在某照射源之下

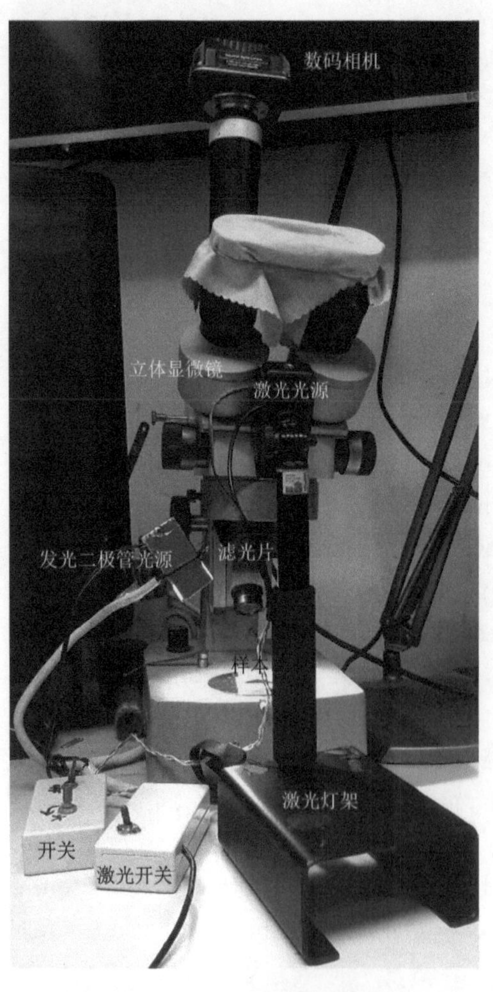

ing, energy is released in the form of a photon, the energy of which is equal to the energy difference between the two orbitals involved. Thus, the material emits radiation, which has an energy characteristic of the atoms present. The term fluorescence is applied to phenomena in which the absorption of radiation of a specific energy results in the re-emission of radiation of different energy (generally lower).

4.12 Laser Stimulated Fluorescence Spectroscopy (LSFS)

This is a very inexpensive setup every paleontological lab can build by themselves. It consists of a high power LED laser light source (>0.5 W) of a given wavelength, such as 405nm, 450nm, and 520nm. These LED lasers are widely and inexpensively commercially available now. The other necessary accessory is the appropriate filter to block out the LED wavelength so that only fluorescence it generated reaches the digital camera. The filters shall/can be 420nm, 470 or 480nm (commercially called Y2 or K2), and 560nm (O2). These filters can be found in the good old day camera stores for black and white photography. Whatever needed can be added.

◀ 图 4.12.1 简单显微型激光荧光设置

◀ Figure 4.12.1 Simple setup of a micro scale Laser Stimulated Fluorescence

看到好影像，固然可以高兴一下，但是有一个严肃的问题："没有看到什么？"，明明有东西存在那里，为什么没有看到？这种现象，姑且称之为"有色眼镜效应"，戴上有色眼镜所看到的景象，只是该色片所能提供给看的，而非全貌；这个问题，应该是所有研究员应该经常放在脑子里面的关键思维。

再看图4.12.3，这四张照片同样是禄丰龙胚胎股骨中央横切面，在不同光源照射之下的影像。左上角为一般发光二极体照明，右上为 405 nm 激光荧光，左下为 450 nm 激光荧光，右下为 520 nm 激光荧光。请注意细微的差异，特别是在骨髓腔内，不同光源波长所产生的荧光反应，各有千秋，应该仔细分辨。

For the microscopic view, as shown in Figure 4.12.1, a dissecting microscope shall be used as shown in the photo. For the macroscopic view, an [I] beam lens shall be attached to the LED laser, and a one-axial automatic movement platform (slider) shall be used in a dark room so that the whole fossil specimen can be scanned and recorded. Furthermore, since the DIY optical spectroscope (for the smartphone) is also available now. The straightforward combination of these two can provide microscopic fluorescence elemental analysis at very meager. This kind of home build equipment does not cost an arm or a leg. Any lab can afford. Using them as the primary screening tool for fossil study can be very useful. For example, if we see a spot on fossil having strong fluorescence response, we can zoom into that spot and conduct a spectroscopic elemental analysis. Elemental spectroscopy is well known and used by such as astrophysics. By this simple combination, we can obtain some ideas on the compositional elements at that spot before conducting further analysis.

Figure 4.12.2 is a pair of μ-CT and neutron scan images of the same *Lufengosaurus* embryonic bones and claw. Please note that in the low image of μ-CT, there are three holes (red). However, these three holes cannot be seen in the neutron scan. This shows that under different

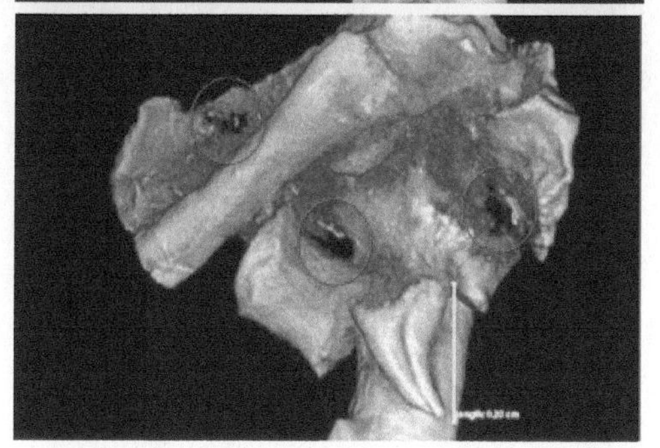

◀ 图 4.12.2　比较 26.7 μm 分辨率中子扫描（上）和 9 μm 分辨率微型计算机断层扫描（下）

◀ Figure 4.12.2　Comparison of Neutron scan (top), at 26.7 μm resolution and μ-CT (bottom) at 9 μm resolution

感谢台湾大学江宏仁教授，提供了手机光谱仪和直接显微镜套筒（图4.12.4），现在看起来，虽然有些地方可能还需要调整调整，但是，至少证明了设备可用可行，我的构想也没错可行。他已经帮忙做好了，他还加装好了一个网络摄影机 (Webcam)，可直接接到计算机，跑 ImageJ。ImageJ 也有现成的“抓网络摄影机外挂 (Webcam Capture

radiation source, it will show different images mentioned previously. Seeing something under one radiation source shall be a great thing. However, there is also a very serious question: What is NOT seen? There is something there, why not see? Let's call this phenomenon as "color glasses effect". When one wears a pair of color glasses the image saw is limited to whatever the tinted glasses can provide, not the whole picture. This a question that every researcher shall keep in

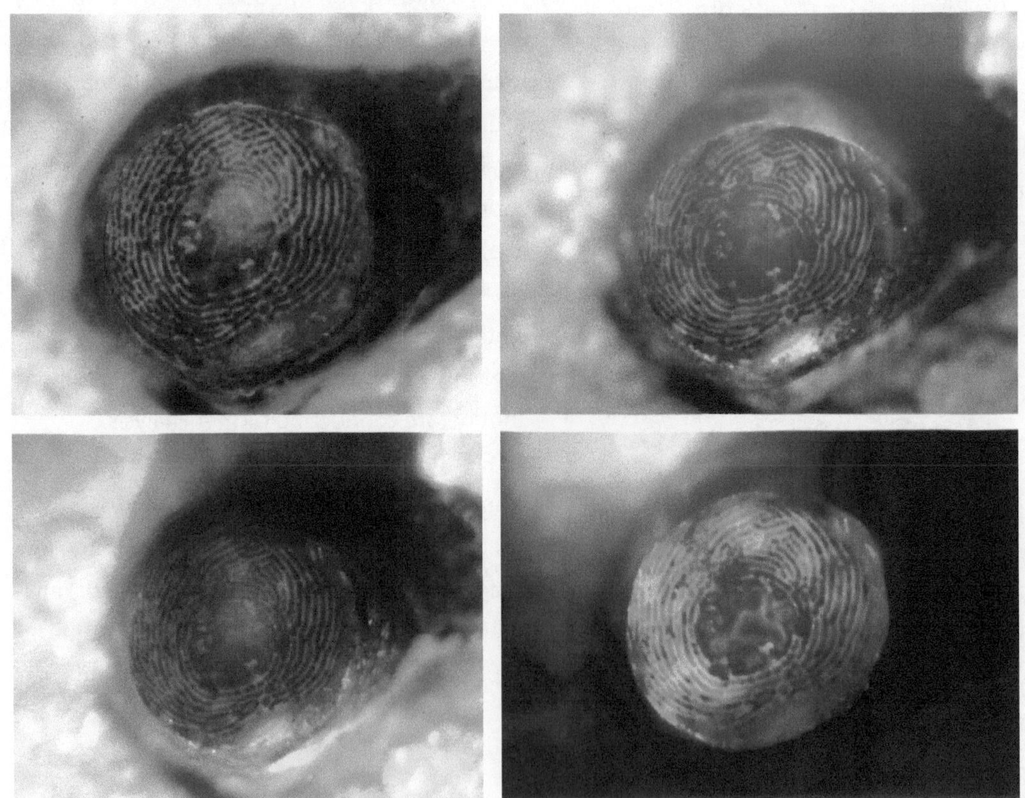

▲ 图 4.12.3　禄丰龙胚胎股骨中央横切面，在不同光源照射之下的影像，左上角为一般发光二极体照明，右上为 405 nm 激光荧光，左下为 450 nm 激光荧光，右下为 520 nm 激光荧光。请注意细微的差异

▲ Figure 4.12.3　Transversal cut of *Lufengosaurus* embryonic femur under different radiation sources. Upper left is a conventional LED light. Upper right 405nm Laser Fluorescence. Lower left 450nm Laser Fluorescence. Lower right 520nm Laser Fluorescence

Plugin)”，人家已经写好免费的后段程序，使用装置如图 4.12.5。

回说一下我的想法：以前说过，我的一个朋友Tom Kaye搞了一套激光荧光(laser stimulated fluorescence, LSF) 设备，近期在几本有分量的学术期刊发表了好几篇轰动“武林”的文章，提供古生物界一种崭新的观测方法，把过去在普通灯光下看不到的化石细节展现出来，如最近他们发表的论文说，在恐龙的脚部分看到了以前看不到的羽毛和鳞片（这个请看官去看看鸡脚和鸡爪部位，是有毛和鳞的）。去年我很受鼓励，就网购了三个不同波长的激光发光二极体 (LED laser)，装到我的立体显微镜，用来观测化石的显微构造，我不像Tom看大件的化石，他需要暗房和自动平移的云台，我只用现成的这台老显微镜，这两者虽有巨观和微观的差异，其实原理是一样的：化石受到强烈激光光源照射，会因成分（不同的元素原子或有机残留物）产生不

◀ 图 4.12.4 江老师的手机光谱仪转接显微镜接头（左上方）和网络摄影机（右下方）

◀ Figure 4.12.4 Professor Jiang's phone spectroscope adaptor to microscope object lens (upper left) and webcam (lower right)

mind all the times.

Look now Figure 4.12.3. These four photos were the very same *Lufengosaurus* embryonic femur mid-shaft taken under different radiation conditions: upper left, regular LED light, upper right 405 nm LSF, lower left 450nm LSF and lower right 520nm LSF. Please notice the minute differences, particularly in the medullary chamber, and different light wavelength produce different fluorescence. They shall be distinguished.

Thanks to Professor Jiang HongZen of Taiwan University, providing the coupling for handheld spectroscope as shown in Figure 4.12.4. As this writing time, some minor adjustments may still be needed; it already proved this gadget and my thoughts work. He even included a webcam which can attach to a computer running ImageJ. There is an available free plugin, Webcam Capture Plugin for download. The whole working setup is shown in Figure 4.12.5.

A bit of my thought: As mentioned a friend of mine previously, Tom Kaye published several intriguing papers showing the Laser Stimulated Fluorescence (LSF) he invented for the paleontology field. It shows the details of fossils not seen under conventional lighting. Their latest paper show hairs and scales of a dinosaur foot (similar can be seen on chicken feet). I was so encouraged and bought three LED lasers of a

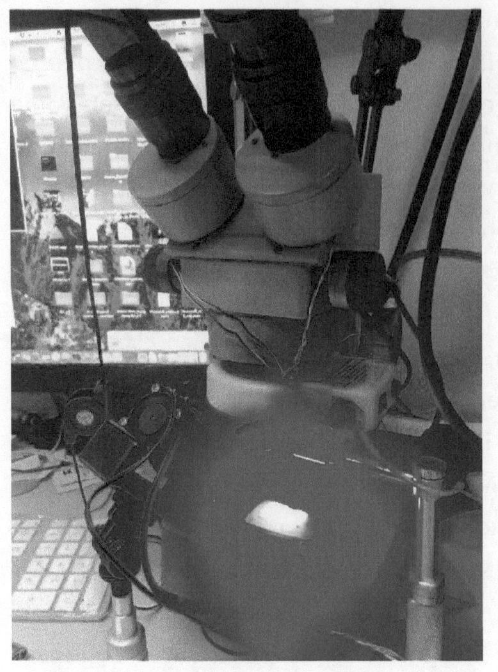

◀ 图 4.12.5 激光显微荧光光谱仪，蓝色光是 450 nm 激光光，照射在样本上（白色），显微镜物镜下方放了一个 470 nm (K2) 滤光片

◀ Figure 4.12.5 Laser Stimulated Fluorescence Microspectroscope. The blue light is the 450 nm laser light shining on the specimen (white). A filter of 470nm is used

同的荧光，这些荧光是一般光源无法提供的。不过不要忘记一个更重要的问题："没有看到什么？"，没有看到的，可能比看得到的更为重要。

所以，在用显微激光荧光的过程中，我就想到，如果激光荧光能和光谱仪结合，那么就可以提供更多的信息，让我们能探讨化石和围岩内的不同元素（成分），对于我提倡的"看进骨头内"研究古生物，这是个关键设备。玩矿物收藏者都知道荧光矿物，在普通灯光下毫不起眼的这些"烂"石头，经由长短波紫外线照射下，竟然发出绚丽的大红大绿，另有一番天地！为何会如此？主要的原因是这些矿物里面含有的微量杂质元素，比方说，很常见的矿物方解石，如果含锰 (Mn) 的话，就会发出深红色，

different wavelength online. I am using them with my dissecting microscope for the observation of small fossils, unlike Tom's setup for big ones. The same principle on both: Under intense radiation, fossil will yield different fluorescence due to different composition (different elements or organic remains). This fluorescence could not be seen under natural lights. Thus, I keep saying as scientists we are, we have to keep in mind a more important question: "what is NOT seen?". Whatever not seen could be more important than what can be seen.

Thus, during the usages of Laser Stimulated Fluorescence Microscopy (LSFM), I thought if the LSF can be combined with spectroscopy, then more information can be gain for us to explore what is inside the fossil and matrix. This is a crucial factor for my promotion of "Look into the Bone" research guideline. Those collectors of fluorescence mineral know under the natural light those "junk" rock can show very vivid colors under short and long wave UV lights as a different world. Why? The main reason is the trace amount of other elements inside the main mineral. For example, if commonly seen calcite contains element manganese (Mn), it will produce deep red color, and it contains mercury

但是如果含汞（水银，Hg）的话，则是蓝白色荧光，主要的成分都是碳酸钙，但是荧光颜色不同，主要是里面所含的微量杂质元素不同！

因此，理论上来说，如果能把手机光谱仪接上激光荧光，看到有趣的荧光反应，在显微镜下调高放大倍率，也就是缩小观测范围到某个很小区域，取得该点的光谱，再进行元素光谱，那么就可以非破坏性地观测得到该点的元素信息了。元素光谱分析，就是天文物理学家判断远处某颗星星是什么种类星星的主要依据，他们从天文望远镜得到某星星的光谱，进而做光谱元素分析，不同的元素会有不同的光谱线，如此就可知道那颗星星的元素组成了，年轻的星星会有较多的轻元素（氢氦等），老年的星星会有较多重（原子序较大）的元素。所以，我的想法是把激光荧光和手机光谱仪结合起来，来日可进行化石的元素光谱分析。

话说回头，初步测试到现在的结果，我很受鼓励，至少看起来我先前的想法没错可行，可以透过这种配合，进行化石的显微激光荧光光谱分析 (laser stimulated fluorescence microscopic elemental analysis, LSFMESA) 了。

现在请看图 4.12.6 到图 4.12.9，在这四张图片的右上角，放着不同光源照射下的普通（图 4.12.6）和荧光照片（图 4.12.7 到图 4.12.9）；图 4.12.6 是用显微镜的发光二极体照射样本的光谱，没有荧光反应，基本上就是发光二极体反射

(Hg), it is bluish-white. The main components are the same, calcium carbonate. The different trace impurities caused different fluorescence colors.

Therefore, in theory, if the handheld spectroscope can be attached to the Laser Stimulated Fluorescence Microscope for the observation of fossil. When and where-ever a strong fluorescent spot shows up, we can then zoom-in to that particular spot to conducting the elemental spectroscopic analysis. So, we can gain some elemental information non-destructively. Elemental spectroscopy is the primary way that astrophysicists use for analyzing stars. From the telescope, the light of a given star undergoes elemental spectroscopic analysis. Different elements will yield elemental spectra line(s). By doing so, they can learn the elemental composition of that particular star. Younger stars will have lighter elements, such as hydrogen and helium, while older starts will have the heavier element (higher atomic number). So, I think combining these two can enable us to conduct elemental analysis on fossils.

Now go back to my initial trials. The results are very encouraging. At least it shows the feasibility of combining these two DIY gadgets to do Laser Stimulated Fluorescence Microscopic Elemental Analysis (LSFMESA).

Now, look at Figures 4.12.6 to 4.12.9. On the upper right-hand corner are the camera images of regular LED light (Figure 4.12.6) and fluorescent photos (Figure 4.12.7–4.12.9). Figure 4.12.6 did not have any fluorescent, just regular LED reflection. Please note that there are three spectrum humps at about 100, 200, and 300 horizontal markers. Figure 4.12.7 was taken

到光谱仪的光谱，请注意到在尺标大约100, 200和300位置，总共有三个波峰；图4.12.7是用405 nm半瓦激光照射样本的荧光光谱，因为这个激光功率最低，计算机荧幕上黄色框内，几乎看不到很黯淡的荧光光谱，其实只要很用力看，还是可以看到一些；在光谱曲线上，可以看到两个，一在尺标约40~80处，另一则在约240~280处；图4.12.8是1.6 W 450 nm照射的荧光光谱，比405 nm的稍微强一些，激光功率是我这三个当中最高的，光波峰出现在约100~150左右，以及约220~280；图4.12.9则是一瓦520

under 405nm, 0.5 W laser fluorescent, and spectrum. Since this is the lowest power laser, inside the yellow box, almost no spectra image can be seen. You have to look at it very very hard to see the very very faint image. However, on the spectrum, to humps can be seen at 40–80 and 240–280. Figure 4.12.8 was taken under 1.6 W 450 nm laser. Two wave humps show up at 100–150 and 220–280. Figure 4.12.9 was taken under 1 W 520nm laser. Fluorescent spectra are the most clearly shown, unlike the other two. However, only one wave hump showed up around 240–280 marker. Please note that the spectra of these three different wavelength laser are very

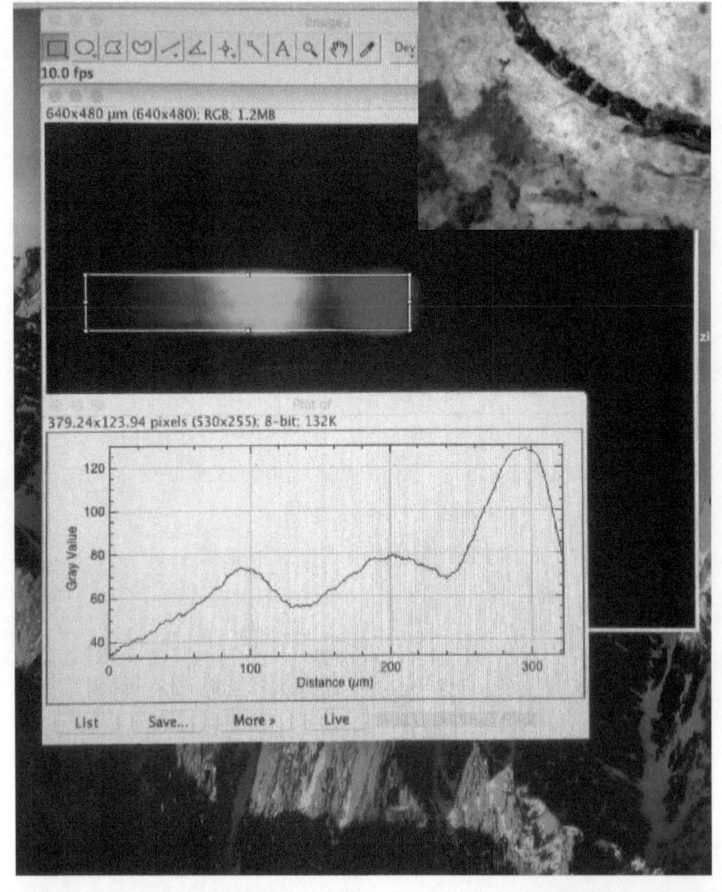

◀ 图4.12.6 普通发光二极体灯源照射在化石样本上的光谱与光谱曲线

◀ Figure 4.12.6 The spectrum and spectra curve of LED on the fossil sample

nm 激光照射样本的荧光光谱，荧光光谱很明确，不像前两者那么暗，它最强的波峰陀，只出现在尺标约 240~280 位置；有没有注意到，三种波长激光灯源所产生的整体荧光光谱曲线，都有所不同？每张图片右上角的实际相片，也显示了这些不同颜色，也就是说，不同波长照射源产生的荧光反应也不一样；图 4.12.10 为同样的样本在四种不同灯源照射之下的光谱和光谱曲线；这一点证明了在不同照射源之下，同样的东西会看到不同的景色！

顺便啰唆一下：自己动手做科研的

different. The photo images at each corresponding corner also show the color differences. Figure 4.12.10 shows the image, spectrum image, and spectra of the same specimen under these four different light source. Again, this proves that under different light source will see a different result.

Using this opportunity, I would like to mention the issue of making our research gadgets and instruments. As far as I know, most of the academics in China do not have a habit of making the tools for their own use. They will buy commercial facilities with public money. Some of them are expensive ones and placed in the

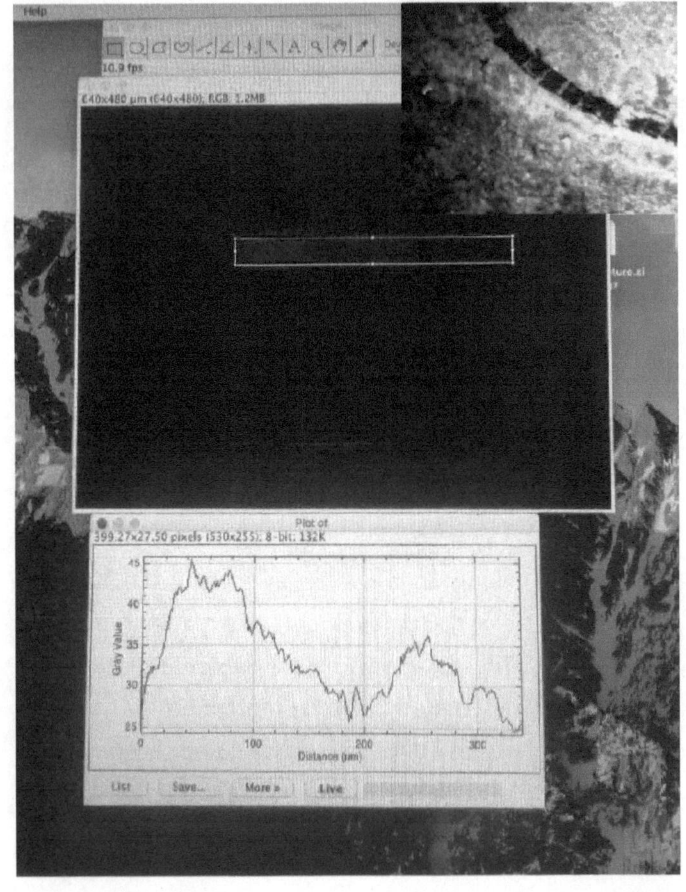

◀ 图 4.12.7 激光 405 nm 灯源照射在化石样本上的光谱与光谱曲线

◀ Figure 4.12.7 Spectrum and spectra curve of 405nm Laser shining on the fossil

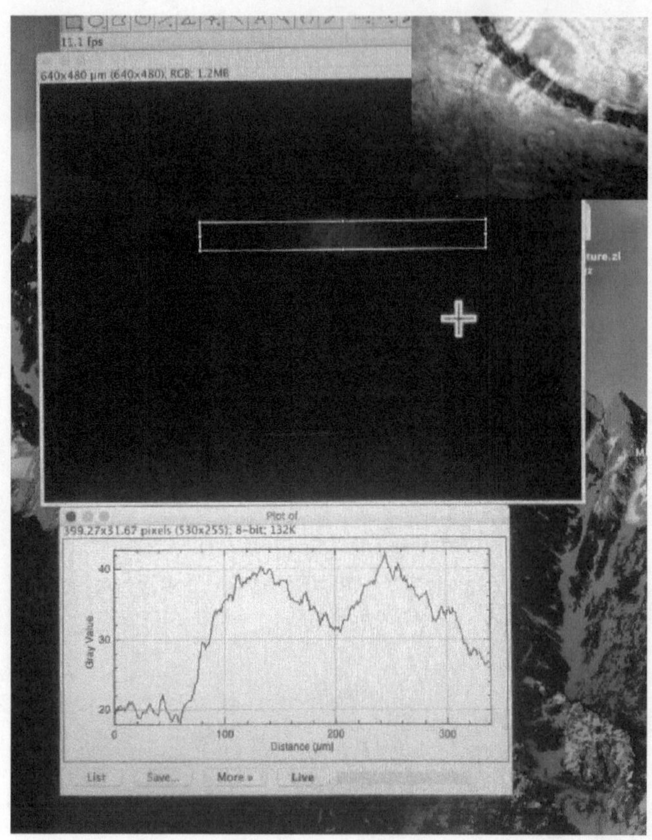

◀ 图 4.12.8 激光 450 nm 灯源照射在化石样本上的光谱与光谱曲线

◀ Figure 4.12.8 Spectrum and spectra curve of 450 nm Laser shining on the fossil

仪器设备，我所知道在国内，好像不是很普遍，通常都是买现成的仪器，买人家做好的设备，那些很昂贵的，还放到“贵重仪器中心”，大家排队轮流用。其实在美国，至少我念书的时候，研究所逼着我们尽量动手做自己需要的仪器设备。反过头来问：台湾有几家大学的化学系，设有玻璃工车间？机械车间？我们的科学教育，应该好好反省了！幸好，有像江老师这么热心的人，愿意几乎免费研发制作基础科学的科研设备，我才能做到今日的成果，我没有几百万的钱去买台不一定适用于我的荧光显微镜，应该给江老师好好鼓掌一下！激光荧光

Precious and Expensive Instrument Center for everybody in the queue to use. When I was in the United States for my graduate studies, the training forced us to do our own specialized gadgets. Let me ask this, how many university chemistry departments in Taiwan have glassware workshop? Mechanic workshop? Some serious reconsiderations shall be done. So, there are few people like Professor Jiang. He developed and made so many of these gadgets for fundamental natural science study. So, I am fortunate to have the result now. Personally, I do not have millions of dollars to buy a possible but not suitable fluorescence microscope. Big claps to Jiang.

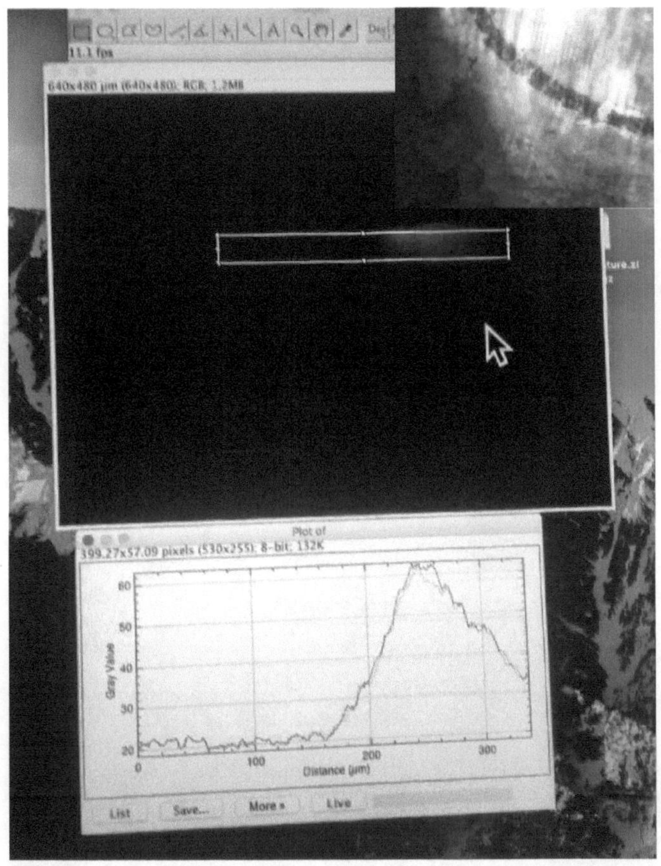

◀ 图 4.12.9　激光 520 nm 灯源照射在化石样本上的光谱与光谱曲线

◀ Figure 4.12.9　Spectrum and spectra curve of 520 nm Laser shining on the fossil

是可以很便宜地做成激光荧光显微光谱分析设备，给我们进一步探讨化石的奥秘！

In conclusion, Laser Stimulated Fluorescence can be very inexpensively expanded to Laser Stimulated Fluorescence Microscopic Elemental Analysis for us to explore the hidden secrets of fossils.

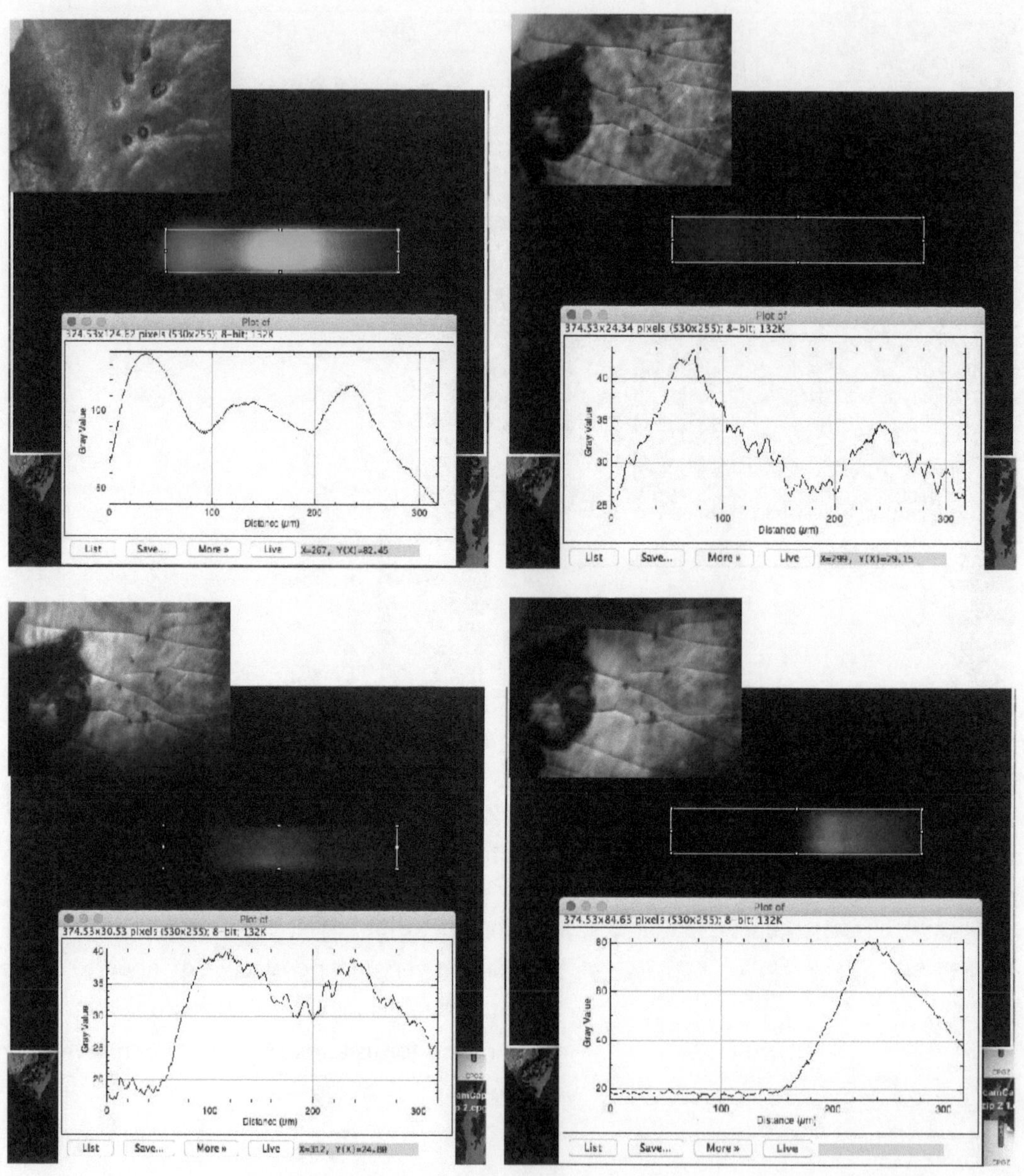

▲ 图 4.12.10 四种不同灯源照射在同样化石（美国蒙大拿州雀鳝目鱼）样本的光谱与光谱曲线

▲ Figure 4.12.10 Spectrum and spectra curve of four different light sources shining on the same fossil (Gar fish scale fossil from Montana, USA)

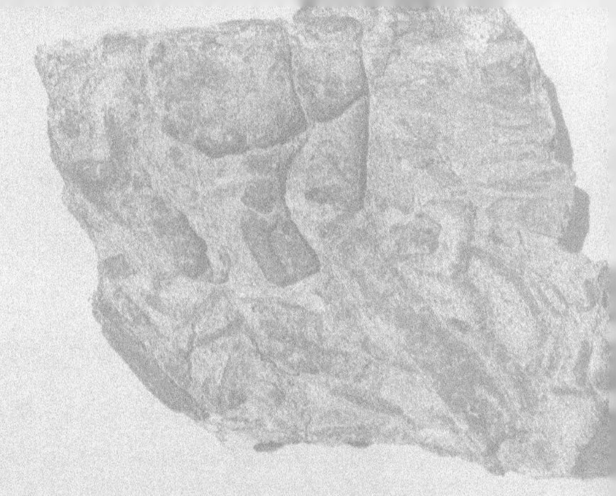

第5章　有机残留物

Chapter 5　Organic Remains

本章介绍古生物化学有机残留物部分与其分类，并强调研究有机残留物的关键三步骤。

首先应该明确的是：化石里外保存着有机残留物，应该是个普遍现象，并非特例，化石不一定是完全无机的硬石头！只要有正确的思路，使用正确精密的仪器设备，应该都可找到或多或少的有机残留物！（下一节图5.1.1和图5.1.2）

在探讨化石里面保存的有机残留物之前，首先应该先考虑一下古代生物体内的化学组成，亦即从生物化学的角度来探讨，这么多不同种类的生物体，到底有哪些主要分类的生物有机物？请注意，在此用的字眼是“主要分类”，而非某特定细分的“生物化合物”，能够做到主要的分类，已经很困难了，要求做到特定细分生物化合物，会不会是过分要求呢？

从大约三十八亿年前地球最早的生命体来说，单细胞双细胞藻类生物体，组成的生物化学成分是什么？到了大约

This chapter introduces the organic side of PaleoChemistry and emphasizes the three key steps.

Preservation of organic remains inside and outside fossils is a common phenomenon, not a special case. Fossils are not just inorganic rocks. With the right thought and using the right precision instrumentations, more or less organic remains can be found! (Figure 5.1.1 – 5.1.2 next section)

Before diving into the preservation of organic remains, consider the chemical compositions of the ancient life, i.e., from the biochemistry point of view first. Among so many organisms, what are the main categories of biomaterial? Please note that the phrase “main categories” is used rather than what particular bio-compound. Be able to tell the main category is hard enough. Asking for a detail bio-compound identification maybe too much.

What are the compositional biochemistry compounds of the single or double cells and algae organisms of about 3.8 billion years ago? Then, at around 2 billion years ago, the multicellular organisms showed up. What was their

二十亿年前古元古代，可能出现的多细胞生物，又是什么有机物质？这么长久可能保存下来吗？或只是铸模 (cast) 呢？到了大约六亿年以后的埃迪卡拉纪生物，全球有二三十处出土了上百个不同埃迪卡拉纪生物的铸模，也从这些铸模化石，确定了埃迪卡拉纪地质年代，但是只有云南中部一处出土了实体化石，因此显得特别珍贵，让科学家们有机会探讨其内外部的形态与结构，还可试着“看进骨头内”探讨这些软组织的生理形态；接着到寒武纪，海洋里面出现了有硬壳的贝壳、有外骨骼的三叶虫等等，还保存了很多原本都是软组织的奇形怪状生物，如怪诞虫 (*Hallucigenia*)，生命的演化开始大幅度的发展，意味着诸多无法计数的生物化学成分出现了，海胆珊瑚甚至有脊椎骨的鱼类，也陆续登上生命的舞台，相对于全身都是软绵绵的生物，这些硬组织，保存成为化石的机会更多，海洋的植物登陆，有些动物也随之登陆，占领了陆地等等，如此的生命演化，也意味着生物体内的生物化学成分的多样性随之增加；古代生物即便本体已经分解，但是在围岩内还保存着当年细胞的分解物，称之为“生物标志 (biomarkers)”、“化学化石 (chemical fossils)”或“分子化石 (molecular fossils)”，用以鉴定远古生物。

我们无法很细很精确地求证每一种生物体的各种有机化学成分，只能以相对粗略的方式来分类，如本章第5.3节所笼统归类的：胶膜/角质 (cutan/

organic material? Could their organic material be preserved for such a long time? Alternatively, just imprints or casts? Around 600 million years ago, the Ediacaran organisms showed up. However, around 20–30 places in the world yielded imprints and casts of these creatures, which helped to establish the Ediacaran Period. There is one, and only one place in central YunNan has Ediacaran body fossils. They are very precious for scientists to study their internal and external morphology and structure. To be able to "Look into the Bone" of them to explore the physiology of these soft bodies is very important. Then to the Cambrian Period, hard shell and exoskeleton such as trilobites were well preserved, including may weird creatures such as the soft body *Hallucigenia*. Life evolution took a big step forward, meaning countless biochemical compounds showed up. Then, sea urchin, coral, and even fish with vertebrae were on the stage of life. In comparison to the soft body creatures, these with hard tissue had more chances to be preserved as fossils. Plants and animals in the water moved to and occupied the land. Such life evolution also means the diversion of biochemical compounds inside the organisms. Furthermore, even the actual body of ancient organisms already decomposed, the matrix still preserved the decomposed remains, and called Biomarker, Chemical Fossils, or Molecular Fossils for identification of Proterozoic organisms.

We can not accurately identify each organic compound in every organism. Rough classified as in Section 3 of this chapter: Cutan/Cutin, Hydrocarbons, Pigments, Fatty acids and esters, Polysaccharides and chitin, Molecular fossil,

cutin)、碳氢化合物 (hydro-carbons)、色素 (pigments)、脂肪酸酯 (fatty acids and esters)、聚糖与几丁质 (poly-saccharides and chitin)、分子化石 (molecular fossil) 和蛋白质类 (proteins)；如果要追究到很细的分子（与构造）层次，举个例子来说，就可知其难：我的硕士论文做青苔的成分分析，因为指导教授说，美国俄勒冈州的印第安人用这种青苔治癌症，所以想找出其内治癌的有效成分，进一步分析鉴定其化学式，我和另外一位研究生跑到森林里采集半天，好不容易采集到够做萃取分析用的量，接着照该做的手段萃取分离等等，结果让我吓一大跳，在这么低等的生物体内，光是倍半萜（sesquiterpene）就有八十多种不同的化学成分（化合物）。较高等的生物，不管是动物或植物体内所含的有机化合物种类数之多，天文数字简直不能想象啊！

所以，从无数古生物体内不同成分数量来说，我们该有心理准备，我们所面临的是个非常复杂的“硬饼干 (tough cookie)”，而且往往混杂在一起，绝非化学实验室内某纯度高达百分之九十九点九九九九的某试剂。接着，孙悟空头上还有一道紧箍咒，相对于研究现生生物的化学成分，我们还必须面对着这些数量难以计数生物化学物质在石化过程中的变化，从活着时候相对来说比较软的古代生物体，变化到硬邦邦的化石，这里面起了什么变化？化石的“化”和化学的“化”，在此成为一体了，原本的物质有所变化！这也是本书一再强调：

and proteins. If we want to look for very fine molecular structural level, my experience below will show how difficult that is. When I was doing my Master graduate study, my advisor asked classmate and me to collect some liverwort in the forest. He knew that the native Indian was using that to cure cancer. We spent a long while to collect enough quantity for extraction. Then separation and related analyses were carried out. The results shocked us. In such a low-class living thing, there are more than 80 close related but different chemical compounds of sesquiterpene! In the higher class living things, no matter whether they are animals or plants, the number of existing chemical compounds inside their bodies is unbelievable astronomical huge.

Thus, we shall keep in mind that the number of different chemical composition is a huge tough cookie. They usually mixed up together, not like the 99.9999% pure reagents in the chemistry lab. Then, there is another monkey wrench to face. Relative to the chemical composition of the currently living beings, the changes during the fossilization process shall be taken into consideration. What took place during the transformation from relatively soft matter to a hard rocky fossil? The “change” of fossil (化石) and the “change” of chemistry (化学) are the same here. Both are dealing with the changes in the matter. This is what repeatedly emphasized in this book. PaleoChemistry is the essence of paleontology. Like it or not! We can not stay at the stage of the just study the morphology after the changes. We need to search for the source of changes and see what took placed in the “bone”!

When the living things die, if the corpses

古生物化学才是古生物学之本的重要原因，我们不能只停留在传统古生物学探讨变化之后的形态而已，更该追溯其源，探讨到底“骨头”里面是怎么“(变)化”的。

想想，某个生物体死亡之后，如果暴露在空气中，通常被各种天候和细菌相对快速分解掉，软组织先被消化，接着硬组织（骨头牙齿）也受不了各种天候和细菌的里外夹攻，都成灰飞烟灭，没有机会变成化石。当然这种情况有例外，如美国蒙大拿那边，曾因为火山爆发毒气杀死了几十万头群居的鸭嘴龙，尸体暴露在空气中腐烂，剩下硬骨头，不知多久之后，硬骨头还没消失之前，山上的天然水库崩解，大量的土石流倾高而下，把这些骨头冲刷到较低处并且埋藏成为化石骨床 (bone bed)，在大约二英里长一英里（1609.344米）宽的范围内，同样方向排列着数不清的鸭嘴龙大骨头，小骨头老早被冲远了或者被土石流磨碎烂了。从大骨头表面的龟裂得知，它们在形成化石之前，曾经被太阳烤到骨头表层龟裂，然后才被埋藏变成化石的。

大部分的化石形成，需要快速掩埋，隔绝空气中的氧气，由厌氧细菌分解躯体内的软组织，不同地层里有各种不同的细菌，它们扮演着重要的消化清道夫角色；地下水中的矿物质则取代置换硬组织内的某些元素，骨头和牙齿内的磷灰石群，也从准稳定 (metastable) 生物性磷灰石 (biogenic apatites)，重新结晶成为热力学上比较稳定的无机磷灰石群矿物；在此过程中，地下水的化学

were exposed in the air, their soft tissues would be decomposed rapidly by weather and bacteria. Then the hard tissues (bones and teeth) would be diminished by the natural elements. Gone, and no chance to become the fossil. Of course, there were exceptions, such as the case in Montana, USA during the late Cretaceous. The volcano poison gases killed a big herd of several deca-thousand hadrosaurus. Corpses were rotten away in the air. Only the bones were left. Then, before the hardy bones were further decomposed, there was a big mudslide from a higher ground lake. The huge current swiped the bones and deposited in an area of about 1 mile wide and 2 miles wide. Only the big bones were in this bone bed with the same orientation. No smaller bones were found in this bone bed. They were destroyed or carried out further away. From the cracks on the surfaces of big bones, it is obvious that these bones were cooked by the sun for a while before they were buried and fossilized.

Most of the formation of fossil need rapid burial and exclude contact with oxygen in the air. Anaerobic bacteria then decompose the soft tissues. Different bacteria exist in the different ground formation. They act as the digestive road sweeper. The minerals in the groundwater would replace some elements inside the hardy bones. Metastable biogenic apatites recrystallized to more stable inorganic apatites. In this process, the chemical components inside the groundwater along with various pH, temperature, pressure conditions get in and out from the bones. For example, in the living bones, there is no Rare Earth Elements (REE). However, two series of Rare Earth Elements (Lanthanide

成分，包括酸碱值、温度、压力等等环境，不停地进出于骨头里外，如通常活生物体骨头内，基本上没有稀土元素，可是在骨头化石内，却可检测到两个系列（镧系和锕系）的稀土元素，有些骨头化石内的总稀土元素浓度，还可超过 20 000 ppm，可以当稀土矿了！这些骨头化石内的稀土元素，肯定是在形成化石的过程中被富集于重新结晶中的磷灰石群晶格内，形成封闭系统的。其他价数的各种阴阳离子，则可能进进出出。

随着时间累积地层掩埋变化，地下水进进出出不知道多少次，有的化石形成不知道多久之后，又可能被暴露在空气中，有的被大自然搬移到别的地方，有可能又重新被掩埋。这些很漫长的时间内，到底发生了哪些情况变化，实在很难精准地说出来，我们无法回到几亿年前，重新走过整个石化的过程，只有手中所拿到的化石，无法回头去做大自然的实验；再者，当化石从地里深处无氧状态被往地表抬升，大约到了距离地表十米的深度，大气中的氧气作用就开始影响到化石，进而与化石里面的有机和无机成分起化学反应。

生物体内的化合物有很多，甚至可说绝大部分都是大分子聚合物。我们来讨论一种可能的情况：蛋白质是由多肽组成的，而多肽 (peptide) 是由更小的氨基酸聚合而成，也就是说，每一条蛋白质是长链氨基酸的聚合物 (polymer)，这个长链的蛋白质聚合物，在某些情况下裂解 (degradation) 成为较小的多肽群，

and Actinides) can be detected in the fossil bones. In some bones, the total La-REE can be as high as over 20,000 ppm! They can be used as REE ores. These REEs in the bones were absorbed from the groundwater, locked and concentrated inside the recrystallized apatite lattices, and formed a closed system. Other ions of various valences can get in and out of the bones.

Along with the burial time increase, groundwater get in and out for an unknown number of times, some fossils were exposed to the air, and some transported to other places, and then reburied again. What happened during such a long time is very difficult to say precisely. We can not go back in time for many hundreds of million years to repeat the whole process. The fossils in our hands are what we can have. No backward experiments can be done. Furthermore, when the deep buried anaerobic fossils were uplift to about ten meters below the ground surface, oxygen in the air starts to interact with the organic and inorganic component inside the fossils.

Inside the body of living things, there are many and even said that they are big molecular of polymers. Let us look at a real case. Proteins are made of many peptides, and peptides are made of many amino acids. So it can be said that proteins are long chain polymers of amino acids. Such a long chain under a specific condition can break down, degradation to smaller peptides or entirely down to amino acid level. That is why the bone or fish soup taste so good, Amino Acids (MSG, Monosodium glutamate)! So, could this happen to the proteins inside the fossils? Of course! There are certainly some

甚至完全分解到诸多组成单体氨基酸的程度，最明显的例子就是厨房所熬的高汤或煮鱼汤，为什么这些汤头味道那么鲜美？就是氨基酸啊！也就是味精啊！那么，原本存在于化石中的蛋白质，会不会也有裂解的可能？根本甭问，化石肯定会有长链蛋白质的裂解产物，至于到底断裂到什么程度，那就难说了。化石里面保存着原本的蛋白质、某程度裂解的多肽群和完全裂解的诸多氨基酸，三者混合完全是可能的。同样的道理，其他长链化合物，如核酸、脂肪酸酯、聚糖(polysaccharides)、胶膜 (chitins) 等，有没有类似的情况，也是个甭问的假问题。

回顾一下，今日的塑料等聚合物，都是把小分子的单体，在某种条件下聚合起来，成为长链的各种塑料，那么在漫长的地质作用年代中，各种化学反应条件下，会不会有上述裂解后小单元的重新聚合作用 (re-polymerization) 呢？这当然是很有可能的，完全不能排除，可是重新聚合产物，会回到原本的长链蛋白质里面氨基酸排列序列吗？难说啊！再者，重新聚合会产生哪些产物？也是难以推测的，其中，有没有水溶性聚合物被地下水带走的？当然有可能！这也同时指出了想通过脱氧核糖核酸 (DNA) 重建基因克隆古代生物（包括电影《侏罗纪公园》恐龙）的荒谬性，一则脱氧核糖核酸非常不稳定，细胞死后，酶就开始分解脱氧核糖核酸主链的组成物质核苷酸（nucleotides）之间的键结，而微生物会加快分解的速度，不过归根究底，

degradation products inside the fossil bones. However, to what degree of degradation is hard to say. So, inside a fossil bone, there is a mixture of original proteins, various degrees of degradation peptides, and amino acids. The same reason applies for other long-chain compounds, such as nucleic acids, fatty acids, and esters, polysaccharides, and chitins, cutan/cutin.

Let us look at the plastics of today. They are polymerized long chain products under some specific conditions from simple monomers. Thus, is that possible that re-polymerization could take place under various chemical conditions during such a long geological time? This cannot be excluded completely. Would they form the long chain of the original amino sequence? Very hard to say. As what repolymerized products were formed? It is also tough to say. During such a process, would some water-soluble products be wash-out? Of course, it is possible. This also points out why to clone an ancient animal, such as dinosaurs in the Jurassic Park movie from DNA is just a joke. For one thing, DNA is very unstable. Once the cells died, enzymes will start to break down the bonding in the DNA to the compositional nucleotides. Microbes will speed up the break down of the long chain. Reaction with water is the main factor. The half-life of DNA is only 521 years. So, how could DNA be preserved after many hundreds of million years? With the most advanced instruments, it is possible to detect nucleotides. However, their order of forming long chain was unobtainable. How in the world to reconstruct the puzzle back to original form?

促成键结分解最主要的因素还是在于和水起反应，脱氧核糖核酸半衰期只有 521 年，几亿年前的古生物，能保存那么久的完整脱氧核糖核酸吗？即便透过各种先进精密仪器设备，可以检测到几种核苷酸，但是，这些组成长链的核苷酸排列顺序完全打乱了，如何重新把这幅没有原稿的基因图谱正确地组合呢？

从保存的可能性和抗拒衰解 (decay) 的角度来说，生物分子通常顺序为：核酸＜蛋白质＜碳水化合物＜脂质＜生物聚合物，也就是说，核酸最容易受到环境影响裂解成碎片化合物；随着长期的细菌和成岩（diagenetic，又称热成熟，thermal maturation）作用导致这些生物分子的裂解和改变；不同复杂生物分子裂解的速率也有很大的差异，核酸保存着原本生物最多生物信息，却是最容易裂解的；蛋白质可提供某生物体的生理和演化关系，比起核酸，可保存得更久，我们在 1.95 亿年前的禄丰龙胚胎和成龙骨头内，找到了被保存的第一类胶原蛋白；被保存在外骨骼的几丁质，可能古老到三四亿年前；黑色素被保存在一亿多年前辽西的龙鸟羽毛内；化石里面的有机物，可能是该生物体内原生 (endogenous) 的，也可能来自外在 (exogenous)，从埋藏环境内碳氢化合物的移入或者两者的混合结果；但是，不能忘记上两段所叙述的裂解和重新聚合，也扮演了相当重要的角色。

总之，化石内所保存的有机残留物，是让我们探索地球生命演化重要的途

From the preservation possibility and resistance to decay points of view to say, the break down of biomolecules is nucleic acids greater than proteins greater than carbohydrates greater than lipids greater than biopolymers. It means that nucleic acids are the easiest to break down into fragmented pieces. Thru the long period of microbial and diagenetic (also called Thermal Maturation) actions, the large molecules would break down. The speed of degradation can vary significantly. The nucleic acids contain the most biological information, but the easiest to break down, unfortunately. Protein can provide information on physiology and evolution. In comparison to nucleic acids, they can be preserved much longer. We reported preservation of collagen type I from 195 million years old *Lufengosaurus* and embryonic bones. Chitin of exoskeleton could be preserved even older, 300, or even 400 million years. Melanin was found in the feather of LiaoXi Cretaceous bird. The organic remains inside the fossils could be endogenous or exogenous moved in hydrocarbons, or mixture of both. However, do not forget what discussed in the previous two paragraphs.

In summary, the organic remains preserved inside fossils is a conduit for us to learn the life evolution of this Earth, particularly on the microbes and animal diversions. By analyzing the organic remains inside and outside the fossils can point to the original biomolecule composition and the close relationship between these ancient creatures.

径，特别是在显微生物和动物的多样化方面，分析化石内外的有机残留物，可能指引原本生物分子生化组成和与这些古老生之间的关系密切程度。

5.1 熬出龙骨汤

找到二亿年前恐龙化石第一类胶原蛋白自白

回到2012年底，我们恐龙胚胎学团队在撰写、后来发表在《自然》(*Nature*)作为封面论文的时候，碰到了一个现在看来很关键的问题：当时用同步辐射傅里叶变换红外线显微光谱分析 (sr-FTIR microscopy)，把 1500~1700 cm^{-1} 这段光谱进一步解析 (deconvolution)并和文献比对，发现在这些快两亿年前恐龙胚胎骨头里面的有机残留物，应该很明确是“第一类胶原蛋白 (collagen type I)”没错！不过，在这个时期，不要说团队外的人，即便是团队内的成员，对于这个重大的发现，还是抱着怀疑的态度，胶原蛋白那有可能保存这么久?

当时想了一想，与其在这个已经很够世界级重大发现的论文中，这么快这么早就提出这么骇人惊闻的成果，还不如玩个文字游戏缓冲一下，免得太多人无法接受这个革命性的发现，心脏病发作；因此，经过反复讨论，决定采用比较笼统含糊的“有机残留物 (organic remains)”这个名词，而先不说“第一类胶原蛋白 (collagen type I)”，因此整

5.1 Brewing Dragon Soup

Monolog of the discovery of collagen type I in almost 200 million years old dinosaur bone

Back to the end of 2012, our dinosaur embryology team was writing and later published on *Nature* cover paper; we encountered one key issue: At that time, we did the deconvolution on the range of 1500–1700 cm^{-1} of the spectra form sr-FTIR Microscopy, and compared with some published papers. We found that in these almost 200 million years dinosaur embryonic bone, they are very clearly collagen type I. This is for sure. However, not to mention from outside non-member but from our team member, doubts on this great discovery arose. How could protein be preserved for such a long time?

So, considering to announce such an astonishing result in our already shocking enough paper, it would be better to tone down a little bit to avoid too many people who can not accept this revolutionary discovery, or causing a heart attack. Thus, after repeatedly discussions, I decide to use more blur term “Organic Remains” instead of “Collagen Type I.” So in the whole paper, the term “Collagen Type I” was never used. Luckily when the paper finally published, the chief editor of Nature realized the importance of this matter, so the second line of our paper title was changed to “with evidence of preserved organic remains.” They recognized the significance and use one whole title line to emphasize.

篇论文中没用到一次“第一类胶原蛋白”这个名词！所幸，在那篇《自然》封面论文发表的时候，期刊主编知道此事的重大性，还给标题做了一点修改，第二行被改成为：“with evidence of preserved organic remains（带着保存有机残留物的证据）”，人家看重，特别用一整行标题标示出来。

不过，这个“妥协”在顶级论文内用语把比较精确小范围的“第一类胶原蛋白”，放大到比较模糊笼统很大范围的“有机残留物”，从这角度和规划来说，其实也是“塞翁失马，焉知非福”啊！从那篇论文被接受后，脑子里面就“将计就计”不停地想，在快两亿年前恐龙胚胎化石内能找到有机残留物（第一类胶原蛋白），难道只是极端太幸运、被台湾这些非古生物界的门外汉成员“瞎鸡啄到米”？或者说，实体化石内保存着有机残留物是个普遍现象？如果有正确的思维，以“看进骨头内”的正确角度、采用适当的实验方法和够灵敏的仪器来探讨分析，在其他的实体化石 (body fossil) 内，是否也或多或少保存着有机残留物呢？

因此，2013年7月14日召集了一个台湾团队四人小组会议，那天，我给了大家一份以下的资料：

有关使用 sr-FTIR 探讨化石内残留有机物的一些想法

目标：

（1）探讨化石内含有机残留物的普遍性；

However, this “compromise” which enlarged the more precise term of “collagen type I” to a term of broader range “organic remains” can also be seen as a blessing! Right after the paper was accepted, I kept wondering that if the fact that we found organic remains (collagen type I) is just a stroke of extreme luck which was accidentally found by our non-paleontologist Taiwan team members? Or, is it a common phenomenon of organic preservation in fossils? If we use the right strategy of “Look into the Bone” with the right methodology and sensitive enough instruments for exploration, can we find more or less preserved organic remains in body fossils?

Thus, on July 14 of 2013 at the MauGeLaDi cafeteria, four of us had a meeting. I handed them the following paper:

Some thoughts about exploring organic remains by sr-FTIR

Goal:

(1) Exploration of generality of organic remain preservation in fossils;

(2) Exploration of the compositional components of organic remains;

(3) Propose a mechanism for preservation of organic remains during fossilization.

Explanation:

(1) Does that organic remain preservation inside fossil only happen in the almost 200 million years old dinosaur embryo fossils of Lufeng area as a particular case we encounter? Or, is it a common phenomenon? This can be checked with body fossils from the different geological times and taphonomic environments of Edi-

（2）探讨有机残留物的成分组成；

（3）提出石化过程保存有机物的机制假说.

说明：

（1）化石内保存着有机残留物，是否只发生于二亿年前禄丰地区恐龙胚胎与成龙化石内的特例，被我们幸运碰上或是一种普遍现象？这可透过测试埃迪卡拉纪实体化石、古生代（爬行类或鱼类）、中生代（恐龙）、新生代（哺乳类）等多种不同地质年代不同埋藏环境的化石，配合现生鳄鱼和鸵鸟骨头为比对。

如果各地质时代不同动物的化石，处于不同地质掩埋环境，都保存着有机残留物，那么，化石保存有机残留物现象，就是一种普遍的情况，而非恐龙化石的特例。

（2）就以恐龙化石来说，包括胚胎、爪子、筋腱等化石，进一步探讨所保留的有机残留物之成分，骨头内的胶原蛋白，爪子里面的角蛋白 (Keratin)，筋腱内的胶原蛋白混合物。正常健康的腱大多是由平行的紧密胶原组成，腱大约 30% 的总质量是水，其余质量组成如下：约 86% 的胶原蛋白、2% 弹性纤维、1%~5% 蛋白多糖和 0.2% 无机成分，如铜、锰和钙，胶原部分是由 97%~98% 的第一类胶原蛋白与少量其他类型的胶原蛋白组成。

本项分析，应该记得，原本的有机物，经过地质作用（压力、温度、细菌、地下水等等），应该有某种程度的分解，成为多肽链或氨基酸，而且这些分解物可能受到微量金属元素催化，而产生二次聚合等作用。故此，本研究应该探讨这些被保存的成分（可能需要用到气相色谱 - 质谱 (GC-MS) 等分析）。

（3）取得以上两项数据信息之后，本研究应该提出一个“为什么这些有机残

acaran Period, Paleozoic Era (reptiles or fish), Mesozoic (dinosaur), and Cenozoic (mammal), and with a comparison of extant crocodiles and ostrich bones.

If the fossils of various geological times, places of different burial environment preserved organic remains, then the preservation of organic remains inside body fossils is a common phenomenon, not a special case for dinosaur fossil.

(2) As far as the dinosaur is concerned, including embryo, claw, and tendon, we can conduct examinations to explore the preserved organic remains, such as collagen inside the bone, keratin inside claw, and collagen mixture inside the tendon. The natural tendon is made of tightly packed parallel collagen. About 30% of the mass is water, and the rest: 86% collagen, 2% elastic fibers, 1%-5% Proteoglycans, and 0.2% of inorganic, such as copper, manganese, and calcium. Collagen portion is made of 97%-98% of collagen type I and a small amount of other type proteins.

While doing this, it shall be kept in mind that the original organic matter went thru geological actions (pressure, temperature, bacteria, and groundwater, etc.). So certainly degrees of decompositions and degradations to smaller molecules of peptides or even amino acids would happen. Moreover, these smaller units could be catalyzed by trace metal and polymerize back to macromolecules. So we shall explore what was preserved. GC-MS may have to be used.

(3) After the two above mention data were obtained, then the question of "why or how these organic remains were preserved?", i.e., Or-

留物/或如何被保存下来”的机制假说。

（以下省略。）

对我来说，这份历史文件，很清楚地叙述了我的“有机残留物计划”想法、设计和应该采取的步骤，就如上述“目标”所列的，这个计划分为三个阶段进行，首先是确认实体化石内残留有机物的普遍性，进而探讨这些有机物是什么东西，最后收整提出有机物保存的机制。

接下来第一阶段的第一步骤，就是我从玩石家博石馆收藏中找了大约一百件各种地质年代的实体化石，从将近五六亿年前埃迪卡拉纪、古生代、中生代、一直到两三万年前新生代台南菜寮溪的哺乳类之各种动物化石，带到同步辐射研究中心，交给江正诚去做试片，做好了试片之后，就几乎天天到同步辐射工作站去“上班”（没有薪水啦！），使用那台非同步辐射光源的傅里叶变换红外线光谱仪 (Bruker FTIR) 进行初步的筛选，看看哪些化石内有可能找到有机残留物的线索。经过两三个月左右的努力，初步发现，其实化石内保存有机物的情况很普遍，甚至可以这么说，红外线光谱中出现烷基波峰 (alkyl peaks, CH_3—, —CH_2—, 2800~3000 cm^{-1}) 太普遍了，几乎不敢相信，只能挑选一些最明显有代表性的样本，计划来进一步做同步辐射红外线光源的光谱扫描分析。接下来把讨论后挑选出来的样本，交给李博士用同步辐射傅里叶变换红外线显微光谱仪扫描，进一步精准地确认（或否定）前面这段的筛选成果。

ganic Remain Preservation Mechanism.

(The rest of this document is omitted.)

To me, this historical document very clearly stated my thought, design, and the steps to take on the "Organic Remains Project". As described in the "Goal" section, this project has three steps. The first is to check the generality of organic remains preservation. Then to find out these organic components. Finally, we need to explore the preservation mechanism.

The first thing of the first step was bringing about 100 specimens of various geological times from my own private museum collection to NSRRC for thin slice preparations by CC Chiang. These specimens were from the Ediacaran Period (~560 mya), Paleozoic, Mesozoic, and 10–30 thousand years old Cenozoic mammals. Once the thin slices were done, I used the Bruker FITR (not synchrotron) to do the initial screening to find out what fossils indicate organic remains. After two to three months of hard works, we found that the preservation of organic remains inside body fossils is very common. The IR peaks of the alkyl group (CH_3-, -CH_2-, 2800–3000 cm^{-1}) were unbelievably common. So, some more representative ones were selected and handed over to Dr. Lee for synchrotron radiation FTIR scans to confirm or reject the initial screening.

Unfortunately, Dr. Lee is a very busy person, and the workstation at SRRC were fully scheduled. We can only use the small spare time slots. Our high-resolution analyses do not go smoothly. However, some results were obtained by narrowing down to the collagen type I in the DaWa, Lufeng. Yunnan *Lufengosaurus* bone, and

无奈，李博士本身工作忙碌，而且同步辐射傅里叶变换红外光谱仪时程排得满满的，只能塞缝这边做一点，那边做一点，因此，整个计划的高分辨率精密分析，一直卡卡不顺，但是还是终于做出了一点点小成果，把原本比较广泛的多种化石内残留有机物探讨，缩减到探讨云南禄丰大洼恐龙骨头内的第一类胶原蛋白，也就是发表在《自然通讯》（*Nature Communications*）这篇“Evidence of preserved collagen in an Early Jurassic sauropodomorph dinosaur revealed by Synchrotron FTIR Microspectroscopy”透过同步辐射傅里叶变换红外线显微光谱仪显示早侏罗纪蜥脚形亚目恐龙保存胶原蛋白的证据。咻！从2013年7月开始算起，总共花费了三年半的时间！

这篇论文提出明确红外光谱分析，证明云南禄丰大洼恐龙骨头化石内，明确保存第一类胶原蛋白的证据，其实这个在那篇《自然》封面论文的增补资料中，已经提供了1500~1700 cm^{-1} 这段光谱进一步解析证明第一类胶原蛋白。这篇新论文中，也提出通过拉曼光谱 (Raman spectroscopy) 观察到很可能是红血球残留物的赤铁矿 (hematite)，推翻了Mary Schweitzer所提出针铁矿 (goethite) 可能是胶原蛋白保存的关键角色。按Mary所使用的分析方法，是透过去钙法将硬骨头溶解掉，因此在样本准备过程中，赤铁矿很有可能被水解 (hydrolyzed) 成为针铁矿，而我们的检测方法完全没有经过化学处理，是在原位

published "Evidence of preserved collagen in an Early Jurassic sauropodomorph dinosaur revealed by Synchrotron FTIR Microspectroscopy" on *Nature Communications*. Whew! Counting from July 2013 to the publication, it is 3.5 years!

In this paper, we clearly demonstrated the preservation of collagen type I inside the DaWa, Lufeng, Yunnan *Lufengosaurus* by IR spectroscopy analysis. But, in the Supplementary of our Nature cover paper, we already provided the deconvolution peak data between 1500–1700 cm^{-1}, showing these organic remains are collagen type I. In this new paper, we also presented Raman Spectroscopy of hematite, the possible remnants of red blood cells. This overturned what Mary Schweitzer stated that goethite plays a crucial role in collagen preservation. The way that her team use is decalcification of the bone. Thus, the hematite was likely hydrolyzed to goethite. Our methodology does not use any chemical process and *in situ* scanning. There is no hydrolization of hematite issue. So, as a reminder, it is possible for the preparation to affect the later analyses result by the precise instrument. It is possible that the results were caused during the sample preparation, not the original state. Be careful on this issue, particularly on thin slice preparation. No (organic) glue shall be used. Do the thin slice by a skillful hand.

Get back to this big project. Finding organic remains of collagen type I in the ~200 million years old dinosaur bone is a big deal. But from the objective and reasonable point of view, this is just one item of the second step. Well, from the first step, we already found that organic remains preservation is a common phenomenon

(*in situ*) 的光谱扫描检测，没有赤铁矿水解的问题。在此提醒一点，样本的制备处理，是会影响到测试的结论，后面用精密仪器检测到的结果，有可能是在前面样本制备所导致，而非原本样本内原有的成分。这一点要特别小心，所以试片制作得非常小心，绝对不用可能造成样本污染的（有机）胶水，只能靠高超功夫，别无他法。

话说回头，此文在整个大规划当中，从快二亿年前恐龙骨头内找到残留有机物第一类胶原蛋白，虽然意义很重大，但是客观理智来说，还只是偏属于第二个大目标里面的一个小项目而已。怎么说呢？从第一个目标的角度来看，实体化石（和围岩内）保存残留有机物是个普遍现象，而古生物种类那么多，即便只限于有血有肉的古脊椎动物来说，骨头只不过是化石的一部分而已，其他如爪子（指甲）、鳞片、羽毛发，甚至尸水（肌肉分解物）、没烂光的树叶等等，也都有保存的纪录，它们的成分原本就很不同于骨头，所以，除了骨头内的胶原蛋白之外，其他古生物体其他构件成分被保存的残留有机物会是什么？再者，如果有注意到，我一直用“第一类胶原蛋白”这个名词，里面的“类”字，表示它是包含很多成员的一大类，而非只是单纯的某一个长链化合物而已。如果上维基百科去查一下“胶原蛋白”，就可发现到目前为止，科学家找出描述的胶原蛋白，就有 28 类，光是纤维状的胶原蛋白，就有五类，第一类胶原蛋白，只

in body fossils (and matrix). However, besides dinosaurs, there are so many other fossils, and even if we limit ourselves to the bone and flesh of paleo vertebrae, fossil bones are just one part of the fossil. Preservations of others, such as claw (nail), scales, hair and feather, even corpse fluid (decomposition of muscle and soft tissue), partially decomposed leaves were recorded. Their chemical compositions are different significantly from that of bone. So, besides the collagen preserved in the fossil bones, what other organic remains preserved inside other parts of the ancient creatures? Furthermore, "Collagen Type I" was repeatedly used. Note that "Type" means there are many many members in this category, not just a simple long chain compound. If WiKi is consulted, there are 28 types of collagens. There are five types of fibrous collagens. Collagen Type I is just one of them.

Some more words about the generality of organic remains preservation in body fossil and immediate surrounding matrix: From the point of view of chemists, all the organic analyses have to rely on modern instruments, Nowadays, the amount of sample needed can be one billionth (ppb) or less. One of these advanced instruments commonly used is infrared spectroscopy. Thus, if a particular functional group can be seen in infrared spectra, we can then be sure that the sample contains specific chemical structures.

From the Chinese medicine dictionary: ① There are two kinds of Chinese dragon bones: White Dragon Bone and Five Flower Dragon Bone. The White Dragon Bone is the Mesozoic dinosaur bones. The later is the Cenozoic mammal bones.

不过是这么多种蛋白质其中的一类而已。

多说几句有关实体化石（围岩）保存残留有机物的普遍性。从学化学者的角度来说，今日的有机分析基本上都得靠相关的仪器来分析，现在的仪器分析，所需要的样本量可小到十亿分之一（ppb），甚至还更少，都可检测得出来，而这其中之一很常用的方法仪器就是红外光谱分析。因此，如果可从红外光谱看到某个化学官能团 (functional group)，就可以断定那个样本里面有某某化学结构成分。

从中药药典可归纳出几点：①中药的龙骨,分为“白龙骨”和“五花龙骨”，前者应该是中生代的恐龙骨头化石，而后者是新生代大型哺乳动物的骨头化石。②传统中医的生药分析认为其成分主要为碳酸钙、磷酸钙，此外还有铁、钾、钠、氯、硫酸根等。不过从现代科学分析的角度来说，骨头化石内的有机残留物，完全被忽略了，过去未从有机残留物的角度思考，也没有适当的仪器和分析方法。能从两亿年前恐龙骨头内找到第一类胶原蛋白，也是我们最近才发现的。③一般以五花龙骨为优，这一点应该是因为五花龙骨年代较轻，保存的未分解胶原蛋白含量较多。

日本汉方“龙角散”的“龙”是指“龙骨”，角指“鹿角霜”，是龙角散的重要成分，也都是中药材，鹿角霜是鹿角处理过后磨成的骨粉。

这段龙角散的描述很有意思，为何要把“龙”骨（化石）和现生的“鹿”

② Traditional Chinese medicine said that the main ingredients are calcium carbonate, and calcium phosphate, containing iron, potassium, sodium, chloride, and sulfates. However, speaking from the modern analytical point of view, organic remains inside the fossil bones were neglected. Such is caused by (a) never thought about the organic remains, (b) no suitable instrumentation could be used. Discovering the collagen type I inside the almost 200 million years old dinosaur bones was done by us recently. ③ The reason why Five Flower Dragon Bone is used should be because they are younger in geological age and hence contains more (un-decomposed) collagens.

There is one Japanese Han formula called "Dragon Horn Power (Long_Jiao_Shan)." The 'Dragon' is Dragon Bone, and 'Horn' is deer horn powder. These two are the main ingredients. Both are Chinese herb medicine. Deerhorn powder was the ground deer horn.

The description of Dragon Horn Powder is facinating. Why mix the fossilized 'Dragon' bone with extant deer horn for sour throat and asthma? Why combine these two? The reason could be that the amount of collagen type I inside the dragon bone is less, have more inorganic, but have certain unique pharmacognosy features. Even if the collagen type I is the same key player, the actual type of collagen from a different individual is different. The extent deerhorn contains more organic collagen. So, these two complements each other, so that this Dragon Horn Power is used for more than two hundred years, and even today.

This paper is the very first presenting *in situ* preservation of collagen type I inside almost

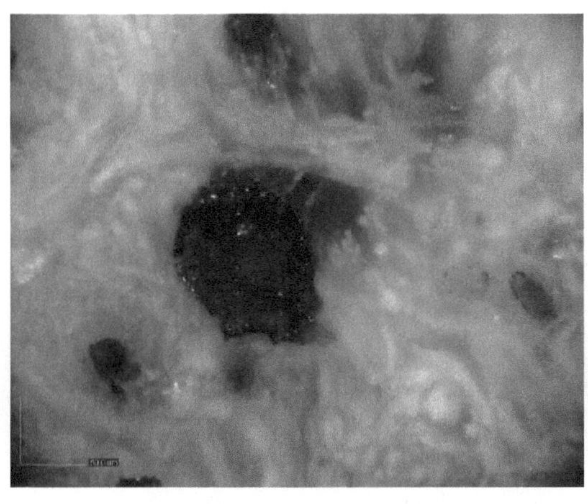

◀ 图 5.1.1　早期的大洼恐龙骨头化石横切面切片，脉管渠道明显可见

◀ Figure 5.1.1　Transversal cut thin slice of a dinosaur bone from DaWa. Vascular cannel can be seen clearly

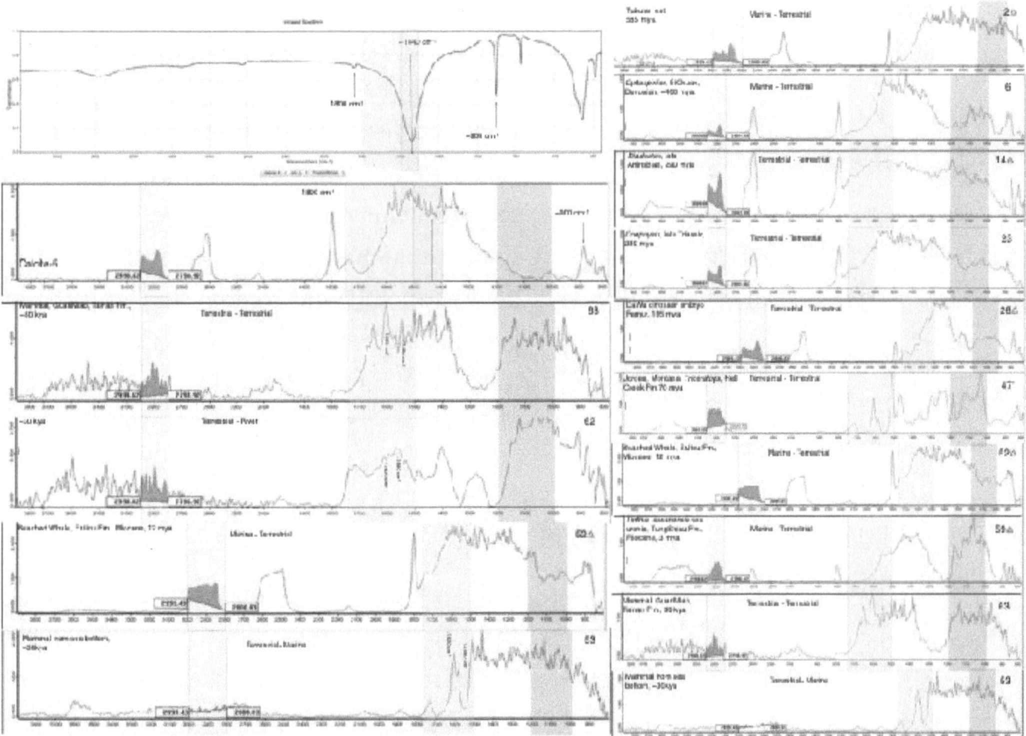

▲ 图 5.1.2　使用 Bruker FTIR 检测出从大约五六亿年前的埃迪卡拉纪到新生代一两万年前澎湖海沟的哺乳类动物化石，在光谱 2800~3000 cm^{-1} 烷基位置都有明确的波峰，显示明确证据，这些化石内还保存着有机残留物。先不要问我是什么有机残留物，那是接下来要探讨的大问题

▲ Figure 5.1.2　Using Bruker FTIR to detect the fossil specimens from Ediacaran Period (~560 million years old) till Cenozoic Era (~10 to 20,000 years old) look at the peaks from 2800–3000 cm^{-1} to see alkyl functional group. These spectra demonstrated that within these fossils, organic remains were preserved. As for what organic remains, that is the question to be explored and answered

角配合在一起？喉咙痛气喘等毛病，为什么需要这两味中药合着用？猜测是，龙骨里面保存着第一类胶原蛋白含量较少，无机矿物较多，但有其独特药性，记得前面提过，即便都是第一类胶原蛋白，不同物种不同个体内的同类胶原蛋白，还是有差异。现生鹿角的有机胶质较多，两者相辅相成，所以两百多年来一直被人们热爱使用着。

我们这篇论文是全球首次提出在原位检测出保存在快两亿年前恐龙骨头内的第一类胶原蛋白，而且发现在骨头内的脉管渠道 (vascular channel) 里面，还有好多赤铁矿微粒聚晶。这些微粒聚晶赤铁矿，应该是来自红血球血红素内的铁离子，这些血红素内的铁离子从二价到三价的变化，把生存必要的氧带到身体各部位、维持脊椎动物的生命。

在追寻此课题的过程中，发生了一些趣闻：照讲既然在初期筛选过程看到大部分的实体化石都有保存着有机残留物（图 5.1.1和图 5.1.2），可是用同步辐射光源的傅里叶变换红外线显微光谱仪来分析横切面试片（图 5.1.3），却找不到胶原蛋白，一下子认为试片太厚，一下子是这个又是那个原因的，后来终于找到了一个主要的原因，骨头化石横切的话，可以看到圆圈状脉管渠道，里面黑黑的，脉管渠道内应该有一层“管壁”，在脉管渠道内有动脉、静脉和神经等三个更小的管子，这些管壁有可能保存下来，以前我们《自然》论文的实验，就指出如此，可是在横切试片圈圈里外找

200 million years old dinosaur bone and, there are many many microspheres of hematite in the vascular channel. These micro hematite aggregates could very likely come from the iron ions of red blood cells. The changes of iron ions from divalent to trivalent brings oxygen to various parts of the body to maintain the life of vertebrates.

In our pursuit of this topic, some interesting things took place. We saw in the most body fossils preserved organic remains in our initial screening, Figure 5.1.1, and Figure 5.1.2. However, when sr-FTIR was used to check the transversal cut thin slice, Figure 5.1.3, we could not find collagen. Perhaps the slice was too thick, or other reasons. Finally, we realized that when the bone was cut transversally, we can see the circular vascular channel well in dark color. Within the vascular channel, there should be a 'tube wall,' and contain smaller tubes of arteries and veins, and nerves. These tube walls could be preserved as we pointed out in our Nature paper. However, we could not see the characteristic amide peaks of collagen. In theory, we should see, but not. Then, we tried to cut the bone longitudinally. If we are lucky enough, we shall have cuts thru parallel vascular channels. By doing so, the chance to find them would be significantly increased. So, many slices were done. However, the results were better but still not ideal. However, while doing so, some transparent fragments were seen falling outside the channels. Ha, we found the collagen preserved inside the bone. Then, we realized that collagen is water soluble. So, during our last grinding process, water was used as the coolant, and that washed out the collagen we want to

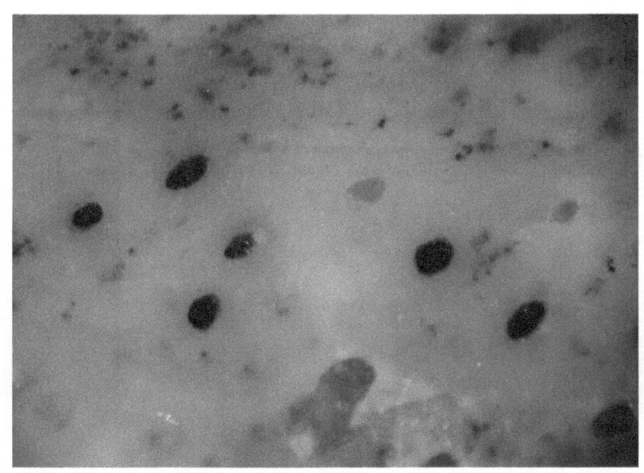

◀ 图 5.1.3 恐龙骨头横切面，这些棕色的圈圈就是脉管渠道的横切面

◀ Figure 5.1.3 Transversal cut of dinosaur bone The dark brown circles are the cross cut of vascular channels

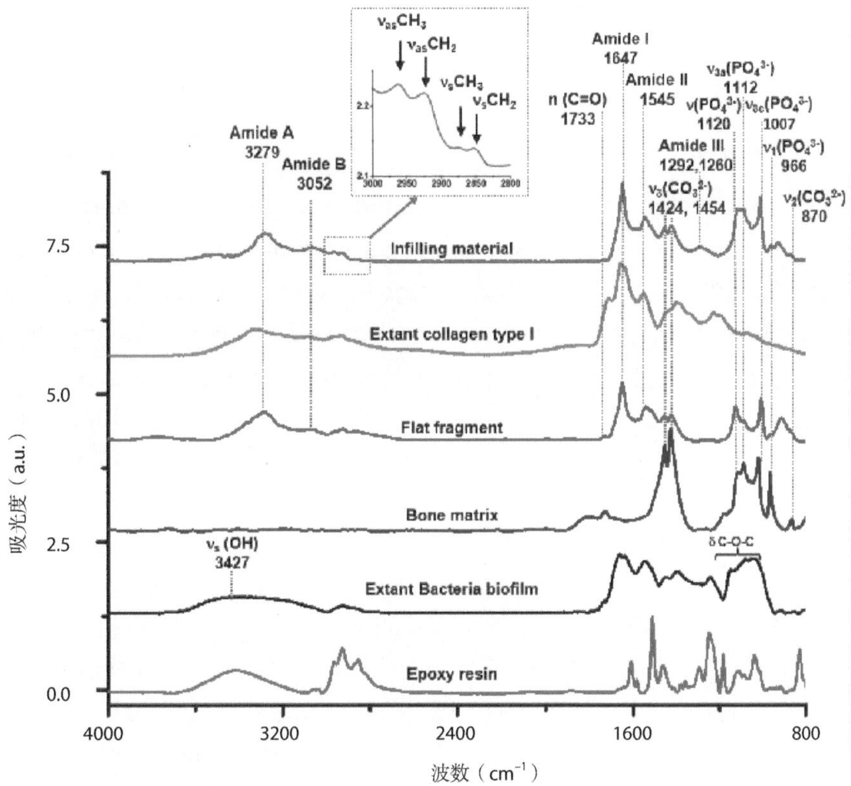

▲ 图 5.1.4 透过同步辐射傅里叶变换红外显微光谱仪测量恐龙骨头化石内胶原蛋白，并以现生已知（牛）胶原蛋白、细菌和可能在研磨过程中污染的环氧树脂做比对

▲ Figure 5.1.4 The spectra did at NSRRC sr-FTIR workstation showing the collagen inside the dinosaur bone, with comparison to extent collagen (calf), bacteria and possible containment epoxy

来找去，就是看不到胶原蛋白的特征诸酰胺 (amides) 波峰，理论上应该看得到，可是实际上没有看到，后来想了想，如果改用纵切，运气好的话，应该可以刚好切到与骨头纵轴平行的脉管渠道，如此一来，找到想要找的东西的机会就大增了，于是又切了不少样本磨了不少试片，结果有改善但还不理想。不过，注意到有些试片上，有些好像透明的小碎块掉出管外，果真不错，终于找到了骨头化石内保存的胶原蛋白！不过，这时候想起：胶原蛋白是水溶性的，因此，如果在最后研磨打薄过程中，用水为冷却剂，那么水会把胶原蛋白溶解掉，所以以前才会看不到东西啊！经过长期痛苦的摸索，终于了解到若要寻找骨头化石内的胶原蛋白，第一，骨头要纵切，而且第二，研磨的冷却剂，不能用水，要改用无水乙醇！第3章图 3.2.1 为最后找到来取得胶原蛋白光谱的试片之一。

图 5.1.4 是历经三年多从错误中摸索出来，最后的红外线光谱分析，胶原蛋白的几个酰胺波峰 (amide I, II, III, A, B) 非常明确，而且将酰胺部分进一步分解的话，可以很明确看到第一类胶原蛋白三螺旋二级构造的波峰（图 5.1.5），证明此物为第一类胶原蛋白。

从蛋白质组成的角度来说：多个氨基酸 (amino acids) 先组成短链的多肽 (peptides)，再由好多多肽组合成长链蛋白质；在形成化石漫长的岁月中，受到各种地质情况（温度、压力、地下水等等）的影响，很难排除长链蛋白质

see! Therefore, we know: ① the bone has to be cut longitudinal, not transversal. ② Water can not be used as the coolant. Change that to absolute alcohol! Figure 3.2.1 of Chapter 3 shows one of the slices we used to get the IR spectra of collagen.

Figure 5.1.4 shows the last results of our 3 years of trials and errors. Several key amide peaks (amide I, II, III, A, and B) were very sure. Further deconvolution also shows the secondary structure of triple helix peak, as Figure 5.1.6. This proves the collagen type I.

Speaking of protein composition: many amino acids polymerized to form short chain peptides, then many peptides to form a long chain protein. During a long time of fossilization, influenced by various geological conditions (temperature, pressure, groundwater, etc.), it is tough to exclude the degradation of long-chain proteins to shorter peptide fragments, or even down to amino acids. The real situation could be even more complicated because these three kinds of fragments could re-polymerized under the right geological condition. This complicates the exploration of collagen preservation even further. All these three possible matters (protein, peptide, and amino acid) contain the amide functional group. So, how to be sure the collagen type I? In short, seeing amide peak means that the dinosaur bone contains preserved matter contains amide functional group. Further analyses of amide peaks are needed to resolve them out. The structure of long-chain proteins has several structural levels. In which the secondary structure is very crucial for the identification of protein type. Such as the collagen type I has a

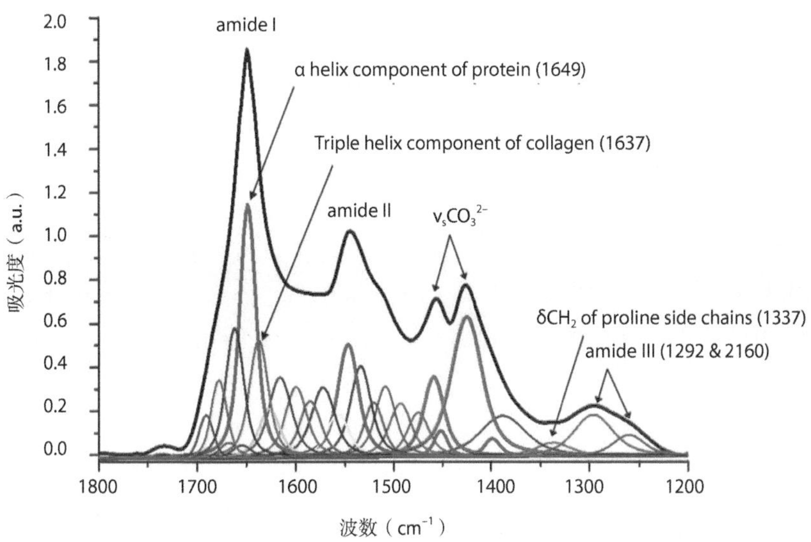

▲ 图 5.1.5　进一步解析几个酰胺波峰 (amide I, II, III）的蛋白质二级结构组成波，在 1637 cm^{-1} 明确出现证明三螺旋的第一类胶原蛋白

▲ Figure 5.1.5　Further deconvolution of protein secondary structure of amide I, II, and III peaks At the position of 1637 cm^{-1}, clearly shows the triple helix of collagen type I

没有被裂解成为短链多肽或甚至个别的氨基酸等碎片段。实际的情况更复杂，因为这三种裂解物质还有可能在适当的地质情况下，又进行了再聚合作用 (repolymerization)，这些情况使得探讨化石内保存胶原蛋白课题，变得非常复杂；这三种可能的物质（蛋白质、多肽、氨基酸），都含有酰胺官能团构造，因此，怎能确认是第一类胶原蛋白呢？简单来说，光是看到诸酰胺波峰，只能说恐龙骨头化石内保存着含有酰胺官能团构造的物质，还不能确定是第一类胶原蛋白。因此需要进一步地分析酰胺的波峰，把它们更细地分解出来；长链蛋白质的构造，可以分成几个构造等级，其中二级

triple helix secondary structure. After deconvolution of the amide I hump, this triple helix has its unique peak at 1637cm^{-1}. From Figure 5.1.5, we can see this peak very clearly. Thus, we can very surely say that we found the preserved collagen type I in the almost 200 million years old dinosaur bone.

About the blood vessel and red blood cells, Schweitzer published paper stated that in the female *T. rex*, she found Heme. Note, she did not say that she found red blood cell (RBC), but the Heme, the remains of red blood cell. These are very different. However, the media people reported red blood cells were found. Of course, it created sever attacks. If real red blood cell could be found inside dinosaur bone, it would be a huge discovery. However, from the practical

构造对于鉴定是哪种蛋白质非常关键，如解析第一类胶原蛋白酰胺I波峰有三螺旋的二级结构，这个三螺旋有它独特的波峰在1637 cm^{-1}，从图5.1.5我们很明确看到这个波峰，因此可以很确定地说，我们找到了二亿年前恐龙骨头化石内保存了第一类胶原蛋白！

关于在脉管渠道内的血管和红血球，以前Mary曾经发表过论文，说她在那头母霸王龙的腿骨内，有找到血基质(Heme)，请注意她不是说找到红血球(red blood cell, RBC)，而是红血球的血基质残留物，这两者有很大的距离，但记者胡乱报道，也因而被人家强烈攻击。毕竟，果真能在恐龙骨头化石内找到真的红血球被保留下来，那可真是一件伟大发现，可是从实际面来说，至少理论上说不通，有一点古生物常识的人都知道。不过，有人认为这些黑点，如果试片磨得够薄，在高倍显微镜下来看，大小、颜色和形状的确是很像红血球。从一开始对此找到红血球念头持着保留的态度，我提醒大家大小颜色和形状类似，并不能也无法证明这些就是红血球，赤铁矿的矿物晶癖有类似的构造。只因看到大小颜色和形状类似，就跳到是红血球的结论，未免太夸张了！后来用拉曼光谱仪来看这些黑点，证明它们是赤铁矿的微小聚晶，为了取得这些赤铁矿的微小聚晶影像，还用同步辐射研究中心的透射X射线显微镜(transmission X-ray microscope, TXM)扫描，让大家看清楚了这些赤铁矿微小聚晶的形态，果真没错（图5.1.6），

point of view, any average paleontologist would know that is not the case. However, someone thought that if the slice can be made thin enough and observed with very high magnification, these dark spots do look like red blood cell size, color, and shape. I told them that I had my reservation. Just because of similar size, color and shape do not and can not prove that these are red blood cells. Jumping to that conclusion is too far-fetched. So, Raman spectroscopy was employed and proved that they are small crystal clump of hematite. Transmission X-ray Microscope, TXM of NSRRC was used. Figure 5.1.6 shows the morphology of a small hematite clump.

Within recent years, the information can be provided by Raman spectroscopy is astonishing, nothing less than infrared spectroscopy. In October of 2015, I went to the Australian Nuclear Science and Technology Organization (ANSTO) in Sydney, Australia to conduct some neutron tomography and visited two experts of the Vibrational Spectroscopy Lab of University of Sydney. We learned that they have the most advanced Raman Spectroscopes and can perform not just a spot or 2D, but also 3D scans so that we can see the distribution of a given component in three dimensions! What Raman spectroscopy can do nowadays can be called as Raman-CT. They are delighted to collaborate with us. So, some *Lufengosaurus* embryonic bone specimens were mailed to them for them to try. When we met again in May of 2016, they gave me the initial results as Figure 5.1.7. In this 3D distribution of Amide III, which is a critical spectroscopic identification of collagen type I, clearly shows the existence of col-

▲ 图 5.1.6 脉管渠道内黑点的 TXM 三维影像，白色的比例尺为 2 μm。证实了它们是赤铁矿而非红血球

▲ Figure 5.1.6 Three dimensional TXM images of the black dots inside the vascular channel The scale bar is 2 μm. This show they are hematite, not red blood cell

就是赤铁矿！

这些年来科技的进展，拉曼光谱仪所能提供的信息，一点也不输给红外光谱仪！我在 2015 年 10 月到澳大利亚核科技组织 (Australian Nuclear Science and Technology Organizarion, ANSTO) 进行中子扫描的时候，顺便拜访了悉尼大学 (University of Sydney)，与该校振动光谱 (Vibrational Spectroscopy) 实验室的两位专家学者讨论，得知他们拥有的最新拉曼光谱仪，不只可以做到单点、二维的光谱分析，更还可做三维扫描，让我们看到物质内某成分的三维分布！拉曼光谱仪如今可做到，也可称为拉曼计算机断层扫描 (Raman-CT)，很乐意合作，因此，回台湾后，就寄了几个大洼恐龙胚胎骨头化石，请他们试试看。2016 年 5 月我们再度碰面的时候，他们给了我初步的成果，如图 5.1.7。在这张显示第一

lagen type I by amide III peak. This is in complete agreement with what I got from μ-CT and other scans. Wonderful. Raman-CT can do nondestructive analysis 200μm down from the surface. Bingo!

Last, but not the least. Let us back to the organic remains. From"Look into the Bone," I proposed several years ago. Although being able to see the internal morphological structure to interpret the secret of ancient life is very advance, but it is not enough, because only very shallow, not deep enough to explore the much richer inner meaning and the changes in the fossilization process, i.e., the goal of taphonomy. Therefore, this originally unintended "Organic Remains inside Fossil" (Figure 5.1.8) topic becomes very important.

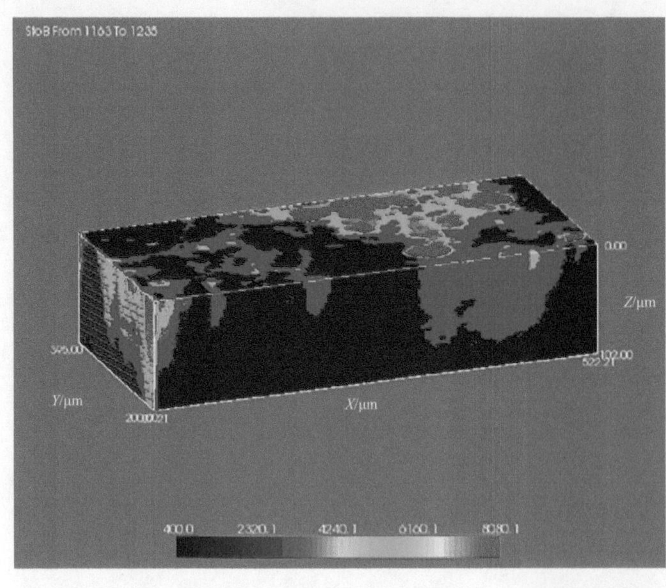

◀ 图 5.1.7 拉曼计算机断层扫描显示大洼恐龙胚胎骨头化石内第一类胶原蛋白酰胺 III 波峰的三维分布

◀ Figure 5.1.7 Raman-CT shows the amide III 3D distribution of collagen type I inside the DaWa *Lufengosaurus* embryonic bone

类胶原蛋白酰胺 III 三维分布的照片中，关键性的酰胺 III 波峰非常明确地证明了第一类胶原蛋白的存在，这影像所显示的酰胺 III 三维空间分布，和我过去用微型计算机断层扫描，以及其他扫描检测所得到的数据吻合，真是太棒了！拉曼计算机断层扫描，可以完全无破坏性地扫描深入表面 200 μm，而且是立体的数据！ Bingo！

最后，回到有机残留物课题来说几句：从我多年前提出“看进骨头内”研究古生物的角度来说，看到骨头（实体）化石内的组织形态构造，用以解读古生物的奥秘，虽然已经很先进，但并不足够，因为只看到浅层结果，而未能深入推论其更丰富的原始内涵以及石化过程 (fossilization) 中的变化，亦即埋藏学 (taphonomy) 所追寻的目标，因此这个误打正着的找寻“化石内有机残留物”（图5.1.8）课题，就显得非常重要了。

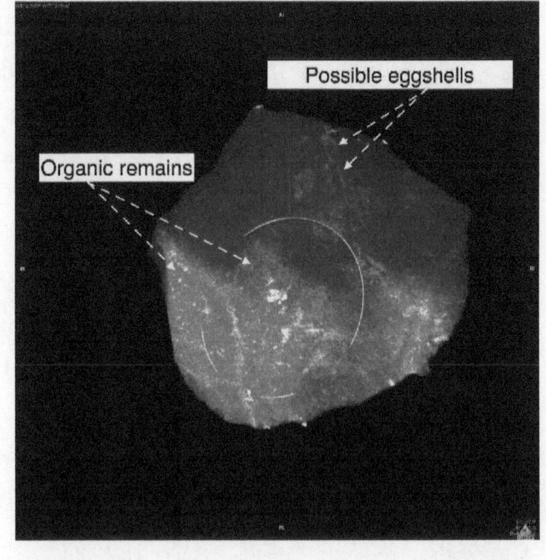

▲ 图 5.1.8 利用中子扫描，看到南非大椎龙胚胎围岩内，似乎还保存着大椎龙蛋壳和很可能是蛋窝内植物树叶的有机残留物

▲ Figure 5.1.8 Seeing the possible eggshell and remains of plant leaves inside the matrix of South Africa *Massospondylus* by using neutron scan

5.2　有机残留物研究三步骤

第一步：有没有有机残留物？

第二步：保存了什么？

第三步：保存机制？

正如前节所述，探讨化石与其紧临围岩内所保存的有机残留物，从理论上来说，有如此的三个步骤，但是在实际情况下，步骤之间往往不是那么楚河汉界泾渭分明，每个化石课题都不尽相同，而多多少少会有些重叠。

当然，探讨化石与其紧临围岩内所保存的有机残留物，第一步“有没有有机残留物？”是最关键的，如果化石与其围岩内根本不含有机残留物，戏就唱不下去了。

那么，怎么知道某样本是否保存了有机残留物？最先最直接的方法就是用肉眼观察。从训练为传统古生物学者的角度来看某块化石样本，首先注意到的，往往着重于该化石本体的形态学，比方说，如果是脊椎动物化石的话，有多少骨头保存下来？绞合程度如何？保存情况如何？可是，从古生物化学的角度来说，我们所要看最主要的要素，不是形态学方面的“死相”如何，而是该样本是否保留着有机残留物！已经清理过的样本，有没有被过度清理，把紧邻化石本体旁边的围岩都清理光光？就拿图 5.2.1 来说，古生物学者所看的，着重于这个样本上该古生物骨骸部分，而古生物化学所着重的，却是骨头旁边那些灰灰的羽毛，而且会看那些环绕骨架白区，在此部分的有机残留物，很可能被过度清理掉了。

5.2　Three Stages of Organic Remains Study

Step I: Is any organic remain preserved?

Step II: What compound preserved?

Step III: Preservation mechanism?

As mentioned in the previous section when exploring the organic remains inside the fossil or the immediate surrounding matrix, there are these steps in theory. However, in reality, the boundary between steps may not be that clear-cut. Each fossil topic is unique and may have overlapping steps.

Of course, in order to explore organic remains preserved inside fossil and the immediate surrounding matrix. The very first step 'Is any organic remain preserved?' is the most crucial. If there are no organic remains, nothing can be done further.

So, how to know if a given specimen contains preserved organic remains? The first and most direct method is to use eye examination. From a trained paleontologist angle, for vertebrae, they will from the morphological point of view look at how many bones were preserved, how the articulation, how the preservation condition? etc. However, from the PaleoChemistry point of view, if the morphological 'dead look' perfect or not is not the primary concern. The key is if any organic remains preserved. Has the specimen been over prepared and all the immediate nearby matrix removed? Take Figure 5.2.1 as an example, the paleontologist would care about the skeleton, but the PaleoChemist would look at the grayish feather next to the bones, and also would look more on the surrounding white

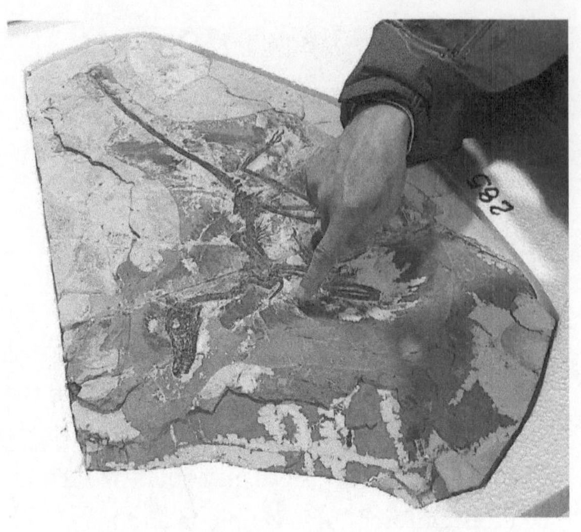

◀ 图 5.2.1 不同的角度，看到不同的景色

◀ Figure 5.2.1 From different angle of view, seeing different scene

但是，肉眼有其天生限制，即便是经验丰富的专家，也经常会碰上看走眼或有看没有到的窘状，何况绝大部分的有机残留物被保存于化石或围岩里面，从外表肉眼看不出来；如第1章图1.3.2所示，这个后肢化石，用肉眼看不出啥名堂，可是在很简单设备的激光荧光下，趾骨周围显然大有文章，呈现明亮的黄色荧光，明确地说出化学成分的不同，造成不同荧光反应的物质，很可能就是有机残留物。仅就这个样本来说，因为荧光部分的形态，很像是原本脚趾肌肉表皮的形态，因此，这些因激光而产生的黄色荧光，非常有可能是趾骨旁边肌肉残留的有机残留物。

再者，除了上述简单的激光荧光设备可用来作为“有没有有机残留物？”的筛选之外，还有借助先进仪器设备的很多方法，其中最令人惊讶的威力设备，就是中子扫描，如图 5.2.2。这块南非大椎龙胚胎的围岩上，肉眼可看出三根很小的肋骨，当然看不到围岩里面有

areas to see if the organic remains in these areas were cleaned out or not.

However, eyes have a natural limitation. Even for experienced experts, misjudgment or not seeing is common, not to mention that most of the organic remains were preserved inside the fossil or matrix, and could not be seen from outside. For example, nothing unusual can be seen of the hind limb, as shown in Figure 1.3.2 of Chapter 1. However, under a very simple Laser Stimulated Fluorescence setup, it is evident that there is something else there not visible by naked eyes as shown by the bright yellow fluorescent color. This indicates that different chemical compositions which generate a different fluorescent response. These are very likely to be organic remains. As for this particular specimen, looking at the shape of the fluorescent areas, they are very similar to the shape of the muscle and skin of the original toes. Thus, the fluorescence generated by laser could be very likely organic remains from the toe muscle.

Furthermore, besides the simple laser stim-

啥东西，然而，在中子扫描之下，围岩表面的肋骨，有看没有到，反而是看见围岩内好多明亮的区域，看来有可能是蛋窝内的蛋壳碎片和腐烂到某种程度的树叶。当然啦！使用中子扫描效果不错，可是这种方法不是人人有机会玩到的尖端科技设施。不过，也不用完全失望，有一些比较“便宜”的检测方法，如傅里叶变换红外显微光谱，拉曼断层扫描和多频显微等等，还有上述非常便宜可以自己做的设施，也都可用来帮忙回答这个最基本初步问题的好工具。

在除了中子扫描之外的其他可用方法中，傅里叶变换红外线显微光谱 (FTIR) 是个好用的筛选工具，事实上，这是我用来检测将近一百多个多种实体化石最早采用的工具，从这些筛选性质的扫描，得到了一个无法否认的结论：其实，实体化石里面保存着有机残留物，不是特别个案，应该会是一个普遍现象。这一点可能会推翻一般大众对于化石已经都是无机物（石头）的认知。

因为当初只先做有机残留物筛选扫描，因此，只用一般实验室可买得到的布鲁克 (Bruker) 公司的商业傅里叶变换红外光谱仪（在此没有任何帮忙推销之意）来进行，所依据的基本原理，从红外线光谱学的角度来说非常基本，我们专注在 2800~3000 cm^{-1} 这段波长范围内，如果出现波峰是因为样本内含有烷基，亦即 CH_3—、CH_2—等官能团，而烷基可以说是绝大部分有机物都有的官能团，所以理论可以这么说，如果某化石样本内

ulated fluorescence, can be used to screen the question of ‘if there are any organic remains?’ There are many advanced instrumentations and methodologies can be very helpful. The most astonishing one is the neutron scan, as Figure 5.2.2. This is a matrix of South Africa Massospondylus embryonic rib bones. Three tiny, tiny embryonic ribs can be seen by naked eyes. However, the neutron scan did not show them well; instead, possible eggshells and decomposed plant material can be seen. Of course, the state of art neutron scan is not accessible for everybody. But, do not be discouraged, other ‘cheaper’ examination ways, such as FTIR, Raman CT, MHGM, etc., and the above-mentioned el cheapo LSF can be used to facilitate the answer to this first and fundamental question.

Besides the neutron scan, the FTIR is an excellent screening tool in the ‘other’ usable tool category. This is the instrument/methodology I used to scan more than one hundred specimens of various geological times and got the undeniable conclusion of Preservation of Organic Remains inside body fossils is not a unique case but is a common phenomenon. This overturned the misunderstanding of the general public on thinking that fossils are just inorganic rocks.

Since the initial works were for the screening of organic remains, thus, commercial FTIR by Bruker was used (without any intention of endorsement). From the most fundamental infrared spectroscopy point of view theoretically speaking, we looked the peaks within the range of 2800–3000 cm^{-1} which are the peaks of alkyl functional groups, i.e., CH_3-, CH_2-. So, it can say

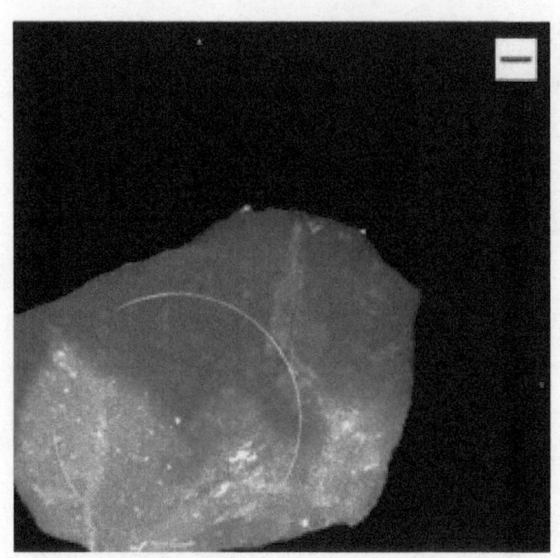

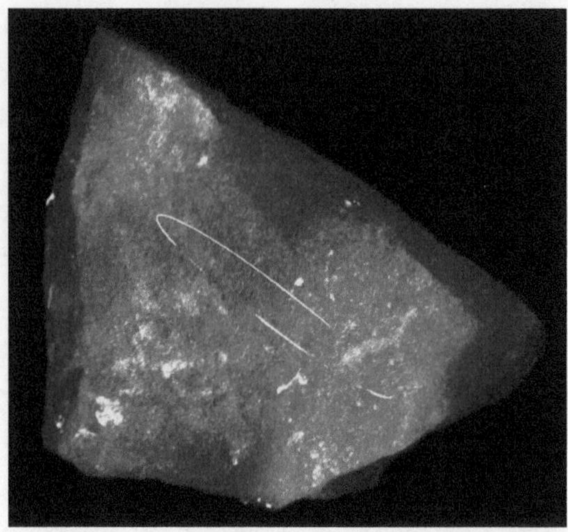

▲ 图 5.2.2 看进围岩里面！上方普通灯光下照片的三根很小的大椎龙胚胎肋骨清晰可见，下面照片是中子扫描所看到的，三个小肋骨几乎看不见，可视围岩里面，却有好多有机残留物

▲ Figure 5.2.2 Look into the Matrix. In the regular light photo on the top three tiny ribs of *Massospondylus* can be seen clearly. However, in the lower photo generated by neutron scan, these three ribs are tough to see. Instead, many organic remains inside the matrix can be seen

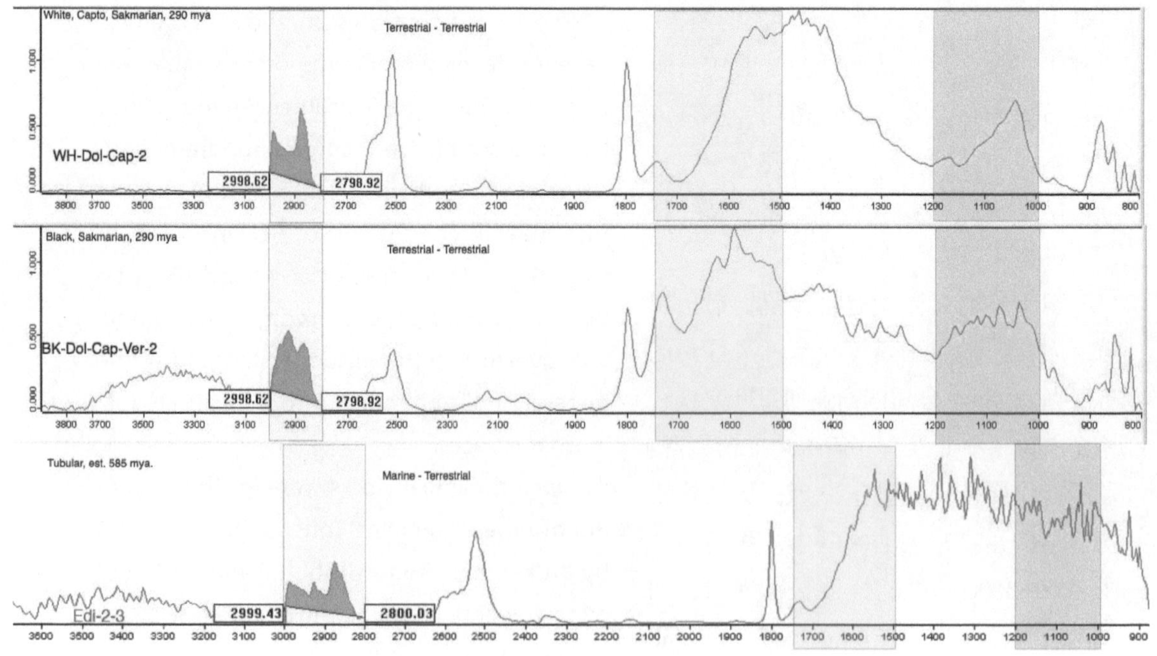

▲ 图 5.2.3 几个典型保存着有机残留物化石样本的光谱比较；请看图 5.1.2

▲ Figure 5.2.3 Some representative spectra of organic remains preservation. Also, please see Figure 5.1.2

保存着有机残留物的话，在这个波段范围内，应该就可以看到烷基特有的几个波峰，反过来说，如果没看到这些烷基的波峰，基本上就可以排除该化石样本内保存了有机残留物，如图 5.2.3。此图的三个光谱，是从扫描中抽样出来的，在光谱的左边，以黄色框标示的，就是 2800~3000 cm^{-1}, CH_3—、CH_2—烷基部分，又因可能会有人说，化石内的碳酸盐（方解石）波峰也可能被误解为有机残留物，此外，请回头看图 5.1.2 的相关光谱，所以特别到美国国家标准相关网站，抓下最权威的纯方解石光谱，以做比较，纯的方解石，根本没有（也不会有）烷基波峰。

在第一步的筛选，如果以上述提到的几种检测方法来看，它们的优缺点如表 5.2.1。

如果某个样本通过了第一关，亦即用了某（些）检测知道，该样本里面还保存了有机残留物，则可开始进行第二步“到底保存了什么？”从表 5.2.1 可看出来，这六种筛选方法当中，目前有用于第二步者，只有后面两种光谱和质谱分析，它们才能提供样本内有机残留物化学成分的信息，进行所需要的分析研判。其他三种方法，目前都没有办法协助进行成分的判读。

回头看看图 5.1.4 和图 5.1.5，以此为例来说说，图 5.1.4 有好几条不同物质的光谱线作为比较，用来证明我们检测的样本里面，不是这些可能的污染（最下面，环氧树脂）或来源（底部第二条，现生细菌生物膜）；也有两条参考比对

that if the given fossil contains preserved organic remains, then we shall be able to see these alkyl peaks. On the other hand, if we can not see these peaks from a given fossil, then we can exclude that, as Figure 5.2.3. In this figure, the top three spectra were from our initial screening. To the left of each spectrum, the yellow box marks the range of 2800–3000 cm^{-1}, CH_3-, CH_2-. Some people may say that carbonate (calcite) peaks could be misidentified as the organic functional group. Please go back to Figure 5.1.2. A most authoritative spectrum of pure calcite by USA national standard agency is attached for comparison. No alkyl peaks can be seen.

In Step 1 screening by the methodologies mentioned above, the advantage/disadvantage comparison of each is as shown in Table 5.2.1.

If a given sample passes the first step, i.e., one or more methods provide positive evidence of organic remains preservation, then, the 'Step II: What compound preserved?' can be started. From Table 5.2.1, it is shown that among these six methodologies, only the last two spectroscopic analyses and mass spectrometry are useful in Step II. They can provide further analyses of the chemical compositions of organic remains inside the sample. The other three can NOT help in this regard.

Look back Figure 5.1.4 and Figure 5.1.5 as an example. There are many spectra of different material to argue that in our experiments, there was no possible contamination (bottom, epoxy resin) or source (2nd from bottom, Extant Bacteria biofilm). There are also two comparison reference spectra: (3rd from bottom) Bone matrix, and (2nd from top) Extant collagen type I. The

表5.2.1 各种筛选有机残留物方法比较
Table 5.2.1 Comparison of various methodology for organic remains screening

设备 Methodology	优点 Advantage	缺点 Disadvantage	注 Remark
激光荧光光谱分析 Laser Stimulated Fluorescence Spectroscopy	设备简单 Simple setup 快速检测 Quick examination 非破坏性 Nondestructive	无法得知化学成分 Do not know chemical composition	元素光谱分析比较有用 Better for elemental analysis 需要进一步探讨 Need further exploration of LSFS
多频显微 Multiple Harmonic Generation Microscopy	快速检测 Quick examination 非破坏性 Nondestructive	无法得知化学成分 Do not know chemical composition 无现成仪器可买 Not commercially available	二倍频三倍频与化学成分，还需探讨 Need research on chemical composition of SHG and THG
中子扫描 Neutron Tomography	非破坏性 Nondestructive 深入整个样本 Deep penetration into sample	无法得知化学成分 Do not know chemical composition 需要申请有限的设备 Need to apply for limited facility	对于检测轻元素很有用 Very useful for detecting light elements
拉曼断层扫描 Raman-CT	非破坏性 Nondestructive 提供完整光谱分析 Provide complete spectroscopic analyses	光谱解读 Spectra Interpretation	有机物光谱分析 Organic compound spectroscopy analyses
傅里叶变换红外光谱 FTIR	提供完整光谱分析 Provide complete spectroscopic analyses	破坏性 Destructive 光谱解读 Spectra Interpretation	有机物光谱分析 Organic compound spectroscopic analyses
质谱分析 Mass Spectometry	提供化学成分 Provide chemical composition	可能破坏性 Could be destructive	选择适当的质谱分析法 Choose the appropriate MS

的光谱：（底下第三条）骨头背景、（上方第二条）现生第一类胶原蛋白；主角则是最上面第一条（填充物质）和第三条光谱线（平坦碎片）；所以，如果专注在从上算下来这四条光谱线，可以清楚看出来，二亿年前恐龙骨头背景内，如果含有胶原蛋白的话，含量非常少，可以看成在硬骨头的背景（底下第三条）内，并没有胶原蛋白，但是磷灰石群矿物质（碳酸磷灰石 A 和 B）却很明显；红色的第一条（填充物质）和蓝色第三条光谱（平坦碎片），则显示了检测区域内所含的官能团：包含碳酸磷灰石 A

main spectra are the Infilling material (the first from top), and Flat fragment (the 3rd from top). So, looking at the top four spectra, it is clear that within the 200 million years old dinosaur bone matrix (3rd from bottom), there is almost no collagen, if any at all. However, the apatite minerals, carbonatoapatite-A, and carbonatoapatite-B are very clear. The first (red, Infilling material) and the third (blue, Flat fragment) spectra show the functional groups of the specimens, including Amide I (1647 cm^{-1}), II (1545 cm^{-1}), III (1292, 1260 cm^{-1}), A (3279 cm^{-1}), and B (3052 cm^{-1}). These only say that the specimens contain Amide functional group, not the collagen type I yet.

和 B 相关的磷酸根和碳酸根、氢氧基以及主角的酰胺基，总共有五个关键的波峰为：酰胺基 I (1647 cm^{-1}), II (1545 cm^{-1}), III (1292 cm^{-1}, 1260 cm^{-1}), A (3279 cm^{-1}), B (3052 cm^{-1})；这些波峰证明了样本内含有酰胺基，到此为止，还不能说这就是第一类胶原蛋白，因为别的蛋白质、多肽或氨基酸，都含有酰胺基，因此必须进一步将酰胺基 I 进行分解成组成波峰，如图 5.1.5 显示出，确认第一类胶原蛋白的关键波峰：α螺旋 (1649 cm^{-1}), 三螺旋 (1637 cm^{-1}), δCH_2 脯氨酸侧链 (1337 cm^{-1})，有了这几个关键波峰，第一类胶原蛋白最终得以确认。

从上述的分析，接下来最重要的第三步骤“保存机制？”这是研究化石与邻近围岩内保存有机残留物课题最重要的关键终极问题。然而因为古生物化学刚刚起步，诸多化石与邻近围岩有机残留物保存，还有待大家努力去发现，目前要说这个大问题的确定答案，为时尚早，在此只能举出两种可能：①铁在磷灰石群骨头化石内胶原蛋白和金属离子对于有机物的催化 (Catalyst) 作用；②碳酸盐的胶合。

铁在磷灰石群骨头化石内胶原蛋白的催化作用，Schweitzer早期的研究指出，针铁矿可能为胶原蛋白的保存催化剂，然而我们团队于2017年1月31日在《自然通讯》论文指出，在原位（恐龙）骨头内存在的，并非针铁矿而是赤铁矿（第1章图 1.2.6 上），这是因为她把骨头用去钙法溶解，在不知的情况下，将赤铁矿水解成针铁矿。不过，虽然如此，

Any protein, peptide, and amino acid contain Amide functional group. Thus, further deconvolution of Amide I is needed as Figure 5.1.5. The key peaks for collagen type I are: α-helix (1649 cm^{-1}), triple-helix (1637 cm^{-1}), δCH_2 of the proline side chain (1337 cm^{-1}). With these key peaks, the last nail for sealing collagen type I is hammered tightly.

From the above discussions, the next most important question is “preservation mechanism.” This is the most important and ultimate question for the study of the preservation of organic remains inside fossil and the immediate surrounding matrix. However, since PaleoChemistry is at its very beginning step, many many preserved organic remains are still waiting to be found, it is too early to say the firm answer to this big question. Here, we list two possibilities: ① the catalyst effect of iron in the apatite bone as well as the metal ions for organics, and ② carbonate cementing.

The catalyst effect of iron in the apatite bone: Previous research of Mary Schweitzer stated that goethite could be the catalyst for the preservation of collagen. However, our paper on Nature Communications of January 31, 2017, pointed out that it is hematite, rather than goethite *in situ* (dinosaur) bone (top of Figure 1.2.6 of Chapter 1). Their misidentification was caused by the decalcification her team used. They did not realize the hydration of hematite to goethite. However, even so, one thing can be sure. Both of these two studies point to the same thing, saying that iron inside the appetites bones plays a crucial role in the preservation of collagen.

Carbonate Cementing: Different from the

可以确定的是，我们两个研究，殊途同归，都指出铁在磷灰石群骨头内，扮演了胶原蛋白保存的关键角色。

碳酸盐的胶合：另一大类的化石，有别于以磷灰石群矿物为主的骨头，就是碳酸盐基质背景的化石，如贝壳、海胆等等；在这类的化石内，前面有提过，也有保存着有机残留物，图5.2.4为红色鲍鱼珍珠母在不同温度六小时加温后的光谱变化，在600℃烘烤六小时之后，烷基波峰(3000~2800 cm^{-1}) 几乎完全消失，在较低波数部分 (1900~600 cm^{-1}) 也有明显变化，表示部分的有机物质被烧掉了。

碳酸盐胶合最明显的例子，如图5.2.5。这是肉食恐龙牙齿小锯齿的二维光谱分布图，上面三排显示三种肉食恐龙之二氧化碳物理吸附 (2345 cm^{-1}) 和化学吸附 (1640 cm^{-1}) 分布；接下来这排，分别显示物理吸附 (2345 cm^{-1}) 和碳酸磷灰石A (879 cm^{-1}) 与碳酸磷灰石B (867 cm^{-1})；最下一排左边为第四排三张分布图的重叠影像，右边则为烷基(3000~2800 cm^{-1}) 的分布图。比较二氧化碳和两种碳酸磷灰石分布重叠影像与烷基分布，非常明显地指出，碳酸盐形成胶合层 (Cementing Layer)，保护了牙本质里面的有机残留物。

这种碳酸盐胶合现象，在以磷灰石群矿物为基质的恐龙牙齿看得最明显，在骨头化石也看到类似的情况，但再以碳酸盐为基质的化石，还尚未有机会深入探讨，因此碳酸盐胶合现象，是否与有机残留物的保存有密切关系有待进一步探讨。

bone fossils using apatites minerals, this is another significant category of fossils base on carbonates, such as seashells, urchins, etc. As mentioned before, in these fossils, there are preserved organic remains. Figure 5.2.4 shows the spectra changes of nacre of red abalone shell after heating with different temperatures for 6 hours. Six hours at 600°C, the alkyl peaks (3000–2800 cm^{-1}) almost gone completely, with significant changes in the lower wavenumber range (1900–600 cm^{-1}). This means some organic matter was burned out.

The most obvious example of carbonate cementing can be seen from Figure 5.2.5. These are the serrations of the theropod tooth. The top three rows are the distribution of physisorption (2345 cm^{-1}) and chemisorption (1640 cm^{-1}). The next row shows physisorption (2345 cm^{-1}), carbonatoapatite-A (879 cm^{-1}), and carbonatorapatite-B (867 cm^{-1}). The left image of the last row shows the overlapped images of row 4. The right image shows the distribution of alkyl (3000–2800 cm^{-1}) group. These two images indicate that there is a carbonate cementing layer protecting the organic remains in the dentin area.

This carbonate cementing phenomenon is the most obvious in the apatite-based dinosaur tooth. Similar also have been seen on bone fossils. We did not have a chance to further explore this on carbonate-based fossils yet. Thus, whether carbonate cementing is closely related to the preservation of organic remains have to be explored further.

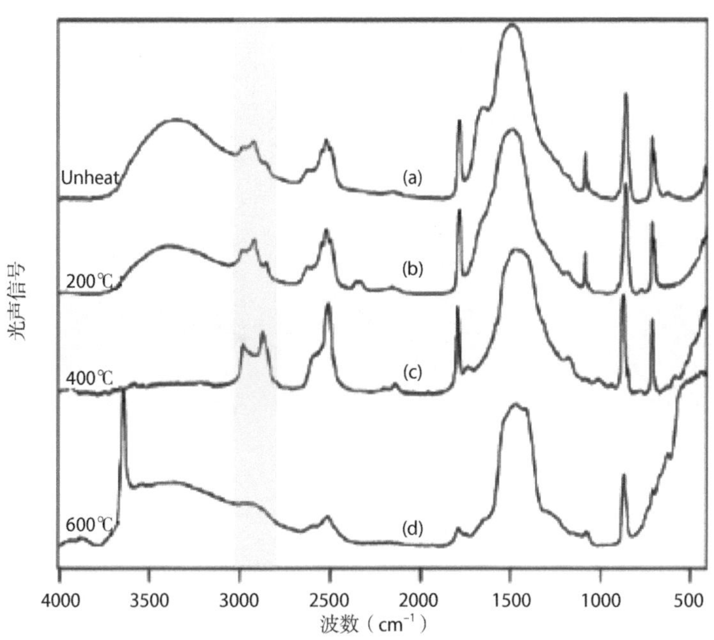

▲ 图 5.2.4 以碳酸盐为基质红鲍鱼珍珠母贝壳温度成分光谱变化，注意烷基波峰随着温度增加变化

▲ Figure 5.2.4 Composition vs. temperature spectroscopic changes of carbonate base nacre of red abalone. Note the change of alkyl peaks changes as the temperature increases

5.3 有机残留物分类

有机残留物分类

胶膜/角质；

碳氢化合物；

色素；

脂肪酸酯；

聚糖与几丁质；

分子化石；

蛋白质分类；

- 生物性功能；
- 溶解度；
- 化学组成；
- 形态；

5.3 Classification of Organic Remains

Classification of organic remains

Cutan/Cutin;

Hydrocarbons;

Pigments;

Fatty Acids and Esters;

Polysaccharides and Chitin;

Molecular Fossil;

Protein Classification;

- Biological Function;
- Solubility;
- Chemical Composition;
- Shape;

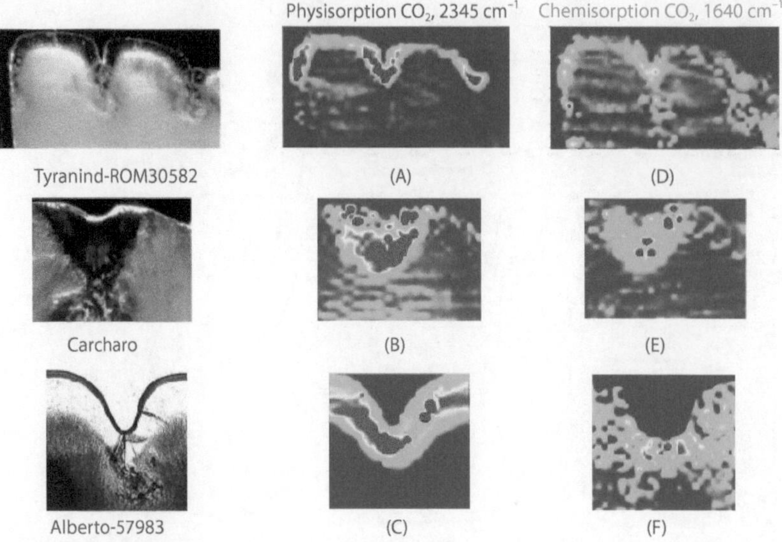

SR-FTIR images of peak height for tooth of Tyranind-ROM30582 (Sample 96),Carcharo and Alberto-57983 (A-C) Physisorption carbon dioxide at 2345 cm^{-1} and (D-F) chemisorption CO_2 at 1640 cm^{-1}

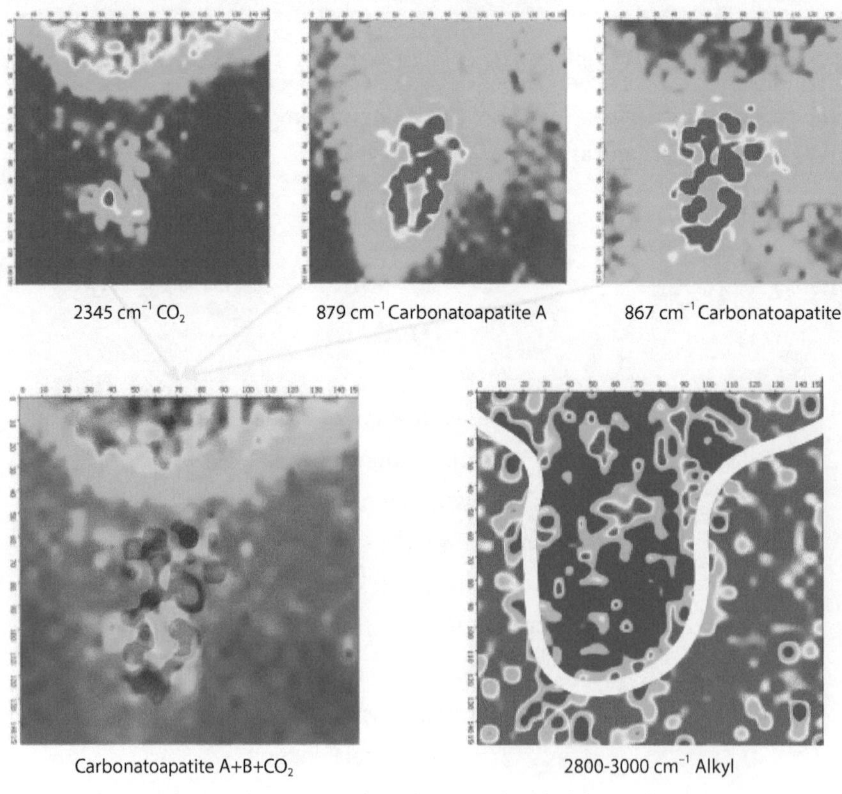

Developmental and evolutionary novelty in the serrated teeth of theropod dinosaurs
K. S. Brink, R. R. Reisz, A. R. H. LeBlanc, R. S. Chang, Y. C. Lee, C. C. Chiang, T. Huang & D. C. Evans

▲ 图 5.2.5 肉食恐龙牙齿上小锯齿的碳酸盐胶合现象

▲ Figure 5.2.5 Carbonate cementing of tooth serrations of theropod dinosaurs

球状蛋白；
丝蛋白；
角蛋白；
弹性蛋白；
纤维蛋白；
第一类胶原蛋白。

这是一个简单的化石或围岩中有机残留物分类,此中不含蛋白质分类中“蛋白质分类/形态/纤维蛋白/第一类胶原蛋白”的其他种类蛋白质。这些不同分类的化学成分，都有文献指出存在于化石或围岩中，当然还有可能其他的有机残留物有待被发现。毕竟化学品的种类与数量太庞大，有太多的可能性了。

5.3.1 胶膜／角质

（维基百科）角质 (cutan)，是某些植物表皮两种聚合物的一种，另一个更常见的聚合物是胶膜 (cutin)，较容易被保存在化石记录中；角质是碳氢聚合物，而胶膜则是聚酯，然而角质的结构和合成未明，它并不普遍如过去所想地存在于很多植物，如银杏就没有它。

角质（图 5.3.1和图5.3.2），最早发现为非皂化成分，不易透过碱性水解去酯，它是蜡聚合物中的一种，覆盖了植物与空气接触的表面。在某些物种，如大花君子兰 (*Clivia miniata*)，随着成熟含量增加，显然取代了在早期成长分泌的胶膜 (Schmidt and Schönherr, 1982)；角质为碳氢聚合物的证据在它的闪燃裂解 (flash pyrolysis) 产物，有成对烷与烯均

Globular;
Fibroin;
Keratin;
Elastin;
Fiberous;
Collagen type I.

This is just a simple classification of organic remains in fossils or matrix, excluding all other proteins except the "protein/shape/fiberous/collagen type I" in which were reported. Of course, it is very likely that others can be found. There are just too many types and numbers of chemical compounds. So, the chance exists.

5.3.1 Cutans/Cutins

(Wiki) Cutan is one of two polymers which occur in the cuticle of some plants. The other and the better-known polymer is cutin. Cutan is believed to be a hydrocarbon polymer, whereas cutin is a polyester, more easily preserved in fossil plants. The structure and synthesis of cutan are not yet fully understood. Cutan is not present in as many plants as once thought; for instance, it is absent in Ginkgo.

Cutan (Figure 5.3.1–5.3.2) was first detected as a non-saponifiable component, resistant to de-esterification by alkaline hydrolysis. It is one of wax polymer that increases in the amount in cuticles of some species such as *Clivia miniata* as they reach maturity, apparently replacing the cutin secreted in the early stages of cuticle development (Schmidt and Schönherr, 1982). Evidence that cutan is a hydrocarbon polymer comes from the fact that it's flash pyrolysis products are a characteristic homologous series of paired alkanes and alkenes (Nip et al. 1986).

The main constituents of cutan are ω-hy-

匀系列的特性 (Nip et al. 1986)。

角质主要由 ω-羟基酸及其衍生物组成，其中单体之间是通过酯键连接，形成大小不一的聚酯聚合物。角质有两种主要的单体家族：C_{16} 和 C_{18} 家族。C_{16} 家族主要包含 16-羟基棕榈酸、9,16- 二羟基棕榈酸或 10,16- 二羟基棕榈酸；C_{18} 家族主要包含 18- 羟基油酸、9,10- 环氧 -18- 羟基硬脂酸和 9,10,18- 三羟基硬脂酸。

droxy acid and derivatives. The monomers are connected by ester bonds forming polyesters of various sizes. There are two main monomers families: C_{16} and C_{18} family. C_{16} family contains mainly 16-Hydroxypalmitic acid, and 9, 16- or 10, 16-dihydroxypalmitic acid. The C_{18} family consists mainly of 18-hydroxy oleic acid, 9,10-epoxy-18-hydroxy stearic acid, and 9,10,18-trihydroxystearate.

5.3.2 碳氢化合物

（维基百科）白脂晶石 (fichtelite) 是玻利维亚木化石内发现的罕见白色矿物，属于单斜晶系，环状碳氢化合物：dimethyl-isopropyl-perhydro-phenanthrene, $C_{19}H_{34}$，很软，摩氏硬度 1，和滑石同样相对密度很低，1.032，只比水重一点点；最早在 1841 年描述，以发现地名命名，德国玻利维亚菲希特尔 (Fichtel-gebirge)，发现于泥炭沼内木化石内和现代含有大量有机物海洋沉积中。

化石燃料是自然形成的燃料，如无

5.3.2 Hydrocarbons

(Wiki) Fichtelite is a rare white mineral found in fossilized wood from Bavaria. It crystallizes in the monoclinic crystal system. It is a cyclic hydrocarbon: dimethyl-isopropyl-perhydrophenanthrene, $C_{19}H_{34}$. It is very soft with a Mohs hardness of 1, the same as talc. Its specific gravity is very low at 1.032, just slightly denser than water. It was first described in 1841 and named for the location, Fichtelgebirge, Bavaria, Germany. It has been reported from fossilized pine wood from a peat bog and in organic-rich modern marine sediments.

Fossil fuels are fuels formed by natural processes such as anaerobic decomposition of

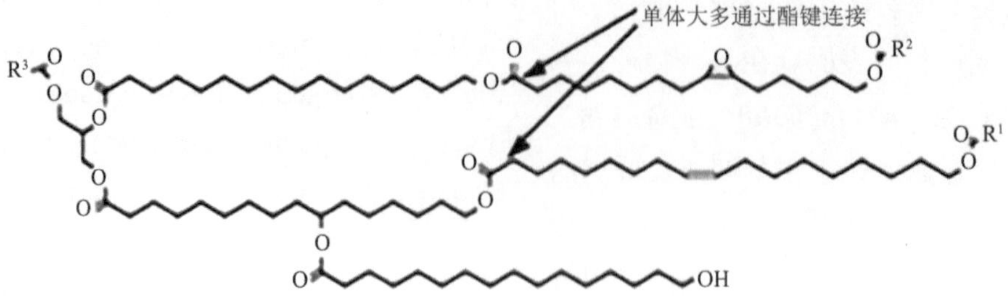

单体大多通过酯键连接

▲ 图 5.3.1 胶膜 / 角质化学结构

▲ Figure 5.3.1 Cutan/Cutin chemical structure

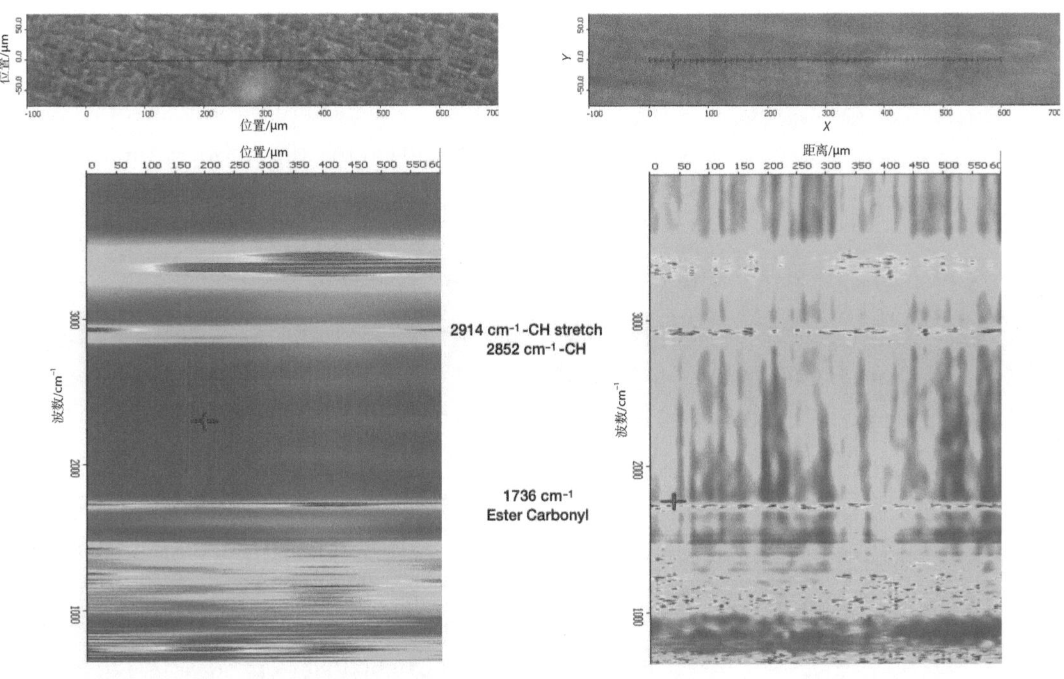

▲ 图 5.3.2 胶膜 / 角质傅里叶变换红外线光谱

▲ Figure 5.3.2 FTIR of Cutan/Cutin

氧死亡埋藏含有古代光合作用能量的生物，这些生物和所产生的化石燃料通常有几百万年，甚至有时候超过 6.5 亿年；化石燃料含有大量的碳，包括石油、煤、天然气，其他常见的衍生物，如煤油和丙烷；化石燃料的范围，从低碳氢比例的高挥发物质，如甲烷，到液态的石油，到不挥发物质，几乎只含纯碳，如无烟煤；甲烷可单独出现于碳氢油田，或与油共存，或甲烷水合物。

化石燃料形成的理论，死亡的植物经由地壳热和压力在几百万年内形成，最早在 1556 年由 Georgius Agricola 提出，后来在十八世纪，又由 Mikhail

buried dead organisms, containing energy originating in ancient photosynthesis. The age of the organisms and their resulting fossil fuels is typically millions of years and sometimes exceeds 650 million years. Fossil fuels contain high percentages of carbon and include petroleum, coal, and natural gas. Other commonly used derivatives include kerosene and propane. Fossil fuels range from volatile materials with low carbon:hydrogen ratios like methane, to liquids like petroleum, to nonvolatile materials composed of almost pure carbon, like anthracite coal. Methane can be found in hydrocarbon fields either alone, associated with oil, or in the form of methane clathrates.

The theory that fossil fuels formed from the fossilized remains of dead plants by expo-

Lomonosov 提出。因为化石燃料是当今的主要能源，相关的研究非常多，古生物化学就不在此方面着笔。

5.3.3 色素

（维基百科）2008 年由中国古生物学者徐星，从侏罗纪（2亿年至1.5 亿年）到晚古新世（Paleocene）和新近纪（Neogene Periods）（6600万年至200万年）岩石内羽毛化石发现的羽毛内找到了被保存的碳。这在过去被认为是分解羽毛组织的微量细菌形成的，然而这些残留物实际上是石化的黑素（melanosomes）的显微有机印模。有些结构还维持着羽毛和软毛组织典型色彩斑斓的颜色。推测这些显微组织可以更进一步研究，探讨显示化石内原本软组织的颜色和质地；耶鲁大学 Derek Briggs 说："发现羽毛内细微的精密构造，打开了探索其他软体生物化石特征的伟大可能，如软毛和甚至内部器官。"

黑素被北京自然历史博物馆用于发现赫氏近鸟龙化石的真正颜色。

刺突鲛

Harpagofututor

Grogan, Eileen D. and Richard Lund. "Soft Tissue Pigments of the Upper Mississippian Chondrenchelyid, Harpagofututor volsellorhinus (Chondrichthyes, Holocephali) from the Bear Gulch Limestone, Montana, USA," Journal of Paleontology, Vol. 71, No. 2, March

sure to heat and pressure in the Earth's crust over millions of years was first introduced by Georgius Agricola in 1556 and later by Mikhail Lomonosov in the 18th century. Since fossil fuels are the primary energy source of today, too many studies already done. PaleoChemistry will not discuss this topic further.

5.3.3 Pigments

(Wiki) Recent (2008) discoveries by Xu Xing, a Chinese paleontologist, include fossilized feathers in rock formations dating from the Jurassic period (200 to 150 million years ago) to the late Paleocene and Neogene periods (66 to 2 million years ago). The feathers contain preserved residues of carbon that were previously thought to be traces of bacteria that decomposed feather tissues; however, these (residues) are, in fact, microscopic organic imprints of fossilized melanosomes. Some of these structures still maintain an iridescent color typical of feather and fur tissues. It is conjectured that these microscopic structures could be further studied to reveal the original colors and textures of softer tissues in fossils. "The discovery of ultra-structural detail in feather fossils opens up remarkable possibilities for the investigation of other features in soft-bodied fossils, like fur and even internal organs," said Derek Briggs of the Yale University study team.

Melanosomes were used to discover the true colors of fossil *Anchiornis huxleyi* by the Beijing Museum of Natural History.

Harpagofututor

Grogan, Eileen D. and Richard Lund. "Soft Tissue Pigments of the Upper Mississippian Chon-

1997, pp. 337–342.

恐龙颜色

https://en.wikipedia.org/wiki/Dinosaur_coloration

近鸟龙

2010 年古生物学者研究一个来自中国义县鸟龙类 (*Averaptoran*) 保存良好的近鸟龙，在保存良好的石化羽毛内发现了黑色素小体 (melanosomes)，它是决定颜色的色素。不同形状的黑色素小体决定不同的颜色，透过研究黑色素小体的形状，他们认为近鸟龙全身有黑色、白色和灰色羽毛，只有在头冠的羽毛是红色的。

中华龙鸟、孔子鸟、尾羽龙和中国鸟龙

布里斯托大学 (University of Bristol) 的 Mike Benton 博士，在 2010 年研究义县的中国鸟龙、孔子鸟、尾羽龙和中华龙鸟遗骸，也发现了黑色素小体；他发现中华龙鸟披覆着橘色羽毛，尾巴有条纹，因此也提出假说，认为有毛但是不会飞的恐龙，利用它们羽毛作为展示，因为其颜色鲜艳。

始祖鸟

研究生Ryan Carney和同僚在2011年，进行了首次始祖鸟的颜色研究，发现了黑色素小体，认为样本羽毛主要是

drenchelyid, Harpagofututor volsellorhinus (Chondrichthyes, Holocephali) from the Bear Gulch Limestone, Montana, USA," Journal of Paleontology, Vol. 71, No. 2, March 1997, pp. 337–342.

Dinosaur coloration

https://en.wikipedia.org/wiki/Dinosaur_coloration

Anchiornis

In 2010, paleontologists studied a well-preserved skeleton of *Anchiornis*, an *averaptoran* from the Yixian Formation in China, and found melanosomes - color-determining pigments- in its well-preserved fossilized feathers. Different-shaped melanosomes determine different colors; by analyzing the shape of the melanosomes, they inferred that *Anchiornis* had black, white and grey feathers all over its body, with only the crest of feathers on its head being red.

Sinosauropteryx, Confuciusornis, Caudipteryx and Sinornithosaurus

Dr. Mike Benton from the University of Bristol, in 2010, analyzed remains of *Sinosauropteryx, Confuciusornis, Caudipteryx*, and *Sinornithosaurus* from the Yixian, and also found melanosomes there. He found that *Sinosauropteryx* was covered in orange feathers and that its tail was stripped. It was hypothesized that feathered dinosaurs which could not fly used their feathers for display because they were brightly colored.

Archaeopteryx

Graduate student, Ryan Carney and colleagues, in 2011, produced the first color study on an *Archaeopteryx* specimen, finding melano-

黑色的。所研究的羽毛可能是隐蔽的，被翅膀的主要羽毛部分遮蔽，他指出这和现代鸟类飞行羽毛一致，黑色的黑色素小体有强化羽毛飞行的构造特性。2013年，一份新的分析在原子光谱分析期刊（*Journal of Analytical Atomic Spectrometry*）发表，指出始祖鸟羽毛可能有更复杂的颜色，全体羽毛有淡色和深色，目前还不知道这些颜色是否作为展示和/或飞行用。

小盗龙

李权国（Quanguo Li）团队2012年的典型小盗龙羽毛研究指出，羽毛是闪亮的黑色，新的样本 BMNHC PH881 是历史上第一次用来决定小盗龙颜色的，它的黑色素小体狭窄堆叠层状，像黑鸟一样。“羽毛闪亮黑色”和认为“小盗龙是夜行性动物在晚间猎食”相互矛盾，现在夜行性鸟类没有闪亮的颜色。认为小盗龙是夜行性的，是从它眼睛巩膜大小得到的结论。

伊卡企鹅

伊卡企鹅（*Icadyptes salasi*）已经灭绝，生存于始新世晚期，大约三千六百万年前，它是南美洲的巨大企鹅，在2008年发现几乎完整的骨骸，包括石化羽毛，这是首次发现企鹅化石的羽毛。研究羽毛内包含色素胞器的黑色素小体，显示为灰色或红棕色，这和现代企鹅不同，现代企鹅的深黑棕色羽毛

somes that suggested a primarily black coloration in the feathers of the specimen. The feather studied was probably a covert, which would have partly covered the primary feathers on the wings. Carney pointed out that this is consistent with the flight feathers of modern birds, in which black melanosomes have structural properties that strengthen feathers for flight. In 2013, in a study published in the Journal of Analytical Atomic Spectrometry, new analyses revealed that the *Archaeopteryx's* feathers may have had more complex coloration, in the form of light and dark colored plumage, with the tips of its flight feathers being primarily black as opposed to the entire feather being dark in color. Whether or not this coloration was integral for display and/or flight is as yet unknown.

Microraptor

The coloration of the feathers of a typical *Microraptor*, according to a 2012 study by Quanguo Li and team was iridescent black. The new specimen - BMNHC PH881 was used to determine the coloration of *Microraptor* for the first time in history. The melanosomes were narrow and arranged in stacked layers, reminiscent of the blackbird. The fact that the feathers were iridescent conflicts with the hypothesis that *Microraptor* was nocturnal, and hunted at night since no known modern birds with iridescent coloring are nocturnal. The belief that *Microraptor* was nocturnal is a conclusion drawn due to the size of the scleral ring of the eye of *Microraptor*.

Inkayacu

Inkayacu is a genus of extinct penguin. It lived in Peru during the Late Eocene, around 36

是从独特的黑色素小体而来，较大和椭圆体的。

始新世伊卡企鹅羽毛内的黑色素小体窄长，类似于其他鸟，它们的形状指出横过伊卡身体有灰色和红棕色羽毛，大部分现代企鹅的黑色素小体长度大约相同，但是更宽，在活体企鹅内有很多，这些黑色素小体形状使得它们为深棕或黑色，这也是现生企鹅大部分是黑白色的原因。虽然伊卡企鹅的羽毛，没有现生企鹅独特的黑色素小体，但还是有很多方面类似。

鹦鹉嘴龙

2016年的一个研究，检验鹦鹉嘴龙与覆盖物样本保存的黑色素小体指出，这个动物身体有反荫蔽颜色，可能是喜欢活在少光的茂密森林中，就像今日活在森林里的鹿或羚羊，它有肢体的条纹和斑点，可能表示杂乱颜色。这个样本同时在肩膀、脸（可能作为展示用）、泄殖腔（可能抗菌用）和连接后肢与尾巴的翼膜上，也有密集的颜色团块。它的大眼显示视力应该很好，在寻找食物和躲开猎食者很有用，请见图 5.3.3。

5.3.4 脂肪酸酯

化石树脂

（维基百科）琥珀是化石树脂，也叫树脂体 (resinite)，来自针叶树和其他树种。柯巴脂 (copal)、贝壳杉胶 (kauri gum)、达马树脂 (dammar) 和其他树脂，也在次化石 (subfossil) 存着处共存。次化石的柯巴

million years ago. A nearly complete skeleton was discovered in 2008 and included fossilized feathers, the first known in penguins. A study of the melanosomes, pigment-containing organelles within the feathers, indicated that they were gray or reddish brown. This differs from modern penguins, which get their dark black-brown feathers from unique melanosomes that are large and ellipsoidal.

The melanosomes within the feathers of the Eocene penguin *Inkayacu* are long and narrow, similar to most other birds. Their shape suggests that *Inkayacu* had grey and reddish-brown feathering across its body. Most modern penguins have melanosomes that are about the same length as those of *Inkayacu* but are much broader. There is also a higher number of them within living penguins' cells. The shape of these melanosomes gives them a dark brown or black color and is the reason why modern penguins are mostly black and white. Despite not having the distinctive melanosomes of modern penguins, the feathers of Inkayacu were similar in many other ways.

Psittacosaurus

In a 2016 study, the examination of melanosomes preserved in the specimen of *Psittacosaurus* sp. preserved with integument indicated that the animal was countershaded, likely due to preferring a habitat in dense forests with little light, much like many modern species of forest-dwelling deer and antelope; stripes and spots on the limbs may represent disruptive coloration. The specimen also had dense clusters of pigment on its shoulders, face (possibly for display), and cloaca (which may have had an an-

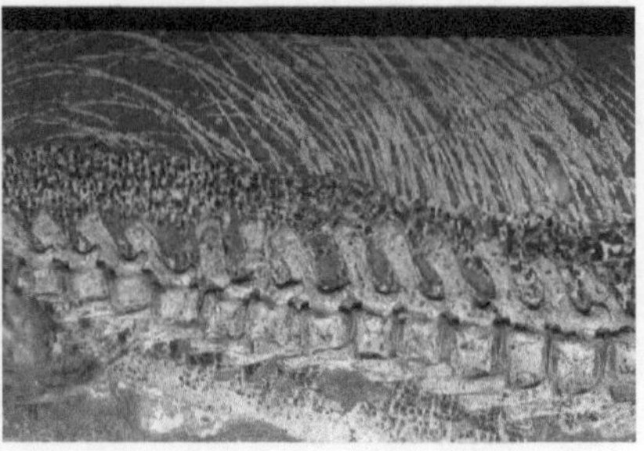

▲ 图 5.3.3 在激光荧光下的鹦鹉嘴龙样本，清楚显示全身体皮肤毛发等的颜色（左），尾部放大显示鬃毛（右），取自 Tom Kaye 网页

▲ Figure 5.3.3 Laser Stimulated Fluorescence images of *Psittacosaurus* specimen. Left: whole animal, Right: Tail portion showing bristles. Credit: Tom Kaye

脂，可以和真正琥珀区分，用一滴丙酮或氯仿滴上，会黏黏的；非洲的柯巴脂和新西兰的达马树脂，也出土于半化石状态。

琥珀分类有几种方法，最基本的是有两种植物树脂可能形成化石：针叶树和被子植物所产生的类萜 (terpenoids)，含异戊二烯 (isoprene, C_5H_8) 的环状结构；另外一类是酚醛树脂 (phenolic resins)，现在只从被子植物产生，也有同样功能；已经灭绝的髓木目(*medullosans*)，产生第三类的树脂，在其脉管内经常发现琥珀。琥珀的成分有很多变化，每一物种产生自己独特的化学成分，可用热解气相色谱质谱仪 (pyrolysis-gas chromatography-mass spectrometry) 来鉴定。整体的化学和构造组成，用来把琥珀分成五类；琥珀当为宝石的分类法，则依照制造的过程。

琥珀有五个分类，依其化学成分而定，都源自软黏树木树脂；琥珀有时候内含有动物和植物的内含物；在煤矿

timicrobial function), as well as large patagia on its hind legs that connected to the base of the tail. Its large eyes indicate that it also likely had good vision, which would have been useful in finding food and/or avoiding predators. See Figure 5.3.3.

5.3.4 Fatty Acids and Esters

Fossil resins

(Wiki) Amber is fossil resin (also called resinite) from coniferous and other tree species. Copal, kauri gum, dammar, and other resins may also be found as sub-fossil deposits. Sub-fossil copal can be distinguished from genuine fossil amber because it becomes tacky when a drop of a solvent such as acetone or chloroform is placed on it. African copal and the kauri gum of New Zealand are also procured in a semi-fossil condition.

Amber can be classified into several forms. Most fundamentally, there are two types of plant resin with the potential for fossilization. Terpenoids, produced by conifers and angiosperms, consist of ring structures formed of isoprene (C_5H_8) units. Phenolic resins are today

脉的称为树脂体 (resinite)，“灰黄琥珀 (ambrite)”这字词是特别应用于在新西兰煤层中找到的灰黄色琥珀。因上层沉积的高压和温度产生分子聚合，首先将树脂转换成柯巴脂，之后的恒常温度与压力，赶出萜烯 (terpenes)，形成琥珀。

要此发生，树脂必须能抗腐化，有很多树产生树脂，但大部分都被物理和生物过程破坏了，日光与微生物（如细菌和真菌）和极端温度，都会裂解树脂。树脂若要持久到变成琥珀，它必须能抗衡这些力量或在没有这些条件下的环境产生。

琥珀的组成非均匀性，但含有多少会溶解于乙醇、乙醚和氯仿的树脂成分和不溶解的沥青质；琥珀由在半日花烷 (labdane) 家族内的前驱体 (precursors)、透过自由基聚合作用而成的巨大分子，包含欧柏酸 (communic acid)、枯茗醇 (cummunol) 和 (biformene)，这些半日花烷是二萜 ($C_{20}H_{32}$) 和三烯 (trienes)，具有三个烯基有机骨架聚合起来，经过年代，琥珀成熟，发生更多聚合、异构化、交联和环化作用。把琥珀加热超过200 ℃ (392 ℉)，就会分解，产生琥珀油，留下称之为“琥珀松香 (amber colophony)”或“琥珀沥青 (amber pitch)”的黑色残留物；当溶解在松节油或亚麻籽油，形成“琥珀清漆”或“琥珀漆”。琥珀是一种很独特的保存方式，保存生物体无法石化的部分，对于重建生物体与其生态环境有所助益，然而树

only produced by angiosperms and tend to serve functional uses. The extinct *medullosans* produced a third type of resin, which is often found as amber within their veins. The composition of resins is highly variable; each species produces a unique blend of chemicals which can be identified by the use of pyrolysis-gas chromatography-mass spectrometry. The overall chemical and structural composition are used to divide ambers into five classes. There is also a separate classification of amber gemstones, according to the way of production.

There are five classes of amber, defined based on their chemical constituents. Because it originates as soft, sticky tree resin, amber sometimes contains animal and plant material as inclusions. Amber occurring in coal seams is also called resinite, and the term ambrite is applied to that found specifically within New Zealand coal seams. Molecular polymerization, resulting from high pressures and temperatures produced by overlying sediment, transforms the resin first into copal. Sustained heat and pressure drive off terpenes and results in the formation of amber.

For this to happen, the resin must be resistant to decay. Many trees produce resin, but in the majority of cases, this deposit is broken down by physical and biological processes. Exposure to sunlight, rain, microorganisms (such as bacteria and fungi), and extreme temperatures tend to disintegrate resin. For resin to survive long enough to become amber, it must be resistant to such forces or be produced under conditions that exclude them.

Amber is heterogeneous in composition but consists of several resinous bodies more or less soluble in alcohol, ether, and chloroform, associated with an insoluble bituminous sub-

脂的化学组成，对于重建产生树脂种系发生学的类同性归类，用途不大。

琥珀有时候会含有动物或植物，被分泌的树脂包裹了，大约一亿三千万年前白垩纪的琥珀内可含昆虫、蜘蛛（可能含其网）、环节动物、青蛙、甲壳类、细菌和阿米巴、海洋微化石、木、花、果实、羽和毛，还有其他小生物。最古老包裹小动物化石的琥珀，来自意大利东部三叠纪喀尼阶 (Carnian)（二亿三千万年前）。

5.3.5 聚糖与几丁质

几丁质（壳质素）

（维基：*http://qdhys.ijournal.cn/hykx/ch/reader/create_pdf.aspx?file_no=19980512*）

几丁质 $(C_8H_{13}O_5N)_n$ 为长链状聚合物，是由约 8000 个葡萄糖的衍生物，*N*-乙酰葡糖胺作为单体聚合而成；几丁质是自然界的一种半透明而坚固的材料，

stance. Amber is a macromolecule by free radical polymerization of several precursors in the labdane family, e.g., communic acid, cummunol, and biformene. These labdanes are diterpenes ($C_{20}H_{32}$) and trienes, equipping the organic skeleton with three alkene groups for polymerization. As amber matures over the years, more polymerization takes place as well as isomerization reactions, crosslinking and cyclization. Heated above 200 °C (392 °F), amber suffers decomposition, yielding an oil of amber, and leaving a black residue which is known as "amber colophony," or "amber pitch; "when dissolved in oil of turpentine or linseed oil this forms "amber varnish" or "amber lac." Amber is a unique preservation mode, preserving otherwise unfossilizable parts of organisms; as such it is helpful in the reconstruction of ecosystems as well as organisms; the chemical composition of the resin, however, is of limited utility in reconstructing the phylogenetic affinity of the resin producer.

Amber sometimes contains animals or plant

▲ 图 5.3.4 激光荧光 (左 450 nm, 右 520 nm) 下看琥珀内的昆虫

▲ Figure 5.3.4 Looking the insect inside an amber under Laser Stimulated Fluorescence, left 450 nm, and right 520 nm

常见于真菌的细胞壁和节肢动物（如虾、龙虾、蟹）或昆虫的外骨骼，软体动物的齿舌及包括章鱼和墨鱼等头足类的喙和内壳，鱼类和滑体亚纲的鳞片和软组织；几丁质与属多醣的纤维素类似，都会构成纳米纤维或细毛状的晶体结构；在实际功能上，则近于构成皮肤的角蛋白，因为具有这些特性，几丁质在医学和工业上具有实用价值。

几丁质可能存在于寒武纪甲壳动物，如三叶虫外骨骼内，最古老保存的几丁质是二千五百万年前渐新世琥珀内的蝎子。鸟类羽毛与蝴蝶翅膀的鳞片上常有由几丁质构成的层状、柱状或三维的纳米晶体结构，能以透过薄膜干涉而产生虹彩光泽。

matter that became caught in the resin as it was secreted. Insects, spiders and even their webs, annelids, frogs, crustaceans, bacteria, and amoebae, marine microfossils, wood, flowers, and fruit, hair, feathers, and other small organisms have been recovered in Cretaceous ambers (deposited c. 130 million years ago). The oldest amber to bear fossils (mites) is from the Carnian (Triassic, 230 million years ago) of north-eastern Italy.

5.3.5 Polysaccharides and Chitin

Chitin

(Wiki, *http://qdhys.ijournal.cn/hykx/ch/reader/create_pdf.aspx?file_no=19980512)* Chitin $(C_8H_{13}O_5N)_n$ is a long-chain polymer of an *N*-acetylglucosamine, a derivative of glucose, and is found in many places throughout the natural world. It is a characteristic component of the cell walls of fungi, the exoskeletons of arthropods, such as crustaceans (e.g., crabs, lobsters, and shrimps) and insects, the radulae of molluscs, and the beaks and internal shells of cephalopods, including squid and octopuses and on the scales and other soft tissues of fish and *lissamphibians*. The structure of chitin is comparable to the polysaccharide cellulose, forming crystalline nanofibrils or whiskers. In terms of function, it may be compared to the protein keratin. Chitin

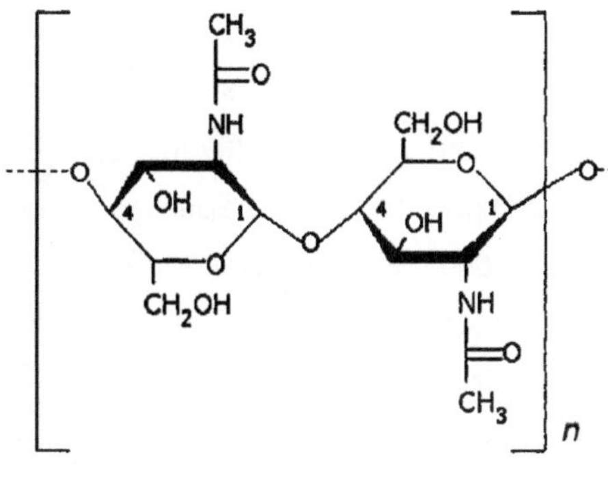

◀ 图 5.3.5　几丁质的分子结构：几丁质是由两个 *N*-乙酰葡糖胺作为单元，于 β-1, 4 位置重复串联而成

◀ Figure 5.3.5　Molecular structure of chitin; using two *N*-acetylglucos-amine as monomer and repeatedly polymerized at β-1,4 position

5.3.6 分子化石

(取自：*http://www.palaeocast.com/life-molecules-and-geology/*)

分子化石或"生物标志(biomarker)"，是能与某个特定生物源头关联而且包括核酸，亦即脱氧核糖核酸和核糖核酸(RNA)、蛋白质、碳水化合物和脂质。在这些当中，脂质最能抵抗降解，能够保存几百万年，因此可用来探讨生命演化和过去生态环境的动荡。固醇—24-isopropyl-cholestane—地质史上复杂动物生命的最早证据；高度分枝的异戊二烯—矽藻出现；*n*-烷(*n*-alkanes)—C_3和C_4光合作用—明确稳定碳同位素签章。

总和起来，这些研究展示使用分子化石，因为它们普遍存在和被保存的潜能，可用它们来重建不同地质时段的生命演化，成为现代古生物学者工具库内的基本要件。

鉴识科学技术协助发现新的分子化石

日本和中国的研究员相信，他们用鉴识科学常用的分析方法，找到了古菌(Archaea)新分子化石。依照微生物学家Carl Woese所设计的系统，在地球上有三种生命领域：细菌、古菌(Archaea)和真核生物(Eukaryota)；到今天，古菌的分布还不清楚，特别是超过二百万年前的地质时代，这是因为除了嗜盐生物(halophilic)、产甲烷生物(methanogenic)、厌甲烷生物等之外，很少发现古菌的分子化石，但是细菌和真核生物的分子化石却很多。

has proved versatile for several medicinal, industrial and biotechnological purposes.

Chitin was probably present in the exoskeletons of Cambrian arthropods such as trilobites. The oldest preserved chitin dates to the Oligocene, about 25 million years ago. It is a scorpion encased in amber. The bird feathers and the scales of butterfly usually have layers, column or 3D nanocrystal structure of chitin. They can produce iridescence by thin film interference.

5.3.6 Molecular fossils

From: *http://www.palaeocast.com/life-molecules-and-geology/*

Molecular fossils (or "biomarkers") are organic molecules which can be tied to a specific biological source and include nucleic acids (e.g., DNA and RNA), proteins, carbohydrates, and lipids. Of these, lipids are the most highly resistant to degradation and can be preserved over millions of years. As such, they can be used to investigate the evolution of life and past ecosystem dynamics. Sterols—24-isopropyl-cholestane—the earliest evidence for complex animal life in the geological record. Highly branched isoprenoids—the rise of diatoms. *n*-Alkanes—C_3 and C_4 photosynthesis—distinct stable carbon isotopic signatures.

Collectively, these studies showcase the utility of molecular fossils. Due to their ubiquity and preservation potential, they can be used to reconstruct the evolution of life over different geological timescales and have become an essential component of the modern paleontologist's toolkit.

Forensic science techniques help discover new molecular fossils

https://www.sciencedaily.com/releases/

研究分子时钟 (Molecular Clock) 指出，古菌出现于大约 38 亿年前，但是古菌分子化石的地质证据更直接结果，却指出只有二亿年（排除 2.5 亿和 27 亿年两个记录），在时间上这么大的差距，可能是因为古菌在早期地质年代只有很低的生物量或者是因为分子化石的易变性，导致分解。

日本东北大学 (Tohoku University) 的Ryosuke Saito博士和Kunio Kaiho教授的研究团队，从中国南方沉积岩样本，利用气相色谱-质谱 (GC-MS)分析法，发现古菌分子化石中的新化石。这是首次研究，以检测使用气相色谱-质谱于成岩产物，气相色谱-质谱是有机地球化学实验室广泛存在的常规仪器，气相色谱-质谱已经被法医鉴定物质和毒品检测、火灾调查、环境分析和炸药调查中视为"金标准"，它也被用来检测地球外的样本。

Saito的研究小组正在研究二百多万年前地球上古菌分布，他们认为通过利用气相色谱 - 质谱法的化石分析，会更容易找到在沉积岩内的古菌。古菌分子化石（特别是产甲烷古菌和厌甲烷古菌）可以存放在超过 2 亿年的冷泉，因为古菌在这样的环境中，保留的生物量庞大。在诸古菌中，较早期的嗜盐古菌分子化石往往被保存在样本中，因为它们的高稳定性。

2017/08/170831091431.htm

Researchers in Japan and China believe they have found new molecular fossils of archaea using a method of analysis commonly used in forensic science. According to a system designed by microbiologist Carl Woese, there are three domains of life on Earth -- Bacteria, Archaea, and Eukaryota. To date, the distribution of archaea remains unclear especially, for geologic periods dating back more than 2 million years. This is because except for halophilic, methanogenic and methanotrophic archaea, molecular fossils of archaea are rarely found, while those of bacteria and eukaryote are commonly found.

Studies of molecular clock suggest that archaea appeared around 3.8 billion years ago, while the results of more direct geological evidence from molecular fossils of archaea indicate a timeline of 0.2 billion years (except for two records of 0.25 and approximately 2.7 billion years). The reason for the big difference in timeframes could be due to the low biomass of archaea in earlier geologic periods, or to the lability of the molecular fossils, leading to decomposition.

A team led by Dr. Ryosuke Saito and Professor Kunio Kaiho of Tohoku University took sedimentary rock samples from southern China and analyzed the organic molecules in them using gas chromatography-mass spectrometry (GC-MS). They found new fossils among the molecular fossils of archaea. This is the first study to detect diagenetic products using GC-MS, a conventional instrument found widely in organic geochemistry laboratories. GC-MS has been regarded as a "gold standard" for forensic substance identification and is used in drug detection, fire investigation, environmental analy-

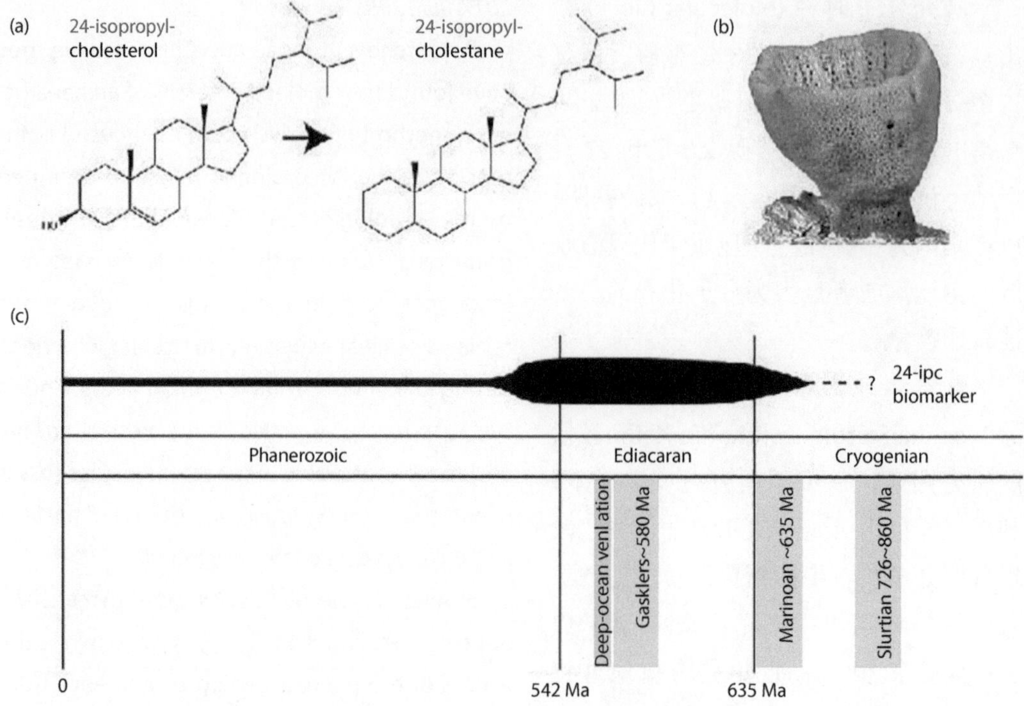

▲ 图 5.3.6 地质记录中多分支的异戊二烯：（a）C25:2 HBI 分解为 C25 HBI；（b）现生硅藻 *Haslea trompii*；（c）在长时间沉积（左）和石油（右）相对丰度的 C25 HBI 群

▲ Figure 5.3.6 Highly branched isoprenoids in the geological record: (a) degradation of C25:2 HBI to C25 HBI, (b) the modern diatom (*Haslea trompii*), (c) relative abundance of C25 HBIs within sediments (left) and petroleum (right) over time (redrawn from Sinninghe-Damsté 2004)

5.3.7 蛋白质分类

蛋白质是生物化学界的一大项目，组成生物体的基本物质，“蛋白质”、“多肽”和“肽”这些名词的含义在一定程度上有重叠，经常容易混淆。(维基百科)蛋白质（旧称“朊”）是大型生物分子或高分子，它由一个或多个由氨基酸残基组成的长链条组成。氨基酸分子呈线性排列，相邻氨基酸残基的羧基和氨基

sis, and explosives investigation. It is also being used to identify extra-terrestrial samples.

Saito's team was now studying the archaeal distribution on Earth more than 2 million years ago. They believe that by analyzing the fossils using the GC-MS method, it will be easier to find archaea in sedimentary rocks. Archaeal molecular fossils (especially methanogenic archaea and methanotrophic archaea) can be preserved in samples deposited in cold seep even older than 200 million years because the biomass of

通过肽键连接在一起。蛋白质的氨基酸序列是由对应基因所编码，除了遗传密码所编码的20种“标准”氨基酸，在蛋白质中，某些氨基酸残基还可以被改变原子的排序而发生化学结构的变化，从而对蛋白质进行激活或调控。多个蛋白质可以一起，往往是通过结合在一起形成稳定的蛋白质复合物，发挥某一特定功能。

“蛋白质”通常指具有完整生物学功能并有稳定结构的分子；而“肽”则通常指一段较短的氨基酸寡聚体，常常没有稳定的三维结构；然而，“蛋白质”和“肽”之间的界限很模糊，通常以20~30个残基为界；“多肽”可以指任何长度的氨基酸线性单链分子，但常常表示缺少稳定的三级结构。

线性氨基酸链称之为多肽(polypeptide)，蛋白质含有至少一个长的多肽；短多肽，含有20~30残基(residues)，很少被认为是蛋白质，通常称之为多肽，有时也叫寡肽 (oligopeptides)；个别的氨基酸残基，经由多肽键邻近的氨基酸残基键合在一起，蛋白质内的氨基酸残基顺序，由基因码的某基因定义；然而，在某些生物，基因码可能包含硒半胱氨酸 (seleno-cysteine)，在某些古菌，则含吡咯赖氨酸 (pyrrolysine)；在合成之后，甚至过程当中，蛋白质内的残基，经常被转译后修饰 (post-translational modification) 化学修改，改变物理和化学性质、折叠、稳定性、活性和最终的蛋白质功能。有时候

archaea in such environments is vast. Among archaea, molecular fossils of halophilic archaea are often preserved in samples from earlier periods because of their high stability.

(Ryosuke Saito, Kunio Kaiho, Masahiro Oba, Jinnan Tong, Zhong-Qiang Chen, Li Tian, Satoshi Takahashi, Megumu Fujibayashi. Tentative identification of diagenetic products of cyclic biphytanes in sedimentary rocks from the uppermost Permian and Lower Triassic. Organic Geochemistry, 2017; 111: 144 DOI: 10.1016/j.orggeochem.2017.04.013)

5.3.7 Protein classification

Protein is one of the most significant topics in biochemistry and is the essential matter for organisms. The words protein, polypeptide, and peptide are a little ambiguous and can overlap in meaning. (From WiKi) Proteins (/ˈproʊˌtiːnz/ or /ˈproʊti.ɪnz/) are large biomolecules, or macromolecules, consisting of one or more long chains of amino acid residues. Proteins perform a vast array of functions within organisms, including catalyzing metabolic reactions, DNA replication, responding to stimuli, and transporting molecules from one location to another. Proteins differ from one another primarily in their sequence of amino acids, which is dictated by the nucleotide sequence of their genes, and which usually results in protein folding into a specific three-dimensional structure that determines its activity.

Protein is generally used to refer to the complete biological molecule in a stable conformation, whereas peptide is generally reserved for a short amino acid oligomers often lacking a stable three-dimensional structure. However, the boundary between the two is not well defined and usually lies near 20–30 residues. The Polypeptide

蛋白质有非多肽基连接，称之为辅成基(prosthetic groups) 或辅因子 (cofactors)，几个蛋白质能合作达成某个功能，通常形成稳定的蛋白质复合体 (protein complexes)。

与其他生物大分子（如多糖和核酸）一样，蛋白质是地球上生物体中的必要组成成分，参与了细胞生命活动的每一个进程。酶是最常见的一类蛋白质，它们催化生物化学反应，尤其对于生物体的代谢至关重要；除了酶之外，还有许多结构性或机械性蛋白质，如肌肉中的肌动蛋白和肌球蛋白，以及细胞骨架中的微管蛋白（参与形成细胞内的支撑网络以维持细胞外形）；另外一些蛋白质，则参与细胞信号传导、免疫反应、细胞黏附和细胞周期调控等；同时，蛋白质也是动物饮食中必需的营养物质，这是因为动物自身无法合成所有氨基酸，动物需要和必须从食物中获取必需氨基酸；通过消化过程将蛋白质降解为自由氨基酸，动物就可以将它们用于自身的代谢。

由于氨基酸的非对称性（两端分别具有氨基和羧基），蛋白质链具有方向性；多肽键有两种共振形式，导致一些双键特性，防止沿着轴转动，因此阿尔法碳大约同平面；另外两个多肽键二面角则依蛋白质主干决定其在地形状。蛋白质链的起始端有自由的氨基，被称为 N 端或氨基端；尾端则有自由的羧基，被称为 C 端或羧基端，蛋白质的顺序，从 N 端由左到右写到 C 端。大多数的

can refer to any single linear chain of amino acids, usually regardless of length, but often implies an absence of a defined conformation.

A linear chain of amino acid residues is called a polypeptide. A protein contains at least one long polypeptide. Short polypeptides, containing less than 20–30 residues, are rarely considered to be proteins and are commonly called peptides, or sometimes oligopeptides. The individual amino acid residues are bonded together by peptide bonds and adjacent amino acid residues. The sequence of amino acid residues in a protein is defined by the sequence of a gene, which is encoded in the genetic code. In general, the genetic code specifies twenty standard amino acids; however, in certain organisms, the genetic code can include selenocysteine and—in certain archaea—pyrrolysine. Shortly after or even during synthesis, the residues in a protein are often chemically modified by post-translational modification, which alters the physical and chemical properties, folding, stability, activity, and ultimately, the function of the proteins. Sometimes proteins have non-peptide groups attached, which can be called prosthetic groups or cofactors. Proteins can also work together to achieve a particular function, and they often associate to form stable protein complexes.

Like other biological macromolecules such as polysaccharides and nucleic acids, proteins are essential parts of organisms and participate in virtually every process within cells. Many proteins are enzymes that catalyze biochemical reactions and are vital to metabolism. Proteins also have structural or mechanical functions, such as actin and myosin in muscle and the proteins in the cytoskeleton, which form a system

蛋白质都自然折叠为一个特定的三维结构，这一特定结构被称为天然状态。虽然多数蛋白可以通过本身氨基酸序列的性质进行自我折叠，但还是有许多蛋白质需要分子伴侣的帮助来进行正确的折叠，在高温或极端 pH 等条件下，蛋白质会失去其天然结构和活性，这一现象就称为变性。生物化学家常常用以下四个方面来表示蛋白质的结构：

一级结构：组成蛋白质多肽链的线性氨基酸序列；一个蛋白质是一个聚酰胺。

二级结构：依靠不同氨基酸之间的 C═O 和 N—H 基团间的氢键形成的稳定结构，主要为 α 螺旋和 β 层叠；因为二级结构是局部的，不同的二级结构的许多区域可存在于相同的蛋白质分子。

三级结构：通过多个二级结构元素在三维空间的排列所形成的一个蛋白质分子的三维结构，是单个蛋白质分子的整体形状；蛋白质的三级结构大都有一个疏水核心来稳定结构，同时具有稳定作用的还有盐桥（蛋白质）、氢键、边链紧密和二硫键；“三级结构”常常可以用“折叠”一词来表示；三级结构控制蛋白质的基本功能。

四级结构：由几个蛋白质分子（多肽链），通常称为蛋白质亚基所形成的结构，在功能上作为一个蛋白质复合体。

因为蛋白质实际上是非常复杂的东西，学术上的分类，也是百家齐鸣，分类的依据大致上有：生物性功能 (biological function)、溶解度 (solubility)、

of scaffolding that maintains cell shape. Other proteins are important in cell signaling, immune responses, cell adhesion, and the cell cycle. In animals, proteins are needed in the diet to provide the essential amino acids that cannot be synthesized. Digestion breaks the proteins down for use in the metabolism.

Due to the asymmetry of amino acids (two terminals of amine and hydroxy group), protein chain has direction. The peptide bond has two resonance forms that contribute some double-bond character and inhibit rotation around its axis so that the alpha carbons are roughly coplanar. The other two dihedral angles in the peptide bond determine the local shape assumed by the protein backbone. The end with a free amino group is known as the N-terminus or amino-terminus, whereas the end of the protein with a free carboxyl group is known as the C-terminus or carboxyl terminus (the sequence of the protein is written from N-terminus to C-terminus, from left to right). Most of the proteins are folded into a specific three-dimensional structure. This special structure is called the nature state. Most of the proteins can fold themselves by the amino acid sequence. However, some proteins need the molecular company to form correct folding. Under high temperature or extreme pH condition, proteins will lose the nature state structures and activities. This is called denaturalization. Scientists usually use the following four to express protein structures.

Primary structure: A linear amino acid sequence that makes up a protein polypeptide chain; a protein is a polyamide.

Secondary structure: Rely on the hydrogen bond between C=O and N-H of different amino

化学组成 (chemical composition) 和形态 (shape)；就以形态分类来说，还可继续分类为：球状蛋白 (globular)、丝蛋白 (fibroin)、角蛋白 (keratin)、弹性蛋白 (elastin)、纤维蛋白 (fiberous)；在纤维蛋白这一小分类中，胶原蛋白质只是其中的一种，胶原蛋白出现在身体很多地方，然而在人类身体超过 90% 是第一类胶原蛋白。到目前为止，有 28 类胶原蛋白已经被描述过，它们可依照形成的构造分成几个族群：

纤维蛋白；

非纤维蛋白；

短链；

基底膜；

其他。

最常见的五类：第一类，皮肤、腱、血管结扎、器官、骨头（骨头内主要有机的成分）；第二类，软骨（软骨胶质的主要成分）；第三类，网状（网状纤维的主要成分），通常和第一类并存；第四类，形成基底膜，基底膜的上皮组织分泌层；第五类，细胞表面，头发和胎盘。

从这简单的进阶分类可以看出来，第一类胶原蛋白 (Collagen Type I) 只不过是无数蛋白质当中的一小分类。

5.4 大问题：有机残留物在生物与演化的重要性

作为这一章的总结，回到古生物化学有机残留物的最终极目标。我们花了

acid to form a stable structure. Mainly α-helix and β-sheet. Since the secondary structure is local, many areas of different secondary structures can exist in the same protein molecule.

Tertiary structure: The three-dimensional structure is formed from the secondary structural elements. It is the overall shape of a single protein molecule. Usually, the tertiary structure of a protein has a hydrophobic kernel to stabilize the structure. Other interactions are salt bridges, hydrogen bonds, and the tight packing of side chains. Tertiary structures usually can be expressed as 'folding.' It controls the basic protein function.

Quaternary structure: The three-dimensional structure consisting of the aggregation of two or more individual polypeptide chains (subunits) that operate as a single functional unit.

Since proteins are a very complicated matter, there are so many academic classifications, base on biological function, solubility, chemical composition, and shape. The shape classification can be further classified as globular, fibroin, α-keratin, elastin, and fibrous. In the subcategory of fibrous, collagen is just one of many. Collagens are in many places of the human body. However, more than 90% are collagen type I. Up to now, there are 28 types of collagen had been described, and can be further classified according to the forming structure into several groups.

FACIT (Fibril Associated Collagens with Interrupted Triple Helices) (Type IX, XII, XIV, XVI, XIX);

Short chain (Type VIII, X);

Basement membrane (Type IV);

Multiplexin (Multiple Triple Helix domains with Interruptions) (Type XV, XVIII);

MACIT (Membrane Associated Collagens with Interrupted Triple Helices) (Type XIII, XVII);

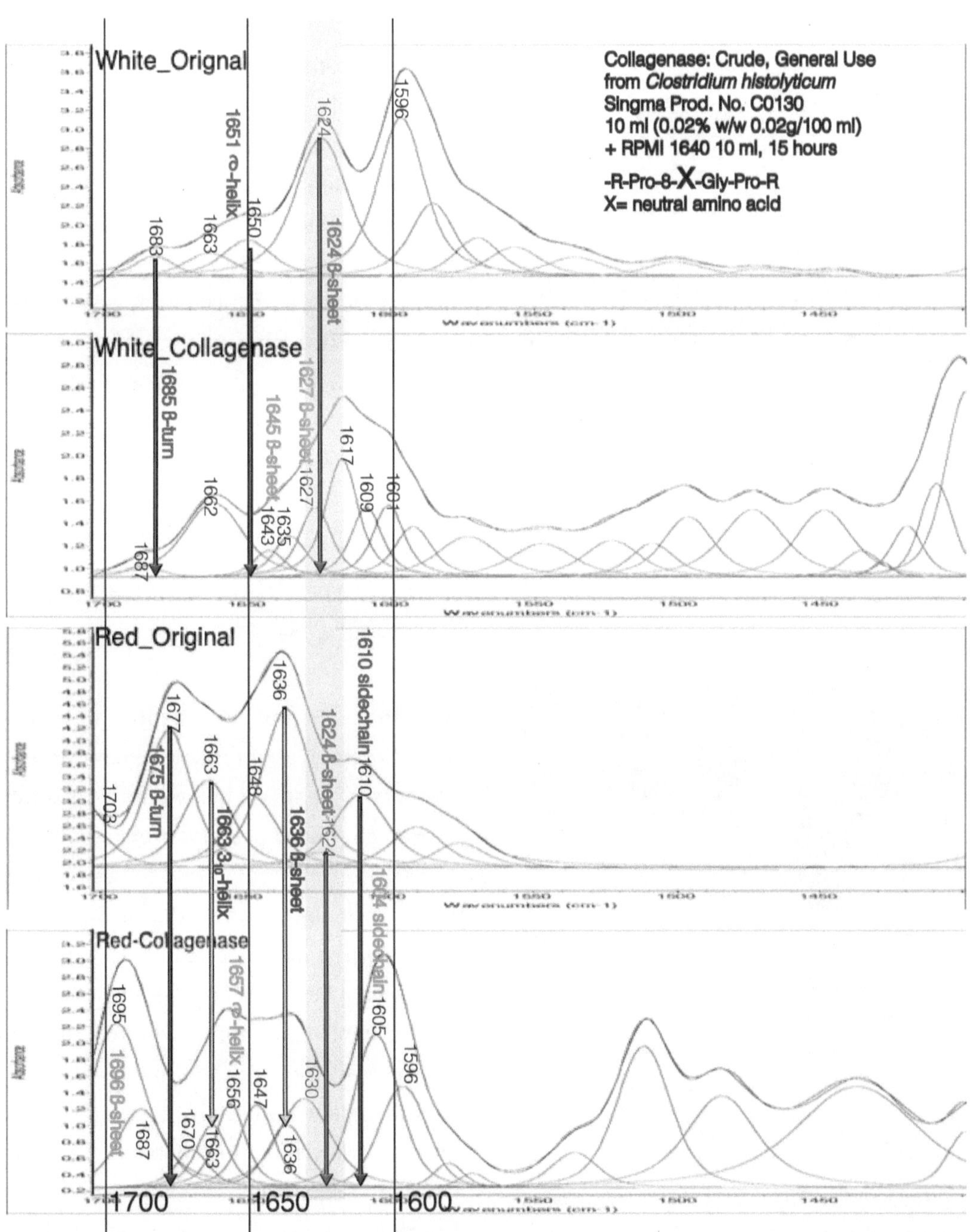

▲ 图 5.3.7 胶原蛋白消化酶处理禄丰龙胚胎骨头红白区域前后红外光谱比较

▲ Figure 5.3.7 Comparison of IR spectra of before and after collagenase treatment of *Lufengosaurus* embryonic bones

很大的代价，得知在某化石内或邻近围岩内，保存了某残留有机物，该做的化学分析都做了，化学式也都有了，甚至保存的机制也都找出来了，很可能也可以发表一篇论文，当然，鉴定出是什么有机残留物和得知保存机制，在科学上有其非常重要的分量，特别是保存机制，对于整个人类的贡献和影响，非常巨大，无法预估。

然而，完成了以上这些，还是尚未达标，没有回答出古生物化学里面最最最重要的问题：这些有机残留物对于生

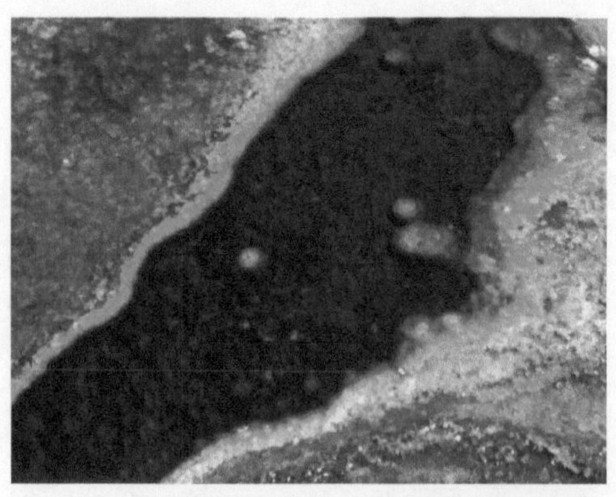

▲ 图 5.4.1　在黄石国家公园高温喷泉水中的颜色，是不同耐温藻类所造成的各种颜色，甚至可以从颜色判断是哪种藻类和水温度。地球早期生命的生活环境可能与此类似

▲ Figure 5.4.1　The colors of high-temperature geyser in National Yellowstone Park are from various high temperature bearing algae and cyanobacteria for the varied colors. From the color, it even can be used to identify what kind of algae and cyanobacteria and water temperature. The Earth early environment for early life could be similar

Other (Type VI, VII).

The most common five types: Type I: skin, tendon, vascular ligation, organ, bone (main organic component of bone); Type II: cartilage (main component of cartilage); Type III: Mesh (main component of mesh fiber), usually co-exist with Type I; Type IV: Forming basement membrane, Epithelial/mesothelial/endothelial tissue: Type V: Cell surface, hair and placenta.

From such a simple further classification, it can be seen that collagen type I is just a small category of many proteins.

5.4 The Big Question: The Importance of Organic Remains in Biology and Evolution

As a conclusion of this chapter and back to the ultimate goal of organic remains in PaleoChemistry, we spend many efforts to know the organic remains preserved inside a fossil or immediate surrounding matrix, we got the chemical formula and even the preservation mechanism. The achieved results can be and shall be well published. Of course, identification of what organic compounds were preserved, and the preservation mechanism are very crucial in science, and particularly the preservation mechanism will yield un-estimable impact and contribution to humanity.

However, even all of the above was finished, there is still one primary shortage. Does it not answer the most critical question of PaleoChemistry: What is the relationship between the preserved organic remains, biology and evolution? For example, both were living in the ocean;

物学和生物演化，有什么关联？比方说，同样都是活在海洋里面，埃迪卡拉纪罕见有软壳或硬壳的生物，可是到了较晚的寒武纪大爆发，几丁质的外骨骼和贝壳碳酸盐硬壳都出现了，这到底怎么一回事？相对来说，埃迪卡拉纪的生物，体形较大，大的可以长到二米，十几二十厘米者很普遍，可是，大部分寒武纪大爆发生物体形都在十厘米以内，很少如奇虾六十厘米者，为什么会有如此的“老倒缩”？达尔文进化假说，如何解释？海水成分改变了生物体内的化学组成？三叶虫的眼睛促成了生存武器竞争？地球暖化？

因此，在此要提醒读者的是，不要赢得战斗，却输掉了战争；我们结合了古生物学和化学成为古生物化学，其主

there were scarce soft- or hard shell animals in the Ediacaran Period. However, at the later time in the Cambrian Explosion, soft chitinous shell and carbonate hard clamshell showed up. What is going on here? In relative terms, the body sizes of Ediacaran creatures are more significant, and some could reach two meters while ten or twenty cm were very common. However, the majority of Cambrian Explosion creature body sizes are less than ten cm and very few big ones such as 60+cm *Anomalocaris*. Why the 'old-shrinkage' of the body took place? How could Darwin's Evolution hypotheses be able to explain this? Chemical composition changes in ocean water, causing the chemical composition changes in organisms? The trilobite eyes started the biological arms race? Global warming?

Therefore, as a reminder to readers, do not win the battle, but lost the war. We combined

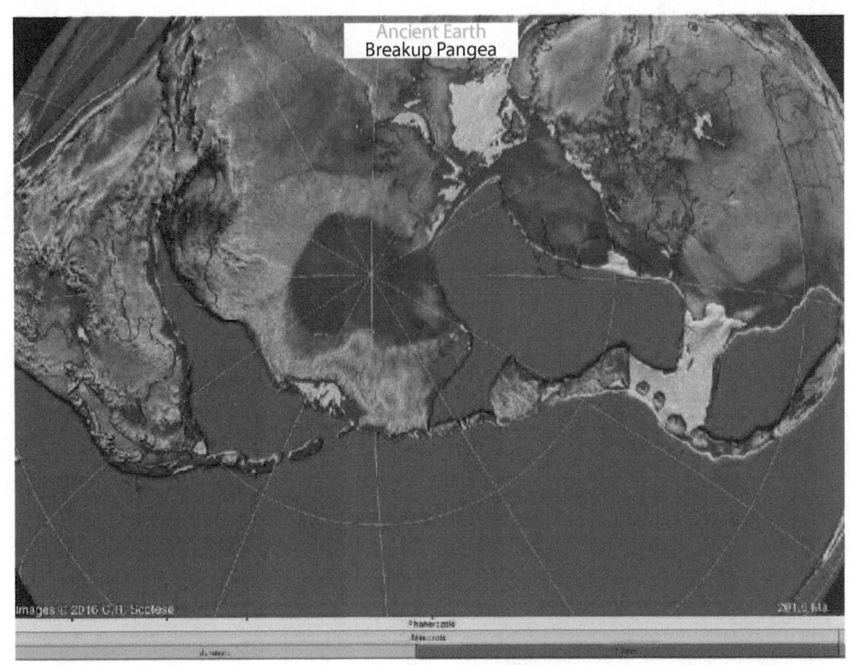

▶ 图 5.4.2　二亿年前（早侏罗纪）的一块相连接地球板块，华北和华南尚未合并，中间有个海洋

▶ Figure 5.4.2　One connected continent of Earth around 200 million years ago (Early Jurassic). A sea separated northern and Southern China

要的一个目标，就是提供更多这个地球的生物学和演化学的资料证据，让我们更认识所脚踏“石”地的这颗蓝色星球来龙去脉。

the paleontology and chemistry into PaleoChemistry. One of the main goals is to provide more information for the biology and evolution information of this Earth so that we can understand more this 'rock-solid' blue planet we are standing on.

第 6 章　无机物

Chapter 6　Inorganics

这一章介绍古生物化学除了探讨有机残留物之外，另外也是很重要的无机化学部分，本章介绍化石里面的几种无机物。

This chapter introduces the other important inorganic side of PaleoChemistry besides the organic remains - the inorganic inside the fossil.

6.1　磷灰石群

6.1　Apatites

骨头不只是羟磷灰石而已

Bones are NOT hydroxyapatite along

一般常认为骨头的矿物为钙羟磷灰石 (calcium hydroxy-apatite, HA)，化学式 $Ca_{10}(OH)_2(PO_4)_6$，最早由W. F. DeJong提出。然而羟磷灰石晶格，六方晶系 $P6_3/m$，可容纳多种额外离子，依序放出Ca^{2+}、OH^-或 PO_4^{3-}离子，而维持电平衡，特别是碳酸根离子CO_3^{2-}可取代氢氧 OH^- 离子或正磷酸离子，成为碳酸磷灰石 (carbonatoapatite)A或B。生物性磷灰石不是纯的羟磷灰石，也不是计量 (stoichiometric)，因此，应该归类为碳酸磷灰石群。碳酸根是骨头矿物首要的次成分，估计5%~8%（质量分数）。以

It is often taken for granted that bone mineral is identical with calcium hydroxyapatite (HA), $Ca_{10}(OH)_2(PO_4)_6$ as first proposed by WF DeJong. However, the HA crystal lattice (hexagonal space group $P6_3/m$) can accommodate various extra ions, releasing in turn Ca^{2+}, OH^- or PO_4^{3-} ions, thus maintaining the electric charge balance. In particular, carbonate ions CO_3^{2-} can replace hydroxyl ions OH^- or orthophosphate ions PO_4^{3-}, leading to carbonatoapatites of type A and B, respectively. Biological apatites are not pure HA and stoichiometric, and should instead be classified as carbonatoapatites. Carbonate CO_3^{2-} is the principal minor constituent of the bone mineral, estimated at 5–8 wt%. Comparative infrared studies of synthetic carbonatoapatites

红外光谱比较合成的碳酸磷灰石与生物性磷灰石，指出生物性磷灰石为B类取代。

已经证明了骨头内的磷灰石缺乏羟基，Cho等估计在骨头矿物内，结构性的羟基化学计量为21% ± 1%（略）。从此角度来说，碳酸磷灰石B最接近骨头矿物（略），这和我们过去的核磁共振研究，指出骨头矿物的化学构造最接近于碳酸磷灰石B。

磷灰石家族矿物群（$A_5(XO_4)_3Z_q$），磷灰石是一群磷酸盐矿物，最普通的四个终端成员，写成 $Ca_{10}(PO_4)_6(OH, F, Cl)_2$，可用通式 $A_5(XO_4)_3Z_q$表达。磷灰石群是牙齿珐琅质和骨头的主要成分，其中大部分缺失羟基，而含很多碳酸根和磷酸盐取代是骨头矿物的主要成分。

（维基百科）计量与非计量

计量比

计量比这个词，通常用于计量比化合物（计量比组成）的元素摩尔比，在 H_2O 内的氢和氧计量比是 2 ∶ 1；在计量比化合物，摩尔比都是整数。

非计量比

非计量比化合物，是那些无法用整数表达元素部分比的化合物，它们几乎都是无机物，几乎都是固体，在这些物质里面，缺失小比例的原子或者有太多原子挤在某完美晶格内。因为这些缺失瑕玼，它们展现特别的电性或化学性质，比方说，若缺失了一些原子，其他的原子在固体内就可以移动通过。很多金属氧化物、硫化物是典型的非计量比，如罕见的计量比氧化铁(iron II)，其化学式为 FeO，而更常见非计量比者的化学式为 $Fe_{0.95}O$；非计量比化合

and biological apatites indicate type B substitution in the latter materials.

It has been already established that bone apatite is deficient in hydroxyl groups. Cho et al. estimated the percentage of structural hydroxyl groups in the bone mineral as 21±1% of the stoichiometric content. … In this respect, CHA-B is most similar to the bone mineral. … This agrees with our previous NMR studies, indicating that the bone mineral is closest in chemical structure to carbonatoapatite of type B.

Efficiency of ^1H-^{31}P NMR cross-polarization in bone apatite and its mineral standards, Solid State Nuclear Magnetic Resonance, 2006 by Agnieszka Kaflak, et al. Anna Slosarczykc, Waclaw Kolodziejski

Apatite Group Minerals, $A_5(XO_4)_3Z_q$, Apatite is a group of phosphate minerals, the four most common endmembers are written as $Ca_{10}(PO_4)_6(OH, F, Cl)_2$, and can be expressed as the general formula,$A_5(XO_4)_3Z_q$. Apatites are the primary component of tooth enamel and bone mineral, in which most of the OH^- groups are absent and containing many carbonates and phosphate substitutions is a significant component of bone material.

Stoichiometric

(from WiKi) Stoichiometric and Nonstoichiometric

The term stoichiometry is often used for the molar proportions of elements in stoichiometric compounds (composition stoichiometry). For example, the stoichiometry of hydrogen and oxygen in H_2O is 2:1. In stoichiometric compounds, the molar proportions are whole numbers.

Nonstoichiometric

Nonstoichiometric compounds are chemical compounds with an elemental composition

物是均匀的而非混合物；蓄电池依赖能以某范围非计量比状态的材料，其他非计量比化合物应用于陶瓷和超导；因为固体整个来说是电中性，离子化合物内的缺失，用固体内其他原子电荷来补偿或是改变其氧化态，或是以不同电价的不同原子取代之。

长期以来人们一直认为骨矿物晶体组成和结构与合成标准计量比的羟基磷灰石(HAP)相似。然而其晶体结构和化学组成不同于纯HAP，骨矿物质是纳米晶体，晶体结构不完善，组成为非化学计量比。它具有特殊的表面，包含不稳定的化学基团，由于其结构中沿六方轴存在一个“隧道”，进入的离子可以充填或置换羟基磷灰石晶体缺陷部位。这些特点使骨矿物晶体的物理、化学、结晶特点乃至生物个体和组织之间的生理环境变得极其复杂。

生物磷灰石相的确切组成结构是什么？碳酸根离子在生物磷灰石中进行具体的替换与占位是怎样？又是怎么维持电荷平衡？这些问题的答案，能够更好地理解生物磷灰石中结合的或被利用的OH^-作用机理；也能更好地解释OH^-与CO_3^{2-}含量的不同是怎样影响天然骨、牙齿和牙釉质中矿物质本身的结晶度、生物相容性、化学活性及力学性能等。

非计量化碳酸磷灰石群

骨头矿物主要成分不是羟磷灰石，而是大约 65% 碳酸磷灰石群、25% 有机成分（主要是第一类胶原蛋白），其余 10% 为水分（质量分数）。非计量化生物性磷

that cannot be represented by usual integral numbers. They are almost always inorganic compounds and almost invariably solids. In such materials, some small percentage of atoms are missing, or too many atoms are packed into an otherwise perfect lattice work. They exhibit special electrical or chemical properties because of these flaws or defects. For example, when atoms are missing, the other atoms can move through the solid more rapidly. Many metal oxides and sulfides represent nonstoichiometry. For example, stoichiometric iron(II) oxide, which is rare, has the formula FeO, whereas the more common nonstoichiometric material has the formula $Fe_{0.95}O$. Nonstoichiometric compounds are homogenous; they are not mixtures. Batteries often rely on materials that can exist in a range of nonstoichiometric states. Other nonstoichiometric compounds have applications in ceramics and as superconductors. Since solids are overall electrically neutral, the defect in an ionic compound is compensated by a change in charge of other atoms in the solid, either by changing their oxidation state or by replacing them with atoms of different elements with a different charge.

For a very long time, it is thought that the bone mineral composition and structure is similar to the stoichiometric synthetic hydroxyapatite (HAP). However, their crystal structure and chemical composition are different from pure HAP. Bone mineral is imperfect nanocrystal structure, and composition is nonstoichiometry. It has a unique surface, including unstable chemical functional group. There is a 'tunnel' along the hexagonal axis. The ions entered can fill or replace defect parts of hydroxyapatite. These features made the physical, chemical,

▲ 图 6.1.1 磷灰石矿物群

▲ Figure 6.1.1 Apatite Mineral Group

灰石群含好几个离子取代，比方说 Na^+ 和 Mg^{2+} 可取代 Ca^{2+} 离子，HPO_4^{2-} 离子可取代磷酸离子，Cl^- 和 F^- 取代 OH^-；再者最多取代者的碳酸离子（3%~8%），在晶格内可占有 OH^-（磷灰石 A）或 PO_4^{3-}（磷灰石 B）位置，骨头的矿物成分通常比较接近于磷灰石 B。

本节的小总结，虽然活体和化石骨头的磷灰石群是那么重要，但是到现在对于骨头内的磷灰石群矿物，还有很多没有搞清楚的奥秘，有待古生物化学研究者来厘清！不过有一点可以明确的，骨头里面的无机矿物，不管是现生或化石里面的，绝对不是只有羟磷灰石那么简单，因此，从现在开始不要继续错误下去说是羟磷灰石！

crystallography, and even causing the physiological environment between the biological individual and tissue to become very complicated.

What is the exact compositional structure of biological apatite? How the carbonate ions in the biological apatite conduct the actual replacement, and where? How is the electrical balance maintained? The answers to these questions can help to understand the mechanism of the combination or used OH^- in biological apatite, and for the explanation of the different amount of OH^- and CO_3^{2-} affect the crystallography degree, biocompatibility, chemical activity, and biomechanics of the natural bones, teeth, and enamel of the teeth.

Nonstoichiometric Carbonatoapatites

The majority of the bone mineral is NOT HO-apatite, but "about 65 wt % mineral component (carbonated hydroxyapatite) and 25 wt % organic component (mainly type I collagen), being the remaining ten wt % of water. (omitted) The nonstoichiometric biological apatites contain several ion substitutions. For example, Na^+ and Mg^{2+} may substitute Ca^{2+} ions, HPO_4^{2-} ions may substitute phosphate ions, and Cl^- and F^- may replace OH^-. Additionally, carbonate ions, the most abundant substitutions (3–8 wt. %), may occupy either the OH^- (type A apatite) or PO_4^{3-} (type B apatite) positions in the crystal lattice. The mineral component of bone is usually closer to B-type apatite."

To summarize this section. Although the apatites in the living and fossilized bones are so important, there are still so many unknowns about the apatites inside the bones and waiting for paleochemists to resolve the issue. However, one thing for sure, the inorganic minerals inside

已知可能生物性磷灰石群（排除 AsO_4, VO_4, CrO_4）
Known possible bio-Apatites excluding AsO_4, VO_4, CrO_4

Alforsite	$Ba_5(PO_4)_3Cl$	(Hexagonal dipyramidal, (6/m), P 63/m)
Carbonate-rich Fluorapatite	$Ca_5(PO_4,CO_3)_3(F,O)$	(Hexagonal)
Chlorapatite	$Ca_5(PO_4)_3Cl$	(Hexagonal dipyramidal (6/m))
Fluorapatite	$Ca_2Ca_3(PO_4)_3F$	(Hexagonal - dipyramidal)
Hydroxylapatite	$Ca_5(PO_4)_3(OH)$	(Hexagonal 6/m – dipyramidal)
Hydroxylapatite-M	$Ca_5(PO_4)_3OH$	(Hexagonal 6/m – dipyramidal)
Hydroxylpyromorphite	$Pb_5(PO_4)_3(OH)$	(Hexagonal - dipyramidal)
Manganapatite	Fluorapatite containing divalent Mn was called maganapatite	
Miyahisaite	$(Sr,Ca)_2Ba_3(PO_4)_3F$	(Hexagonal, 6/m - Dipyramidal, P 63/m)
Mn-bearing Fluorapatite	$(Ca,Mn_2+)_5(PO_4)_3(F,Cl,OH)$ or $Ca_5([P,Mn_5+]O_4)_3(F,Cl,OH)$	
Oxypyromorphite	$Pb_{10}(PO_4)_6O$	(Hexagonal, 6/m - Dipyramidal, P 63/m)
Pieczkaite	$Mn_5(PO_4)_3Cl$	(Hexagonal, 6/m - Dipyramidal, P 63/m)
Pyromorphite	$Pb_5(PO_4)_3Cl$	(Hexagonal, 6/m - Dipyramidal, P 63/m)
Stronadelphite	$Sr_5(PO_4)_3F$	(Hexagonal, 6/m - Dipyramidal, P 63/m)
Unnamed (F-analogue of Pyromorphite)	$Pb_5(PO_4)_3F$	
Belovite-(Ce)	$NaCeSr_3(PO_4)_3F$	(Trigonal - Pyramidal, H-M Symbol (3) Space Group: P3)
Belovite-(La)	$NaLaSr_3(PO_4)_3F$	(Trigonal, Class (H-M):3 - Rhombohedral, Space Group:P3)
Carlgieseckeite-(Nd)	$NaNdCa_3(PO_4)_3F$	(Trigonal, Class (H-M):3 - Rhombohedral, Space Group:P3)
Deloneite	$(Na_{0.5}REE_{0.25}Ca_{0.25})(Ca_{0.75}REE_{0.25})Sr_{1.5}(CaNa_{0.25}REE_{0.25})(PO_4)_3F_{0.5}(OH)_{0.5}$ (Trigonal)	
Fluorcaphite	$SrCaCa_3(PO_4)_3F$	(Hexagonal to Pyramidal)
Fluorstrophite	$SrCaSr_3(PO_4)_3F$	(Hexagonal, 6/m - Dipyramidal, P 63/m)
Kuannersuite-(Ce)	$NaCeBa_3(PO_4)_3F_{0.5}Cl_{0.5}$	(Trigonal - RhombohedralH-M Symbol (3) Space Group: P3)
Britholite-(Ce)	$Ca_2(Ce,Ca)_3(SiO_4,PO_4)_3(OH,F)$	
Britholite-(La)	$Ca_2(La,Ce,Ca)_3(SiO_4,PO_4)_3(OH,F)$	(Hexagonal, 6/m - Dipyramidal, P 63/m)
Britholite-(Y)	$Ca_2(Y,Ca)_3(SiO_4,PO_4)_3(OH,F)$	
Fluorbritholite-(Ce)	$Ca_2(Ce,Ca)_3(SiO_4,PO_4)_3(F,OH)$	
Fluorbritholite-(Y)	$Ca_2(Y,Ca,Ln)_3(SiO_4,PO_4)_3(F,OH)$	(Hexagonal, 6/m - Dipyramidal, P 63/m)
Fluorcalciobritholite	$(Ca,Ce)_5(SiO_4,PO_4)_3F$	(Hexagonal, 6/m - Dipyramidal, P 63/m)
Fluorphosphohedyphane	$Ca_2Pb_3(PO_4)_3F$	(Hexagonal prisms with pyramidal terminations)
Phosphohedyphane	$Ca_2Pb_3(PO_4)_3Cl$	(Hexagonal, 6/m - Dipyramidal, P 63/m)

指定六方晶系是因为大部分磷灰石的六方对称，然而在氯磷灰石和羟磷灰石可能的氯原子或羟自由基可导致单斜对称

The hexagonal crystal system designation is due to a hexagonal symmetry of most Apatite. However, positional ordering of chlorine atoms or the hydroxyl radical in Chlorapatite and Hydroxylapatite can lead to monoclinic symmetry

6.2 稀土元素

镧系——蜘蛛线用于相对定年；

锕系——U/Pb 同位素用于绝对定年；

“赖利的诅咒”。

相对于成龙，胚胎骨头有约 1/11 的镧系稀土 (≈P) 比例，大约 1/5 的锕系稀土比例。

在此说的稀土元素，只包括镧系十四个天然元素：镧 (La)、铈 (Ce)、镨 (Pr)、钕 (Nd)、钐 (Sm)、铕 (Eu)、钆 (Gd)、铽 (Tb)、镝 (Dy)、钬 (Ho)、铒 (Er)、铥 (Tm)、镱 (Yb)、镏 (Lu)，其中钷为人造元素以及与镧系的 14 个元素相关的两个元素钪 (Sc) 和钇 (Y)，没用于讨论之内；此外，第 Ⅲ 族副族元素锕系元素也列入讨论范围。

在脊椎动物活着的时候，骨头内基本上没有稀土元素，骨头化石内会有稀土元素，是在死亡后一万年内骨头里面生物性磷灰石，重新结晶转换成无机磷灰石群的过程中，从地下水中吸附而来的。一旦稀土元素被磷灰石吸附，就被锁在晶格内，不会流失或被置换，形成一个封闭系统 (closed system)，因此稀土元素可用来作为该脊椎动物埋藏地层的定年。将镧系的十四个天然元素成员正常化之后取对数值画出的“蜘蛛线图”（图 6.2.1），可提供很多信息：比方说，图中两组曲线中央部分都向上突，轻稀土 (LREE) 和重稀土 (HREE) 浓度较低，而中稀土 (MREE) 浓度较高，虽然浓度上相差一个数量级，但是两组曲线形状类似，

the bones, no matter if it is living or fossil bone, it is absolutely NOT as simple as just hydroxyapatite. Thus, from now on, do not keep using the mistake and say hydroxyapatite!

6.2 Rare Earth Elements

La Series — Spidergram for relative dating;

Ac Series — U/Pb isotopes for absolute dating;

“Larry’s curse”.

As comparison to adult bones, embryonic bones have ~1/11 of La-REE (≡P) ratio, ~1/5 of Ac-REE ratio.

Here the discussion of Rare Earth Elements only limited to the 14 natural elements of lanthanide series: La, Ce, Pr, Nd, Sm, Eu, Gd, Tb, Dy, Ho, Er, Tm, Yb, and Lu. Since Pm is a non-natural produced element, it is excluded along with two other commonly referred to as REE members, Sc and Y. However, the natural elements of Group III Actinides are included.

When the vertebrae were alive, there is no REE inside the bones. The bone containing REE took place within 10 K years after the death of animals while the biogenetic apatites transformed to more stable inorganic apatites. They were absorbed from underground water. Once apatites absorbed the REE, they were locked inside the crystal lattice and will not be washed out or replaced and formed a closed system. Therefore, REE can be used as stratigraphic dating for such vertebrae animals. The Spidergram from 14 La-REE after normalization (Figure 6.2.1) can provide a great deal of information. For example, in the figure, two central convex curves indicate the mid-REEs have a higher concentra-

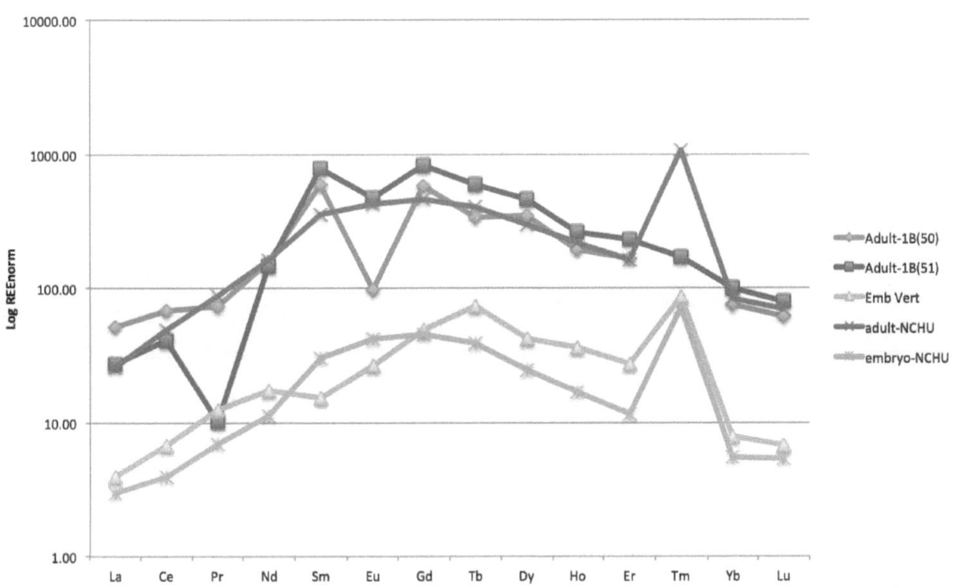

▲ 图 6.2.1　云南禄丰大洼出土禄丰龙成龙与胚胎化石镧系稀土元素比较；图内上方三条蜘蛛线为成龙骨头内的稀土元素分布，下方两条蜘蛛线为同一地点禄丰龙胚胎的，两者相差一个数量级

▲ Figure 6.2.1　Spidergram of La-REE of DaWa Adult and Embryonic *Lufengosaurus* Bones

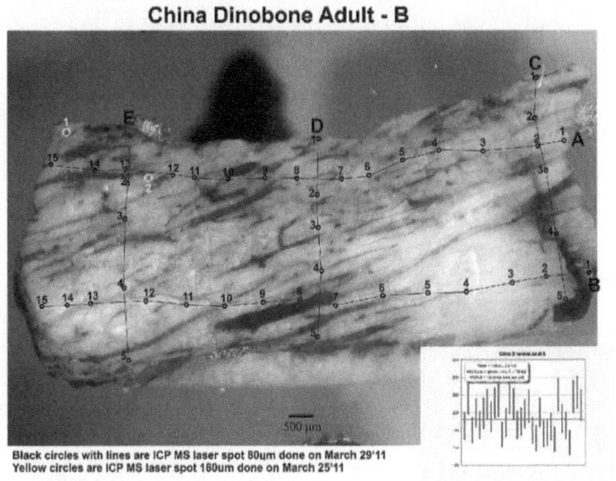

◀ 图 6.2.2　云南禄丰大洼出土禄丰龙成龙化石激光剥蚀电感耦合等离子体质谱图

◀ Figure 6.2.2　LA-ICP-MS of DaWa Adult *Lufengosaurus* Bones

它们应该来自同样的地层，如果来自不同地层的话，两组曲线基本形状会有较大的差异。

在周期表同样属于第 Ⅲ 族副族元

tion than the light-REEs and heavy-REEs. The log position of these two curves shows one order of magnitude difference, but the curve shapes are similar. This means there were from the same stratum. If they were from different strata, then

素以锕为首的一系列锕系元素，是原子序数第 89 元素锕到第 103 元素铹，共 15 种放射性元素，在周期表中占有一个特殊位置，天然界元素只到铀，以后的元素都是在 1940 年后用人工核反应合成的，称为人工合成元素。

锕系元素的铀，是所有大自然出产原子序最大的元素，也会在骨头内生物性磷灰石重新结晶过程中被吸附锁在晶格内，形成一个封闭系统。铀和铅的同位素比例，是地质界惯用的绝对定年元素同位素对，是我们用来进行骨头化石绝对定年的元素。

在我们探讨云南禄丰龙胚胎过程中，发生了一件有趣的事情，姑且称之为“赖利诅咒 (Larry's Curse)”，他在加拿大亚博特大学有一台先进的激光剥蚀电感耦合等离子体质谱，曾经发表一篇论文说在 K/T 界线之上方而非界线之下方，发现了鸭嘴龙恐龙化石，亦即不会飞的恐龙，并没有随着六千五百万年前那颗打到墨西哥湾陨石，全部死光光，还有少数存活下来了，多活了 70 万年。先暂且不讨论到底这个发现能否站得住科学界的严苛检验，当我看到此文自己打了一下脑袋瓜，同样的（锕系）稀土元素，我怎么没想到过呢？能够使用骨头化石内锕系的铀来做绝对定年，而且这种方法可说是几乎没有破坏性，如果没有刻意说出来，在骨头上用激光所烧的洞，肉眼很难看出来，真是太好了。因此，我就寄了几块大洼禄丰龙成龙和胚胎骨头，请他用他的设备帮忙做大洼

the curve shapes would have more significant variations.

Log normalized (by CBA) individual La-REE values were plotted as the vertical axis and the individual rare earth element as the horizontal axis. It can be seen clearly that these curves are in two distinct groups. The top 3 curves are the La-REE of DaWa *Lufengosarus* adult bones, and the bottom two curves are the DaWa *Lufengosarus* embryonic bones. These two groups of curves are well separated. However, they are from the exact same place (same formation).

The Group III Actinides, are from atomic number 89 to 103. These 15 radioactive elements occupy a unique position in the Periodical Table. The elements from nature are up to Uranium. All elements after U did not occur naturally on Earth and were created artificially after 1940.

The Uranium of the Ac-REE is the biggest atomic number element from nature. It can be locked inside the crystal lattice during the re-crystallization of apatites and formed a closed system. Thus, the commonly used U/Pb isotopes ratio by geologist can be used for absolute dating of fossil bones.

While we were conducting the study of Yunnan *Lufengosaurus* embryo, an interesting event took placed, I called Larry's Curse. He has an advanced LA-ICP-MS in University of Alberta, Canada. He published a paper stating that *hadrosaur* dinosaur was found ABOVE, not below the K/T boundary. That means the dinosaurs did not completely wipe out by that 65 million years ago meteorite; some survived for another 700 thousand years after that great event. *(Direct U-Pb dating of Cretaceous and Paleocene dinosaur bones, San Juan Basin, New*

禄丰龙的定年，他也很爽快，没有多久就给我了禄丰龙成龙的绝对定年，(192.4 ± 2.8）百万年，我就等着他给我同一地点出土的恐龙胚胎化石绝对定年，可是等啊等的，也写过几封Email催他，他不是用这个借口，就是说仪器维修中等等，后来终于告诉我，找来找去，他在胚胎骨头中，很难找到磷灰石，处处都是方解石，无法取得可用的数据。

哼，这就有趣啦！为何禄丰龙成龙骨头很快就可取得可用可信赖的数据结果，而胚胎骨头就是得不到可信赖的数据？它们都是来自同样地点的禄丰龙，只是一个是成龙，一个是胚胎，埋藏的环境完全相同，怎么会有如此大的差异？这个问题困扰了我好长一段时间，从理论上来说不通啊！后来，我把同样的禄丰龙成龙和胚胎骨头，送到台湾的两个不同研究单位，再度使用电感耦合等离子体质谱检测镧系稀土元素与铀／铅同位素，计算出来的绝对定年是 1.98 亿年，因此就取两次绝对定年的平均值，说云南禄丰大洼恐龙山的“大洼层”绝对定年为 1.95 亿年。另外一方面，我重新检查过去做过几次电感耦合等离子体质谱总镧系稀土元素含量和铀含量，哇啦！发现一个很有趣的情况，大洼恐龙山出土的禄丰龙成龙骨头内的总镧系稀土元素量，是胚胎的 11~12 倍！整整一个数量级多！而铀的含量也是五倍多！因为总稀土含量和磷（灰石）有绝对密切的关联性，因此可说，在胚胎骨头里面，磷灰石群的量，远低于成龙的，意

Mexico, James E. Fassett, Larry M. Heaman and Antonio Simonetti, Geology 2011; 39; 159–162 doi: 10.1130/G31466.1). Let us not to argue if this can stand critical academic review or not as now. The point is that when I read this paper, I hit my head and said to myself why I did not think about this. The actinides are REEs too. Uranium in this Ac-REE group can be used for absolute dating, and this method is almost non-destructive. The tiny holes burned on the bone surface were almost invisible. So, several fragments of *Lufengosaurus* adult and embryonic bones were sent to him. It did not take long for him to give me the dating figure of 192.4 ± 2.8 million years old for the adult bones. I was waiting for the absolute dating of my embryonic bones. Wait and wait. Emails sent, one reason after the others, he did not give me the embryonic REE result. Finally, I was told that in the embryonic bones, it is very challenging to find apatite, calcite is everywhere. So no useful data could be obtained.

This is very interesting. Why were reliable data obtained very quickly for the adult bones, but not for the embryonic bones? They came from the exact same spot with the same taphonomic condition. The only difference is adult and embryonic. Why this big difference? This issue bothers me for a long while. It could not be explained theoretically. Then, more specimens were sent to two different institutes in Taiwan using ICP-MS. The resulting absolute dating is 198 mya. Thus, I use the average value of these two datings as 195 mya for DaWa Layer. Furthermore, I re-examined the data from the past several ICP-MS on total La-REE and U concentrations. Wala! An exciting finding showed up. The total La-REE amount of the *Lufengosaurus* adult

味着胚胎阶段的骨头，并没有像成龙那样多的硬骨头，即便刚刚孵化的禄丰龙，它们骨头的机械强度，无法支撑它们自由行动自己觅食。

Bingo！这就解释了“赖利诅咒”的原因了，因为在胚胎阶段的禄丰龙骨头内，硬邦邦的磷灰石群矿物，还正在形成当中，而且因为恐龙蛋内的总磷含量是固定的，并非全部用来建构胚胎的骨头，有一部分的磷被用于其他生命关键的组织细胞去了，用于建构骨头的磷量有限。这也导致了另外一个非常有趣的科研课题：恐龙为何需要喂养它的宝贝？或说，恐龙是晚成性 (altricial) 而非早成性 (precocial) 的动物。

bone is about 11–12 times more than that of the embryonic bones. One order of magnitude more. Moreover, the U concentration is five times more. The total REE concentrations have a perfect correlation with phosphorus P (apatites); thus, it can be said the amount of apatites in the embryonic bones were far less than that of the adult bone. There were no that many hardy bones in the embryos or even in the hatchling. The *Lufengosaurus* neonates did not have enough mechanical strength for them to feed themselves.

Bingo! This explains the reason for Larry's Curse. During the embryonic stage, the apatitic hardy bones were just starting to form. The total amount of phosphorus inside a dinosaur egg is limited and had to be used for not only the bones but also for many other vital organs and cells. The amount of P for the bones is limited. This leads to another very fascinating research topic: Why the dinosaur had to feed their babies? In other words, dinosaurs were altricial, not precocial animal.

附　透过稀土元素分析探讨云南恐龙地层

黄大一[1]，黄春兰[2]，林启灿[3]，Mike Knell[4]，杨文质[5]

[1] 天龙科普国际基金会总监，中兴大学客座教授，timd_huang@yahoo.com

[2] 高雄海洋科技大学水产养殖系

[3] 高雄海洋科技大学海洋环境工程系

[4] Department of Earth Science, Montana State University

[5] 云南楚雄彝族自治州博物馆

摘要

稀土元素在石化过程中从生物性转变为自生性磷灰石动物所吸附，并固定于晶格内，同样地层内的化石，会有相似的稀土元素指纹，因而可作为化石地层之比对；云南中部以出土中生代早期恐龙著名，本文试图透过化学分析恐龙化石与围岩内所含的稀土元素指纹，推论云南楚雄州内主要恐龙出土地点之三个恐龙地层（暂称为："大洼层"、"姜驿层"、和"川街层"）资料，做将来本地区内恐龙研究之重要参考；同时，也对楚雄市内饱满街恐龙脚印地层提出探讨。

关键词：恐龙，地层，稀土元素，蜘蛛线图，楚雄州，磷灰石，下禄丰组

Appendix　Exploring Dinosaur Strata of Yunnan *via* Rare Earth Element Analyses

Timothy Huang[1], ChunLang Huang[2], ChiTsan Lin[3], Mike Knell[4], WenJi Yang[5]

[1] DinoDragon International Research Foundation, Visiting Profssor of ChongHsing University, timd_huang@yahoo.com

[2] Department of Aquaculture, KaoHsiung Marine University

[3] Department of Marine Environment Engineering, KaoHsiung Marine University

[4] Department of Earth Science, Montana State University

[5] Yunnan ChuXiong Prefecture Museum

Abstract

Rare earth elements (REE) are absorbed and fixed into the lattice of biogenic-to-authigenic apatite within the remains of organisms during the fossilization process. Fossils within the same formation have similar REE fingerprints; thus, they can be used to correlate different fossil-bearing strata. Central YunNan is famous for early Mesozoic dinosaur discoveries. This paper presents findings on rare earth element fingerprints using chemical analyses of dinosaur fossils and matrix from this area, and establishes a database profile for three distinct dinosaur-bearing strata (tentatively named as DaWa Layer, JiangYi Layer, and ChuanJie Layer), providing reference information for future research. An examination of the stratigraphic placement of the dinosaur tracks at BaoManJie, ChuXiong

1. 绪论

云南恐龙的困惑

云南的中生代早期，亦即晚三叠纪到中侏罗纪的恐龙，闻名世界，可是对于古生物的业余爱好者来说，一直存在着某些困惑，其中之一就是云南恐龙相关资料的缺乏和不易取得，特别是不容易找到准确的恐龙地层资料和绝对年代；如，楚雄州禄丰县大洼恐龙山是中国人自己研究恐龙的发源地，此处出土的许氏禄丰龙（*Lufengosaurus huenei*）是杨钟健纪念他德国老师许氏而命名的，当时他认为是晚三叠纪下禄丰组地层，大约二亿多年前（百度百科，2009）；但是听闻，后来有研究者认为没有那么久远，“只是”早侏罗纪，一亿八、九千万年“而已”；更后来又有学者指出，毕竟杨没错，应该是晚三叠纪到早侏罗纪的地层；到底大洼的恐龙是那个地质年代的？又如，禄丰县川街乡恐龙谷的地层，如何说是 1.6 亿年中侏罗纪而非早侏罗纪的地层？

不论从科研或科普的角度来说，云南恐龙地层的定年，似乎也是云南恐龙迷人要素之一。

稀土元素地层对比

在动物死亡后，原本生物骨头内的生物性磷灰石 (Biogenic Apatite)，含有相对浓度较高的碳酸根、钠与其他离子的准稳定 (Metastable) 碳酸羟磷灰石

City is also discussed.

Keyword: Dinosaur, Strata, Rare Earth Elements, Spidergram, ChuXiong Prefecture, Apatite, Lower LuFeng Group

1. Introduction

Puzzles of Yunnan Dinosaurs

The dinosaurs of YunNan from late Triassic to mid-Jurassic are world famous. But for paleontology amateurs, there are some puzzles. One of them is the difficulty of getting information about, mainly the precise dinosaur stratigraphic information and absolute date (Luo et al.). For example, the DaWa Dinosaur Hill of LuFeng, ChuXiong Prefecture is the birthplace of Chinese Dinosaurology. *Lufengosaurus huenei* came from here. CC Yang named it in honor of his tutor Friedrich von Huene. At that time, Yang thought the stratum was late Triassic Lower Lufeng group (Staron et al., 2001) of more than 200 million years ago (Baidu, 2009). However, it was heard that place is not that old, 'just' early Jurassic, about 180–190 million years ago. Then, some scholars said that Yang was right. It should be late Triassic to early Jurassic. So, what geological age is it? Another example is how to say the Dino Valley at ChaunJie is 160 million years old mid-Jurrasic, not early Jurassic?

No matter from scientific research or science popularization point of view, the dating of YunNan dinosaur strata is one of the attracting myths.

Comparison of Rare Earth Element Strata

After the animal died, the biogenic apatite,

(Carbonate Hydroxylapatite)，此化合物具有较大溶解度和反应性，因而比较不稳定，而在形成化石的重新结晶过程中，它们会转变成无机、热稳定性的氟磷灰石 (Fluorapatite)，生物性磷灰石内的钙、钠、碳酸根、和羟根等等，被岩层内溶解在水中的氟、稀土元素和微量元素等离子取代，被吸附于磷灰石的晶格内，此过程大约在动物死亡后几千年到一万年内完成；这些稀土元素一旦被固定于化石后，非常稳定，不会再从化石中溶解出来、流失掉、或被其他元素置换，因而一直保留着在石化过后生物遗骸内；稀土元素被吸附的比率，取决于当时地层内化学组成与孔隙水情况，因而保留了该地层特有的"指纹"(Trueman, et al., 2003)，亦即，化石内的化学成分，记录了当时的成岩流体 (Diagenetic Fluids) 成分，可用来探索古环境的情况 (Trueman, 1999)；再者，相同地层之化石，不论生物品种，只要属于相同地层，都有相同或类似的"稀土元素指纹 (Rare Earth Element Fingerprint)"，亦即，透过分析和比对这些化石里面的"稀土元素蜘蛛线 (Spider-gram)"，亦即化石内所含、经正常化后的稀土元素相对比例曲线，可用来反推某块化石属于那个地层，即使曾经被迁移过或再埋藏的化石，也不受影响，故透过本项分析，可让某化石和原本地层对比；到目前的研究指出，透过此法判定相对地层，一旦石化过程完成后，不只不受不同生物种的限制，适用的范围，也可相距数百公里，

which contains a metastable higher concentration of carbonate, sodium, and other ions carbonate hydroxylapatite. These compound have higher solubility and reactivity, and hence relatively unstable. During the fossilization process, they recrystallized to inorganic and more stable Fluorapatite. The biogenic calcium, sodium, carbonate, and hydroxyl groups were replaced by fluorite, rare earth, and trace elements, and locked in the crystal lattice of apatite. This took place around several thousand to ten thousand years after the animal died. Once these REE been fixed in the fossil, there are very stable, will not be dissolved out, lost, or replaced by other elements, and stay inside the bones of the animals. The ratio of REE absorption depends on the strata chemical composition and groundwater. So, it preserved the unique 'fingerprint' of that stratum (Trueman et al., 2003), i.e., the chemical composition inside the fossils recorded the composition of the diagenetic fluids of that time and can be used to explore the paleoenvironment (Trueman, 1999). Furthermore, as long as the fossils were from the same stratum, no matter what species, they will have the same Rare Earth Element Fingerprint. That is after analysis and compare the fossils "Spidergrams," which are the curves of the normalized relative ratio, it can deduct what stratum a given fossil belongs to, even if it was moved or re-buried. Thus, by this analysis, a given fossil can link to its original stratum. Up to this point, using this method to compare strata shows that once the fossilization process was done, not only limited by animal species, but also the range can be several hundred kilometers apart. As long as they were from the same stratum, they would

亦即，只要是来自相同地层内的生物化石，都会有相同或类似的稀土元素指纹(Martin, et al., 2005)。

举例来说，在某山沟找到一块从山坡上被冲下来的恐龙化石，山坡上有好几个恐龙层位，那么要怎么判断，到底手中这一块山沟里捡到的化石，是从那一层掉下来的恐龙呢？以往只能靠经验（如骨头形态）来猜测，如今可借着分析比对稀土元素指纹特性 (Trueman & Benton, 1997)，就可确定地说，这块骨头化石，原本是山坡上那一地层的恐龙；或者，另外一种情况，假设在山沟捡到好几块恐龙化石，分析出现好几种不同的稀土元素指纹，那么，就可推论出山坡上的恐龙地层数目。

骨头化石内吸附稀土元素，提供了一种有效的地层对比方法，虽然此法不像同位素或其他地层定年法可提供绝对年龄的信息，然而却可透过它了解不同地层之间的相对关系，或用来判断不同地点出土的恐龙化石是否来自相同的地层 (Patrick et al. 2004; Trueman et al. 2003)。本文仅就中国云南省楚雄州内主要早中生代恐龙出土地区的样本做分析，推论出楚雄州内三个主要地区恐龙层位：暂称为：大洼层、姜驿层和川街层。

2. 研究方法

分析样本描述

云南楚雄州为世界著名之早期恐龙发现地，州内九县一市，除了两个县之

have the same or very similar REE fingerprints (Martin et al., 2005).

For example, a washed down from hill dinosaur bone was found in a ditch, and there are several dinosaur strata on the hill. How to know which stratum this dinosaur bone came from? In the past one can only rely on experience to guess, such as the shape of the bone. Now we can use the analysis and comparison of REE fingerprints (Trueman & Benton, 1997) to be able to identify the stratum it came from. Alternatively, the other scenario, assuming many dinosaur bones were found in the ditch and showed many different REE fingerprints, then, it can say there is x number of strata in the hill.

REE absorbed inside bone fossil provides an excellent strata comparison method. This method does not provide absolute dating as isotopes or other strata dating. However, the relationship between different strata can be understood, or to use it for determining if dinosaur fossils from different places are from the same strata (Patrick et al. 2004; Trueman et al. 2003). This paper discusses the major dinosaur sites in ChuXiong Prefecture of YunNan Province of China. We concluded there are three major dinosaur layers, temporarily called as DaWa Layer, JiangYi Layer, and ChuanJie Layer.

2. Method

Description of Specimens

ChuXiong Prefecture of Yunnan is the world famous early dinosaurs. There are nine counties and one city in the Prefecture. Only two counties did not report having dinosaurs or

外，均有恐龙化石或恐龙脚印出土记录；然而，州内这些恐龙地层的信息，并不完整，甚至相突矛盾；本文就从楚雄州内某些恐龙点所采集到的化石，透过化学分析稀土元素与统计分析，试做本州内恐龙地层的探讨。

本文的绝大部分化石样本，采集于2006年底到2007年初，第一作者带着美国蒙大拿州立大学博士生麦可·坎内尔（Mike Knell）做地质考察，主要为姜驿 (JY) 地区（38个样本），也包括禄丰县大洼 (DW)、禄丰县川街乡细细坡 (SSP)、武定县 (WD)、牟定县 (MD)、双柏县 (SB) 等地区少量样本（共13个）；又第一作者往日于禄丰县举办“百战天龙恐龙营”所采集之两块恐龙化石样本，亦列入比对分析。

在表6.2.1中，序号1至40者，为姜驿地区所采集的恐龙化石样本，编号41至54者，则来自楚雄州内其他地点：包含禄丰县大洼恐龙山、武定县、牟定县猫街和禄丰县细细坡等地。基本上这些样本涵盖了楚雄州内主要的恐龙地点，建立楚雄州内不同盆地内恐龙地层的重要基本资料。

稀土元素分析

这些化石样本之稀土元素与其他元素的化学分析，系本文第二、三作者遵循台湾环境事务主管部门NIEA M104.01C标准方法（2003），使用电感耦合等离子体发射光谱法ICP-OES (Inductively Coupled Plasma, Optical

dinosaur tracks. However, the information is not complete, and even contradictory to each other. This paper uses the dinosaur fossils collected in several sites for REE analyses and statistics for the exploration of dinosaur strata of this Prefecture.

Most of the specimens for this paper were collected during the end of 2006 and early 2007. The first author took a Ph.D. student Mike Knell of Montana State University for a survey in mainly JiangYi (JY, 38 specimens) area, and also include DaWa (DW), ShiShiPao (SSP) of Lufeng County, WuDing (WD) county, MoDing (MD) County, and ShuanBao (SB) County (13 specimens in total). Also, two specimens collected by the first author during DinoDragon Project were included.

In the Table 6.2.1, the serial number 1 to 40 were the specimens collected in JiangYi area, number 41 to 54 were from various places in ChuXiong Prefecture, including DaWa Dinosaur Hill, WuDing County, MauJie of MoDing County, and ShiShiPo of Lufeng County. These cover the major dinosaur sites in the ChuXiong Prefecture for the establishment of the important necessary information of dinosaur strata in various basins.

Rare Earth Element Analyses

The second and third authors did the experiments of Rare Earth Elements and other elements following NIEA M104.01C 2003 using ICP-OES (Inductively Coupled Plasma, Optical Emission Spectrometer) Perkin-Elmer Optima 2100DV. The samples were done as 0.1 g of representative specimens, 10 ml concentrate nitric acid, and 1 mL hydrogen peroxide in Teflon bea-

表6.2.1 云南楚雄州恐龙地层恐龙化石稀土元素

Table 6.2.1 REE of Dinosaur Strata of ChuXiong Prefecture of YunNan

	La	Ce	Pr	Nd	Sm	Gd	Dy	Ho	Er	Tm	Yb	Lu
(1) JY-1-TWD-1B	1.347	8.348	13.545	6.009	14.882	11.270	16.795	7.704	10.191	40.976	6.157	4.544
(2) JY-1-TWD-2B	1.146	8.157	12.912	5.064	13.809	10.200	15.849	6.839	8.870	32.720	5.295	3.979
(3) JY-1-TWD-3B	2.752	4.713	12.400	10.375	37.547	38.436	38.242	26.022	30.700	64.740	16.418	13.639
(4) JY-1-TWD-4B	2.493	26.869	45.427	14.830	45.058	35.834	39.706	18.941	16.218	55.616	10.301	7.389
(5) JY-4-1803-1B	2.359	17.552	30.169	10.807	31.390	24.761	28.660	17.203	18.081	53.619	12.819	9.915
(6) JY-4-1803-2B	0.703	3.101	6.245	2.627	5.186	3.777	6.574	3.374	5.220	41.254	4.096	3.054
(7) JY-4-1803-3B	0.655	3.437	7.014	2.622	5.345	3.915	6.660	2.998	4.645	41.440	3.634	2.729
(8) JY-5-Hill-1B	2.930	2.955	10.307	12.483	37.800	56.742	39.023	26.216	22.094	142.315	9.812	3.388
(9) JY-6-Clam-1B	-	0.124	-	-	-	-	-	-	-	-	-	-
(10) JY-8-Verta-1B	0.488	1.366	2.777	2.630	8.692	9.517	11.014	6.115	7.742	-	3.498	-
(11) JY-10-Zeng-1B	1.783	3.918	-	6.930	17.325	24.045	24.683	12.952	14.057	13.917	6.540	4.908
(12) JY-10-Next-1B	1.715	4.166	-	10.856	40.028	51.924	41.806	25.811	25.728	23.958	12.769	8.989
(13) JY-11-Next1-1B	0.546	1.102	-	1.144	2.297	2.161	3.709	0	3.716	30.099	1.965	2.098
(14) JY-11-1Next2-1B	1.240	3.731	-	8.247	26.805	36.973	27.251	17.183	18.588	48.460	4.430	6.425
(15) JY-13-Roadside-1B	2.013	5.991	0.983	10.683	28.908	37.458	31.453	18.113	17.635	17.848	7.675	4.324
(16) JY-14-4Leg-1B	0.974	2.190	-	3.226	0.162	6.584	7.667	1.943	5.971	34.212	3.334	1.452
(17) JY-16-AboveHill-1-1B	7.673	10.856	9.420	13.772	28.390	31.395	26.877	15.706	17.842	-	9.243	6.993
(18) JY-15-AboveHill-1B	0.263	0.455	-	0.302	-	-	2.016	-	1.130	-	0.792	-
(19) JY-17-2Drangons-1B	32.536	96.749	-	180.609	380.936	422.261	634.437	128.727	447.013	-	255.549	175.833
(20) JY-18-Scapula-1B	0.808	11.475	9.050	5.027	15.212	17.555	14.087	4.765	6.364	-	3.803	2.666
(21) JY-20-BigDrangons-1B	1.003	2.802	-	7.320	23.904	30.913	24.098	14.217	15.144	-	6.495	4.577
(22) JY-21-SmallSite-1B	2.446	1.412	-	-	-	6.942	7.359	2.321	5.571	-	-	-
(23) JY-22-1Leg-1B	0.296	4.150	-	2.267	8.608	9.466	9.944	3.699	7.180	-	3.652	2.593
(24) JY-24-FireMtn-1B	0.447	1.804	-	2.401	5.690	5.640	6.424	1.037	5.086	-	2.770	2.037
(25) JY-25-ScapLegTooth-1B	2.177	6.378	5.410	20.532	95.528	134.323	86.692	55.355	45.227	43.945	15.516	10.966
(26) JY-26-SmalDragon-1B	1.036	3.667	-	9.942	40.491	54.380	38.050	21.617	22.198	16.528	9.180	6.598
(27) JY-27-SmallDragonLeg-1B	1.538	4.201	-	11.485	33.003	49.769	40.581	25.956	27.199	54.133	12.413	7.959
(28) JY-28-Random-1B	1.613	3.495	-	6.825	15.797	19.026	20.307	12.507	14.334	-	6.051	4.984
(29) JY-28-Random(shell)-2B	-	0.076	-	-	-	-	0.135	-	0.655	-	0.039	-
(30) JY-29-Wang-1B	0.814	2.857	-	6.958	20.954	27.398	23.062	12.967	15.538	-	6.343	4.896
(31) JY-30-OnWall-1B	0.814	2.857	-	6.958	20.954	27.398	23.062	12.967	15.538	-	6.343	4.896
(32) JY-31-AcrossTWD-1B	0.485	10.870	4.921	2.151	5.383	3.184	6.331	-	5.070	6.095	5.106	5.136
(33) JY-32-Complete-1B	-	2.874	-	6.160	17.558	20.428	19.488	10.543	12.542	-	5.138	4.431
(34) JY-33-Skull-1B	1.858	3.673	-	5.722	11.404	12.134	10.577	3.766	7.032	-	3.462	1.789
(35) JY-34-TenVerte-1B	0.985	13.222	9.399	6.466	20.721	24.598	21.176	7.649	9.926	-	5.421	4.086
(36) JY-35-NoTail-1B	14.623	148.424	265.853	70.775	216.446	246.111	239.074	152.847	161.606	499.190	102.311	78.269
(37) JY-36-MinorVerte-1B	19.171	21.901	-	46.881	122.232	187.485	165.159	110.598	112.189	315.345	58.099	43.355
(38) JY-37-LotOfVerte-1B	53.139	62.138	87.729	185.900	683.957	877.626	570.440	403.082	372.325	664.436	189.974	155.472
(39) JY-39-TopPick-1B	9.582	-	644.282	79.472	199.868	235.794	405.655	216.536	201.554	1040.796	199.311	175.453
(40) JY-41-Dynamited-1B	20.975	23.784	-	16.876	18.214	29.758	61.458	13.728	46.731	-	32.185	14.500
(41) MD-1-pelvet-1B	6.611	18.258	-	31.715	62.575	89.279	72.994	22.562	41.522	-	20.269	3.971
(42) MD-2-Birthday-1B	5.339	20.067	-	46.079	134.660	173.333	135.925	59.873	76.533	3710.750	1356.636	16.268
(43) WD-2-Another-1B	55.232	121.620	309.843	206.984	460.766	359.478	339.792	218.445	201.790	592.981	91.187	63.672
(44) WD-1-Top-1B	123.665	186.425	385.198	145.496	210.477	173.429	171.057	82.726	70.067	-	48.534	34.668
(45) WD-3-NextToskull-1B	73.294	173.866	439.634	323.147	741.937	605.003	512.012	338.123	287.849	805.189	132.561	100.552
(46) WD-5-OnTheRoad-1B	28.113	68.781	86.067	191.106	689.820	780.828	517.305	314.380	264.600	965.941	116.185	92.758
(47) WD-6-Road-1B	18.870	54.902	56.046	171.672	676.531	688.441	438.618	249.887	188.641	203.727	66.065	40.055
(48) WD-8-Road2-1B	23.240	48.012	39.604	112.078	238.912	238.401	211.841	116.800	105.042	53.393	38.641	19.938
(49) WD-10-StrateFish-1B	23.381	44.801	-	40.614	50.174	14.093	42.533	-	19.934	116.623	17.877	8.593
(50) DW-1-2ndHall-1B	51.106	68.383	74.424	159.582	591.438	575.967	348.907	193.571	168.419	-	75.252	62.192
(51) DW-2-Hill-1B	27.667	40.924	10.097	145.163	786.764	832.350	460.200	265.828	230.688	184.072	100.691	80.639
(52) SB-2-ShuanBao-1B	3.887	52.200	-	23.299	57.709	60.058	90.555	39.860	71.302	-	59.045	48.569
(53) SSP-1-ShiShiPo-1B	1.539	2.943	3.395	3.978	9.177	9.193	11.482	4.401	7.616	34.901	3.547	0.569
(54) SSP-2-3Verte-1B	3.059	6.618	12.607	11.636	28.028	26.629	24.745	14.535	16.346	56.155	7.103	3.352
CBA**	39.70	68.40	7.17	26.40	5.22	4.60	4.13	0.87	2.54	0.04	2.44	0.360

* 标示格式：: AA-n-Text-1B（mm），AA为某地区名称代码：JY姜驿，MD牟定，WD武定，DW大注，SB双柏，SSP细细坡，CBA全国土壤（A）稀土元素背景值基本统计量：n为在该地区内不同采集点代号，Text-1B为该采集点简称，最后的（mm）为化学分析样本序列编号。

** 除了CBA之浓度单位为百万分之一（ppm）外，所有各元素之数值，均为对应的CBA元素浓度做正常化，故没有单位。

Emission Spectrometer) Perkin-Elmer Optima 2100DV 检测所得。样本之处理过程，系将约 0.1g 的代表性样品，加 10 mL 浓硝酸和 1 mL 过氧化氢于铁氟龙容器中，进行四阶段微波加热消化，收集上层萃取液体，重复三次，再将所收集到的萃取液，以去离子水定量到 100mL 后，利用 ICP-OES 进行 48 种元素的测定。仪器测定原理及参数设定，依照 NIEA M104.01C 标准方法所述，检测相关之品管项目，包含检量线 (r^2>0.995)、检量线确认（回收率介于 90%至110%之间）、样品空白（<2 倍 MDL）、样品重复（RPD<20%）、添加样品/添加重复样品（回收率介于 75%~125%）等。

3. 结果与讨论

化学元素分析结果列于表 6.2.1，表内的数据都已经过归一化 (Normalized)，因此这些数字是比率数值，而非具有浓度单位的原始数据；做稀土元素数据之统计分析，首先要将化学分析所得的原始数据归一化 (Normalization)，用以除去奥多–哈尔金斯规则 (Oddo-Harkins Rule) 效应。此效应为：在自然界土地内镧系（原子序数 57~71）稀土元素中，原子序数为偶数者的元素丰度值，比相邻两个原子序数为奇数者的丰度高；亦即，若要能从化石化学分析出来的各稀土元素浓度，找出有意义的信息，必须把原本化学分析所得的数据，先除以某种标准的正常化数值，用来排除天然背

ker and use a microwave for heat digestion. Upper layer extract was collected, repeated three times. Then the extract was diluted to 100 mL with deionized wat er. Then ICP-OES was used to measure 48 elements. The instrument was set according to the standard method described in EPA NIEA M104.01C 2003. The quality control items are r^2>0.995, recovery 90%-110%, blank sample <2 MDL, RPD<20%, spike recovery 75% -125%.

3. Results and Discussion

Chemical element analysis results are in Table 6.2.1. The data in the table was normalized. Thus these figures are ratios, not the initial concentration raw data. For the statistic analyses of REE, the chemical raw data has to go thru normalization to remove Oddo-Harkins Rule effect. That effect is that the within the La-REE (atomic number 57–71), the richness of even atomic number element is higher than the two adjacent odd atomic number elements. That is, in order to have meaningful information from the analysis of each element concentration, the raw data must be normalized to a particular standard value set to remove the nature background influence to see the relative concentration ratio. This ratio is different from every stratum. So, it is the characteristic indicator for each stratum. Without removing such effect can be normalized too. However, it is easier to understand after normalization.

The standard normalization values used in other papers usually is the North American Shale Composite (NASC) (Gromet et al. 1984) or Post Archean Australian Shales (PAAS) (Le.cuyer

景的影响，才能看出稀土元素之间的浓度相对比例；而这个比例，每个地层都不同，也就是每个地层的特征指标；不除去该效应，也可正常化，但正常化更容易了解结果。

正常化的稀土元素标准数值，欧美相关的论文，通常采用“北美组合页岩 (North American Shale Composite, NASC) (Gromet, et al., 1984)”，也有使用“后太古界澳大利亚页岩 (Post Archean Australian Shales, PAAS)(Le.cuyer et al., 2003)”者，但本分析采用中国“全国土壤（A 层）稀土元素背景值基本统计量 (Chinese Background Average, CBA)（王云、魏复盛，1990）”为准则，虽然此三套标准数值相差不大，但因中国已有此项基础研究数据，故以中国的土壤所含的稀土元素浓度来作为正常化的准则，应该比采用北美或澳大利亚者，更为接近当地实际情况。若需与使用其他正常化基准者做精密对比，可将正常化基准改换即可；但若不需要精密比对，蜘蛛线图形状相似，就足够了。

镧系稀土元素群中原子序 61 的钷 (Pm)，是一个人造元素，从原子炉里面得到的，在大自然界中并不存在。因此，在本次化学分析中和另外两个浓度大半低于分析仪器极低灵敏度之铕 (Eu) 和铽 (Tb) 元素，都被排除在外，因此统计分析所用稀土元素的数目，只有十二个。

群 聚 分 析

群聚分析 (Clustering Analysis) 是一

et al. 2003). However, in this paper, we used Chinese Background Average (CBA) by Wang, Sun and Wei, FuShen, 1990. Although no significant differences between these three standards, we think since China already has this basic research data, it would fit better to use the Chinese standard than using the North American or Australian standards. If it is needed to use another standard for an accurate comparison, change the normalization standard. However, if an accurate comparison is not needed, a similarity of spidergram shall be enough.

The atomic number 61 element Pm is a man-made element in the La-REE series. It does not exist in nature. Thus, it is excluded from our chemical analyses along with two other very low concentration elements Eu and Tb. Thus, only 12 elements were used in the statistics.

Clustering Analysis

Agglomerative Nesting (Hierarchical Clustering) is a statistical method to help the user to extract the structure of big data set and to simplify the complication so that the information meaning behind the data can be understood. The objective is to extract meaningful groups. Thus, analyzing the normalized REE in dinosaur fossils can reveal the relationship between them, ie. dinosaur sites clustering together shall have identical or very similar REE fingerprints, and could be the same dinosaur strata.

From the Dendrogram (Figure 6.2.1), looking from lower left to lower right, three clusters can be seen. The leftmost ones to constitute Cluster 1, and the middle ones Cluster 2. These two

种统计分析的方法，它可以协助使用者从大量的资料中，挖掘出资料之间的结构，并简化资料的复杂性，进而能够了解隐含在资料背后的信息，其目标就在一堆未知资料中粹取具有意义的群体。因此，透过本项统计分析各恐龙点所采集恐龙化石内正常化之后的稀土元素群，可得知彼此之间的相关性，亦即聚合在一起的恐龙点，应该有相同或类似的稀土元素指纹，也可能是相同的恐龙地层。

从此群聚的树状图 (Dendrogram)（图 6.2.3）来看，自底左边往底右看，总共有三个聚集群。最左边诸样本组成"团聚一 (Cluster 1)"和中间诸样本组成"团聚二 (Cluster 2)"，这两个团聚又组成较高层次的聚合，故合并称"第一族群 (Group 1)"，最右边的那些样本另成一个团聚集群，称之为"第二族群 (Group 2)"。

clusters form yet another higher level cluster. Thus, they are combined and called Group 1. Those on the right constitute Group 2.

The majority members of Group 1 are from JiangYi area with three exceptions: SSP-1 (number 53), SSP-2 (Number 54) from ShiShiPo and MD-1 (number 41) from MoDing. However, in Group 2, the majority are from areas other than JiangYi, but includes 5 of JiangYi specimens: 19, 36, 37, 38, 39.

From this statistic analysis, it can be said that there are three major dinosaur layers in ChuXiong Prefecture and named as DaWa Layer (Group 2), JianYi Layer (Group 1), and ChuanJie Layer. According to the known information, the relative ages of these strata as DaWa Layer is the oldest (about 200 million years old), and ChuanJie Layer is the youngest, about 160 million years old. In the future, if isotope or another dating can be done, further information of ChuXiong dinosaur strata

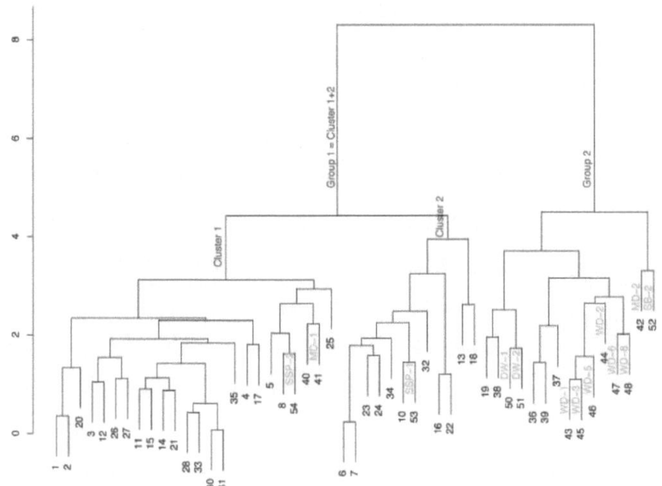

◀ 图 6.2.3 云南楚雄州内恐龙化石群聚树状图（序号请见表 6.2.1）

◀ Figure 6.2.3 Dendrogram of the ChuXiong Dinosaur Fossils (See Table 6.2.1 for numbers)

第一族群的成员，主要是姜驿地区的恐龙，但包含三个例外：即两个禄丰县川街细细坡 SSP-1（序号 53）和 SSP-2（序号 54）者和一个牟定恐龙 MD-1（序号 41）。相对地在第二族群内，大部分都是楚雄州内非姜驿地区的，但还包含有五个姜驿地区的恐龙点，化验分析序号为：19, 36, 37, 38, 39。

经由此统计分析研判，楚雄州内恐龙地层，主要有三，本文暂以代表性地名称之："大洼层"（第二族群）、"姜驿层"（第一族群）和"川街层"；依目前的资料显示，这些地层的相对年代，以"大洼层"被认为最古老（约 2 亿年），而"川街层"最年轻（约 1.6 亿年）。来日若能进行同位素或其他绝对年代定年，将可提供进一步楚雄州内恐龙地层信息。

从图 6.2.3所显示的信息初步判断，或可作如下的解释：

1. 暂称为"大洼层"的第二族群，分布于禄丰县大洼恐龙山、武定县、双柏县和牟定县等地域。因为相对来说，禄丰县大洼恐龙山的地层研究资料比较早也比较多，地质学者认为是"下禄丰组地层"，年代在晚三叠纪到早侏罗纪之间，故以大洼地名称之。

2. 楚雄州姜驿地区，含有两个恐龙地层，其中一个涵盖大部分地区，亦即上述暂称"姜驿层"的第一族群，此族群含有两个次族群；虽然目前本文暂时推论将它们合并为一个恐龙地层，但不能排除为两个相邻地质年代的恐龙地层，来日需更多的分析，才能做出更精

can be obtained.

From Figure 6.2.3, perhaps the following explanation can be given.

1. The DaWa Layer Group 2 spread among DaWa Dinosaur Hill of Lufeng County, WuDing County, ShuanBo County, and MoDing County. Relatively speaking, since more studies and researches were done for DaWa Dinosaur Hill of Lufeng County as the Lower LuFeng Group Formation of age between late Triassic to early Jurassic. Thus, we use DaWa as the name for this.

2. In the JiangYi area, there are two dinosaur formations. One of which covered the most area and called as JiangYi Layer Group 1. This Group has two sub-Groups. In this paper, we put them together as one dinosaur strata, but we cannot exclude they are two adjacent dinosaur formations. More analyses are needed for more precise judgment. However, there are several sites in this area belong to DaWa Layer.

3. Two specimens were collected at MoDing area. The higher place Huang's Birthday Site, MD-2 (number 42) belongs to DaWa Layer, while the lower site MD-1 (number 41) belongs to JiangYi Layer. This means that in the MoDing dinosaur area, there are two dinosaur strata.

This is interesting. According to superposition law, the older formation should be at the lower place. In other words, dinosaur strata at higher elevation should be the younger one, and at the lower elevation should be older. However, at this place, no formation reversal was seen, then, how could the older formation at a higher place? This could be caused by the inclination of the formation, N 80°

准的判断；但本地区内也有较为少数恐龙出土地点，属于“大洼层”的。

3. 牟定地区采集的两个样本，在海拔较高处“黄大一生日点 (Huang's Birthday Site)”者 MD-2（序号 42），属于“大洼层”，海拔较低的 MD-1（序号 41）样本，则属于“姜驿层”；亦即牟定恐龙地区也有两个恐龙地层。

这一点很有趣：依据叠置定律，比较古老的地层，应该是被埋在比较下面，也就是说，在海拔比较高处的恐龙地层，应该是地质年代比较年轻的，在海拔比较低处者，则为比较古老的；但是在这里，既然没有观察到地质学上所谓的地层反转，怎么会有比较古老的地层，反而在比较高的地方呢？可能原因为地层倾斜所致（层态为 N 80° E/10° SE）。

4. 禄丰县川街细细坡梁子高速公路边的恐龙地层，属于“姜驿层”；但本处山顶也有恐龙，未含于本分析之内，山顶恐龙是否属于相同地层，有待来日进一步分析；又本恐龙点距川街恐龙谷只有 4 公里，但两者属于不同恐龙地层，川街恐龙谷为“川街层”，本处为较古老的“姜驿层”，姜驿和细细坡的直线距离约为 120 公里。

5. 武定县、双柏县、禄丰县大洼等地的恐龙点，都属于“大洼层”。这些地点间的直线距离约在 50~100 公里。

蜘蛛线图比对

经由群聚分析的树枝图所得到的信息，将本次化学分析每个恐龙点的 12 个

E/10° SE.

4. The dinosaur formation of the ShiShiPo in Lufeng County next to the freeway belongs to JiangYi Layer. We did not analyze the dinosaurs from the top of the hills here. Whether they belong to the same formation is awaiting for further analyses. Furthermore, this site is only 4 kilometers away from ChuanJie Dinosaur Valley. They are not the same formation. The one of ChuanJie Dinosaur Valley is called ChuanJie Layer. However, here is the older JiangYi Layer. The direct distance between JiangYi and ShiShiPo is about 120 kilometers.

5. The sites in WuDing, ShuanBo, and DaWa of Lufeng County belong to DaWa Layer. The direct line distances between them are 50–100 kilometers.

Spidergram Comparison

The information obtained from the dendrogram of cluster analysis is used to make the spidergrams with the log of 12 normalized REE concentration ratio as the vertical axis, and the atomic number of REE as the horizontal axis.

Figure 6.2.4 is the spidergrams of the three dinosaur layers of ChuXiong Prefecture. From this further exploration of the correctness of the cluster analysis. Therefore, spidergrams of Cluster 1 (green line in the upper middle) and Cluster 2 (green line in the bottom middle) were made for comparison. All the individual spidergram of Cluster 1 almost overlaps completely. They are identical. The curve lines of Cluster 2 are the same and similar to Cluster 1, with a slightly less on the concentration axis. Thus, these two Clusters can be combined as Group 1

稀土元素正常化后之比值，画出“蜘蛛线图”；这种图的纵轴为正常化后稀土元素浓度比例之对数值，横轴为稀土元素，按原子序数小者由左往右增大。

图 6.2.4为云南楚雄州内三个恐龙层的稀土元素蜘蛛线图。从此图可更进一步探讨比对群聚分析分类的正确性，故此将团聚一（中上边的绿色曲线 Cluster 1）和团聚二（中下边的绿色曲线 Cluster 2）所属的各恐龙点之蜘蛛线图制作出来比较；团聚一所属样本的

(thick black line). These are two sections of the same formation.

The spidergrams of top blue curve Group 1 are similar to Group 1, except with higher concentration. Thus, it is a different formation. The bottom red curve came from the trial run of these chemical analyses. The first author collected these two bone fragments in the previous trips for REE analysis. Exact location information is not available. Their data and curves are almost identical. They must come from the same Layer. However, they are different from

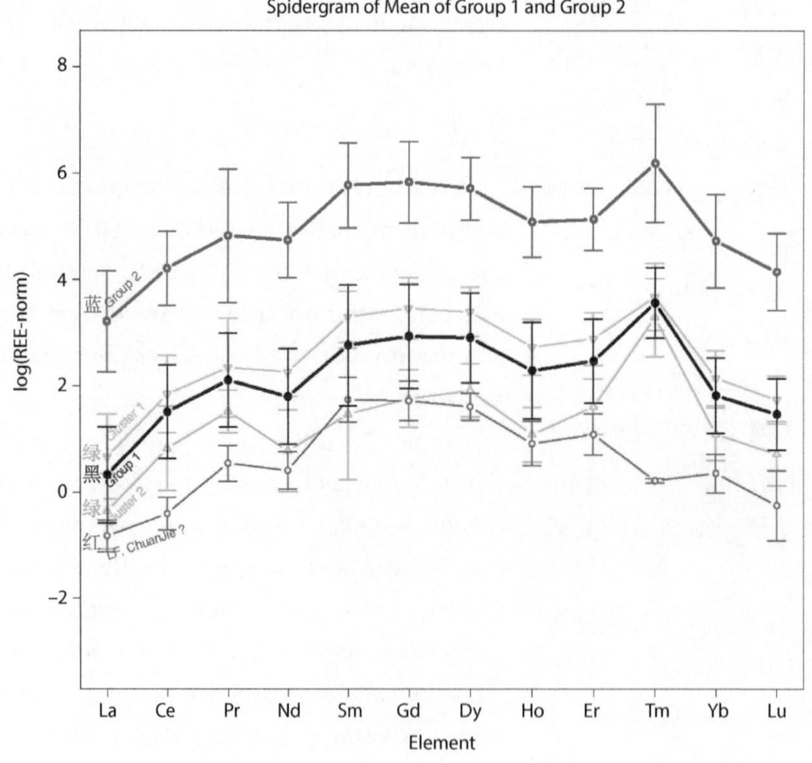

▲ 图 6.2.4 云南楚雄州内三个恐龙地层内恐龙化石的稀土元素蜘蛛线图以全国土壤（A 层）稀土元素背景值基本统计量 (CBA) 为正常化基准

▲ Figure 6.2.4 REE Spidergrams of Dinosaur Bones in ChuXiong Prefecture show 3 Dinosaur Strata Normalized with REE Chinese Background Average (CBA)

个别曲线，几乎都重叠在一起，可说是完全相同，团聚二所属样本，曲线形状和团聚一者也都很类似，唯在浓度轴上稍低于团聚一者，因而两个团聚可以合并，如上述称之为第一族群（粗黑曲线 Group 1），这应该是相同地层的两段。

第二族群（最顶端蓝色曲线 Group 2）所属者的蜘蛛线形状，与第一族群者类似，但在浓度较高，因此应该是另外一个不同的地层。图 6.2.4内最底端的红色曲线，来自本次化学分析之前的测试实验 (Trial Run)；本文第一作者从过去多年来于禄丰县川街境内所采集的零星恐龙碎骨头，挑出两块确实采集地点已无法考证者，进行化学稀土元素分析，拿其数值来和本次实验数据比对，发现这两块的数值与曲线几乎相同，应该是来自相同地层，但与上述的两个地层（大洼层、姜驿层）都不相同，曲线形状稍微有异，稀土元素的浓度也较低，故假设为禄丰川街恐龙谷大坟场者，暂称为“川街层”，依照相关文献报道（方晓思等人，2000），此地层被认为是中侏罗纪，约为一亿六千万年前。

在图6.2.4中，五条曲线均为所属聚集各恐龙点正常化后稀土元素浓度之平均值 (Mean)，而非各恐龙点的个别曲线。若用各恐龙点的个别曲线展示本图，则会因为曲线太多交互重叠，过于繁杂，反而不容易看清楚这些恐龙地层的相互关系。再者在每个元素的数据点，并以“工 (I Beam)”字符号标示标准误差 (Standard Deviation, σ) 范围。

the other two layers (DaWa and JiangYi Layer). The curve shape is a little bit different, and REE concentrations were lower. So they were assumed to belong to the dinosaur graveyard of Lufeng ChuanJie Dinosaur Valley and temporarily called ChuanJie Layer. According to the paper published (Fang et al. 2000), this should be mid-Jurrasic, about 160 million years old.

In Figure 6.2.4, the five curve lines represent the mean REE concentration value of each cluster of dinosaur sites, not the individual of each site. If the individual lines are used, the graph will have too many overlaps causing too much complication to see the relationship among these strata. Furthermore, at each element's data point, I-beam mark is used to indicate the Standard Deviation, (σ) range.

If a comparison of Cluster 1 and Cluster 2 green lines containing standard deviation, it can be found that the degree of overlap is large, so they were combined and call as Group 1 (black line). Look the standard deviation range of this Group 1, it has a high degree of overlap with the two green lines. So, it is reasonable to combine them and call it JiangYi Layer.

If a comparison of the spidergrams and standard deviation of the blue DaWa Layer and black JiangYi Layer, it can be seen that there is no overlap. Thus, they are two different Layers. If a comparison of the spidergrams and standard deviation of the black JiangYi Layer and red ChuanJie Layer, it can be seen there is no overlap, except the standard deviation of three elements. Thus, they are two different Layers.

Furthermore, in the curve lines of blue DaWa

若比对含有标准误差两条团聚的绿色曲线（团聚一和团聚二），它们重叠程度很大，故而推论且合并为一，并以第一族群（黑色线）称之；再观合并后第一族群之标准误差范围，与两条绿色曲线标准误差范围的重叠度极高，因而认为此合并合理，而暂称为“姜驿层”。

若以蓝色线“大洼层”和黑色线“姜驿层”的蜘蛛线和标准误差范围来比较，可见两者所有稀土元素曲线，并无任何重叠之处，因而推论为两个不同的地层。比对黑色线“姜驿层”和红色线“川街层”的蜘蛛线与标准误差范围，虽有两三个元素的标准误差范围重叠，但基本上它们是两条没有重叠的曲线，故推论为两个不同的地层。

除此之外，蓝色“大洼层”和黑色“姜驿层”的曲线，在铥 (Tm) 元素的位置往上凸出，而在红色“川街层”此元素位置的曲线，却是略为往下凹，造成明显的曲线形状差异。这是这三个地层有趣之处，但这些凸出和凹陷的含意如何，目前并不清楚，或许值得进一步探讨。

整体来看，这三条曲线（三个恐龙地层）的稀土元素浓度，显示出粗略的钟形曲线，指出“轻稀土元素群 (Light Rare Earth Elements, LREE)”和“重稀土元素群 (Heavy Rare Earth Elements, HREE)”，比“中稀土元素群 (Medium Rare Earth Elements, MREE)”少，亦即这些恐龙化石的中稀土元素群浓聚效应较为强烈，此中稀土元素群浓缩效应，也可从 Nd-Gd-Yb 三元素（分别代表轻、中、

Layer and black JiangYi Layer, there is an upward kink at element Tm, where at this position of the red ChuanJie Layer line is a downward kink, causing a noticeable curve shape difference. This is an interesting point of these three formations. But what this means is not clear, pending further investigation.

As a whole to see this three curve lines (three dinosaur layers), the REE concentrations roughly show as a bell shape of less Light Rare Earth Elements (LREE) and Heavy Rare Earth Elements (HREE) than that of Medium Rare Earth Elements (MREE). Which means within these dinosaur fossils, MREE has higher concentration effect. This MREE concentration effect can be seen from triagram of Nd-Gd-Yb (representing LREE, MREE, and HREE) in Figure 6.2.4. Most of the dinosaur sites located at the upper center position.

From the papers published (Trueman & Palmer, 1997), we do not have the relationship between the precise age and concentration ratio. But in general, it can be said that the higher REE concentration, the older the layer. The vertical positions of these three curve lines in this graph when compared with known information of dinosaur formations inside ChuXiong Prefecture, DaWa Layer is the oldest (at the very top) and the youngest ChuanJie Layer at the very bottom with lowest REE concentration. The age of DaWa Layer is around 200 million years old, and the ChuanJie ~160 million years old. Then, can we estimate that JiangYi as 170 million years old? Further study will be needed.

重稀土元素群）浓度比例的三角图中（图6.2.4），大部分恐龙采集点坐落在中央偏上的位置看出来。

从国际上已经发表的相关论文（Trueman & Palmer, 1997）得知，目前虽然没有精确年代与浓度比例关系的数据，但大致上来说，化石内稀土元素含量浓度越高，化石地层年代也越古老；这三条主要曲线所处纵轴位置，与楚雄州内已知相关地层年代来看，最古老“大洼层”者在最上方（浓度最高），最年轻的“川街层”者在最底端，恐龙骨头内稀土元素浓度最低，前者估计年代为二亿多年，后者为一亿六千万年，那么是否可以粗略推论说，“姜驿层”者约为一亿七千万年前的地层？这一点，还有待进一步研究。

楚雄苍岭恐龙脚印地层的困惑

在楚雄市，离市中心大约 16 公里东边，有一个小村落，名为苍岭黄草坝，又名饱满街的小山头上，在 1990 年代，曾经发现了五百多个恐龙脚印，分散在一片广大的山坡上，在其中，有一小块用铁丝围起来，面积为7.8 m x 17.15 m = 134 平方米，算是对比较精华地区的保护。第一作者与蒙大拿州立大学维立其欧博士 (Dr. D. Varricchio) 和他的博士生肯内尔等人，于 2006 年春到此考察的时候了解到，似乎自从这个地方被发现之后，除了原本的地质调查，曾在《云南地质》发表过两篇地质学（非古生物学）者们的一些看法（陈述云和黄晓钟；

The formation confusion of the dinosaur tracks at BaoManJie, ChuXiong City

A small village about 16 kilometers from the ChuXiong City center is called Cangling, also called BaoManJie. More than five hundred dinosaur footprints were found in a big slope area. In which a small section, 7.8m x 17.15m = 134 m^2 was fenced for protection. The first author and Dr. D. Varicchio and his Ph.D. student Mike Knell visited the site in the Spring of 2006. We learned that since this place was discovered, except the original geological survey published two geological (not paleontological) papers in YunNan Geology (Chen, SuYun and Huang, 1993, 1994), stating this is a Cretaceous formation. No one else did any works since then, so we decided to do a systematic study.

In October of 2016, the fourth author made the first initial report and published that in the annual meeting of Society of Vertebrate Paleontology (SVP).

1. A recently found dinosaur footprints site, ChuanJie Formation, used to be called as the bottom of upper Lufeng Formation, or upper Lufeng Formation, about early the stage of mid-Jurrasic

2. Footprints on more than two meters thick dark red clay. This small layer ranges from 1 to 5 cm, with multiple cross-over cracks and water wave marks. The areas have footprints is think, about 5 cm, dark red mud and cross-over cracks and water wave marks.

3. The area containing dinosaur footprint is about 105 m^2, with at least six tracks and densely packed overlapping belt, indicating at least

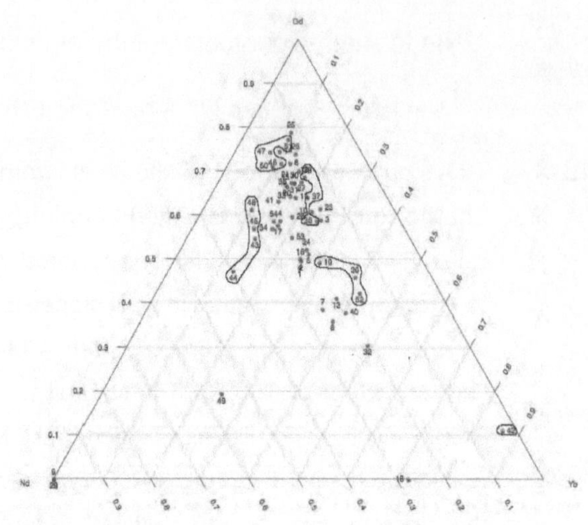

◀ 图 6.2.5 轻中重稀土元素 (Nd, Gd, Yb) 三角图。封闭曲线包围者为第二族群，其余属第一族群

◀ Figure 6.2.5 Triagram of Light, Medium, and Heavy REE. Enclosed belong to Group 2, the rest Group 1

1993, 1994)，认为这是白垩纪的地层之外，未曾有人认真地做过本处的地层研究。因此，我们在此做一些系统性的考察。

2006年10月，第四作者整理出如下的初期报告，并于当年美国的古脊椎动物学会 (Society of Vertebrate Paleontology, SVP) 年会中发表：

1. 一个最近在中国云南楚雄州内发现的川街层恐龙脚印，川街层相当于以前叫做上禄丰层的底部或上禄丰层，年代可能是中侏罗纪早期；

2. 足迹平面在于 2 米多厚的暗红色泥岩序列，这些小层面厚度从 1 到 5 厘米，并显示多重交互的泥土龟裂和水波纹，有脚印的单位薄（约 5 厘米），暗红色泥岩，同时有龟裂和水波纹；

3. 有恐龙脚印的平面，大约有 105 平方米，至少含有 6 条足迹路径，另有紧密重叠的足迹带，表示最少还有另外两条足迹路径；

two more tracks.

4. Each narrow track contains paired mixed front and hind legs. The front leg's footprints are new moon shape, but changes according to the front and hind limb distance, and overlapped by the hind limb footprint. The hind limb footprint shape is less triangular to oval with a side toe pointing out. Few show four small forward-pointing toes. The distance between the front and hind limbs is between less than 1 meter to 1.7 meters, corresponding to the size of the hind limb (length).

5. In one single track, the footprint of the front and hind limb of the same side is still clear, not overlap with the overlapping length. However, along with the increase of the distance between the front and hind limb, the hind limb footprints started to overlap with the footprint backside of the front limb. The posture of four legs, half risen hind leg, walking with elevated legs and reduced size digits point out these footprints belong to sauropods.

6. In this formation, there is only one sauro-

4. 每条宽度狭窄的足迹，含有成对交互的前肢和后肢；前肢印痕新月状，但其外观随着前后肢距而改变，又被后肢脚印重叠；后肢足迹呈次三角形到椭圆形，侧面指向突出，少数显示四个小而向前的爪痕；前后肢距离，大约从少于1米到1.7米,对应于后肢大小(长度)；

5. 在某单路径中，同一边的前肢和后肢路径，仍然明显，也没有和短的前后肢距长度重叠；然而，随着前后肢距的增加，后肢脚印开始与前肢脚印的后端重叠，四脚的姿态，半提脚后肢，和提脚走路以及缩小的脚趾骨，这些足迹是蜥脚类恐龙的；

6. 在本地层，仅有一种蜥脚类恐龙，近亲未详的川街龙，其前后肢只有一个单一的爪趾，这些足迹的狭窄宽度和指向前方的前后肢趾，可能是更原始的蜥脚类恐龙，形态上更接近于蜀龙或火山齿龙；这种说法与最近的基本蜥脚类恐龙演化时间解释一致。

从我们实地测量考察以及恐龙脚印的形状和路径（图6.2.6）等资料，我们认为这是中侏罗纪或更早的恐龙所留下来的遗迹化石。进一步论文搜索，出现了两篇日本和西方人合写发表的论文 (Lockley et al. 2002; Chen et al. 2006)，都引用原始《云南地质》地质调查者的文章，认为这些恐龙脚印和路径，属于白垩纪的恐龙，但未就地层的部分深入讨论。这真是一个莫大的困扰，从实地观察，不论从恐龙脚印的形状来说，或从其他相关的证据来说，我们认为是侏罗纪地层却和文献上所认为的

pod dinosaur. Unclear relative *Chuanjiesaurus* have a single claw in front and hind legs. This narrow and forward pointing track show the front, and hind legs could belong to the more primitive sauropod, morphologically closer to *Shunosaurus* or *Vulcanodon*. This saying agrees with the latest sauropod evolution time.

From our measurement at the site and the shape of footprints and track, we think the mid-Jurrasic or even earlier dinosaurs left these. After searching, there are two papers by Japanese and Westerner (Lockley et al. 2002; Chen et.al. 2006). They directly quoted the YunNan Geology papers and thought these footprints and tracks are Cretaceous. They did not explore the strata. This is big trouble. From our observation at the site, not only just the shape of the footprints but also other related information, we consider this is Jurassic formation, millions of years apart from Cretaceous. They belong to different Periods of the Mesozoic Era.

Within the whole Yunnan Province, the main dinosaur area is the ChuXiong Prefecture, and neighboring KunMing and YiMen and ErShan County of YuXi Area. In the past 70 or more years, in this vast area, numerous dinosaurs were found. All of them are earlier than mid-Jurassic. No a single piece of Cretaceous dinosaur bone ever unearthed. However, at this particular site, the Cretaceous dinosaur footprint showed up. This is odd and unexplainable. Plant-eating dinosaurs would move along with the vegetation growth. So, there was a possibility that a herd of the Cretaceous dinosaur walked thru this muddy earth and left their footprints. However, thousands of migrating dinosaurs with the

白垩纪，相去好几千万年，属于中生代不同的纪，难以吻合。

在整个云南出土恐龙地区，主要是楚雄州和邻近的昆明市以及玉溪地区易门县、峨山县等地也都有恐龙出土的记录，七十多年来在此广大地区内发现了无数只恐龙，全都是侏罗纪中期以前的恐龙，但从来没有一块白垩纪恐龙的零星骨头出现过，可是，竟然会在这个地方出现白垩纪恐龙的脚印，的确有些奇怪，也难以解释。恐龙有随着植（食）物生长而迁移的习性，因此，有一种可能，这些被认为是白垩纪的恐龙，刚好走过这一片泥巴地面，留下了这些珍贵的脚印。然而，整群恐龙随着气候季节迁移，在整群成千上万只恐龙当中，总会有老弱病死或落单被猎食性恐龙杀死遗留的尸体，应该或多或少有些残骸骨头遗留下来才对，就像北美晚白垩纪的鸭嘴龙三角龙等，从美国蒙大拿到加拿大的北极地区都有发现。然而在此地却只找到白垩纪恐龙脚印，但没找到任何白垩纪的恐龙化石，相当奇怪，而且不

seasonal changes, there must have sick and old, or single slower. The predatory dinosaurs would kill them, and their corpses remain in this vast area. Just like the North American hadrosaurs or triceratops. From Montana of the USA thru Canadian Arctic region. So, just the Cretaceous dinosaur footprints found without any bone fossil in this area is very strange. Furthermore, not only no Cretaceous dinosaur ever found in YunNan province along, in the much bigger areas of neighboring QuiZhou and Sichuan provinces, no report of a Cretaceous dinosaur ever found. So, where were these Cretaceous dinosaurs left these footprints went?

In this geological investigation, besides the collection of dinosaur fossils, we also collect the matrix, including the BaoManJie and HeMeng-Ko dinosaur footprints site. HeMenKo dinosaur footprints were discovered by us in the Spring of 2006 when we were "escaping (from stuck by rain)" from ShuanBo County. The geological map clearly shows this site as mid- and early Jurassic. No question about it.

Figure 6.2.7 redraws the analyzed spidergrams of the dinosaur fossils. DaWa Layer is on

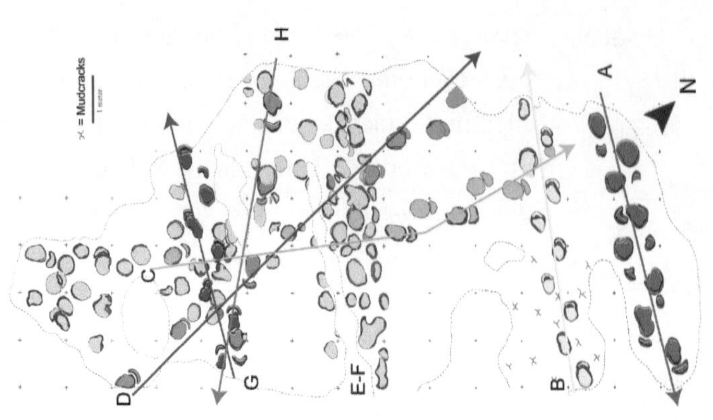

◀ 图 6.2.6 楚雄市苍岭饱满街恐龙脚印路径

◀ Figure 6.2.6 Dinosaur Tracks at CangLing of ChuXiong City

仅云南没有白垩纪恐龙出土的记录，邻近的贵州省和四川省等更广大的地区，也都没有报道。那么，这些留下密密麻麻脚印的恐龙，到底跑到那里去了？

在本次地质考察的过程中，除了采集恐龙化石样本之外，也采集了化石围岩，含楚雄市苍岭饱满街和双柏县河门口恐龙脚印的围岩。河门口的恐龙脚印是我们于2006年春从双柏县“逃难”回程中无意间最早发现的，此脚印地点在地质图上标示为中早侏罗纪，没有争议。

图6.2.7把前面所分析出来的恐龙化石蜘蛛线，再度画出来，“大洼层”，在最上面，以蓝色—B—表示之(B代表Bone)，“姜驿层”者在中间，以黑色—B—表示之，“川街层”则以绿色线—C—表示之(C代表ChuanJie)；在图左边中间部分，前面几个元素的曲线几乎重叠、有一条蓝色—M—的曲线(M代表Matrix)，此蜘蛛线，是第二族群“大洼层”恐龙化石围岩的平均值曲线；在此图的最下方，另有一条黑色—M—的蜘蛛线，这是“姜驿层”恐龙化石围岩的平均值曲线。

从恐龙化石和其围岩的蜘蛛线来说，在围岩里面的稀土元素浓度，在对数值的纵轴上，数值大约相差3~4，亦即围岩里面的稀土元素浓度，只有化石内含的千分之一到万分之一而已，亦即诸稀土元素在化石的形成过程中，有被浓缩集中的现象。这一点，“大洼层”和“姜驿层”都差不多。

再看纵坐标大约在0到1之处，这里有两条较细的蓝色—▼—线条，部分

top and marked as blue -B- (B means bone), JiangYi Layer in the middle as black -B-, and ChuanJie Layer with green -C- (C means ChuanJie). To the center of the left side, the curves of several front elements almost overlapped. Then, there is a blue -M- (M means Matrix) spidergram curve. This is the mean matrix REE of the Group 2, DaWa Layer. To the bottom of this graph, there is another black -M- spidergram curve. This is the mean matrix REE of the JiangYi Layer.

Looking at the spidergrams of dinosaur bone and matrix, it can be said that the REE concentration of matrix along the vertical axis has a difference of 3–4, which means that the matrix REE concentration is about one of one thousand or ten thousands. That means that various REE was not concentrated during the fossilization process. For this, both DaWa Layer and JiangYi Layer are about the same.

Now, look at the 0 to 1 on the vertical axis, there are two thinner blue -▼- lines, with portions as dash-line, and one red dash-line -▲-. The two blue lines were from the BaoManJie dinosaur footprint site and the water wave marks. There were almost identical and in the range of DaWa Layer matrix. The red line is the matrix of HeMengKo dinosaur footprint site, also within this range. The HeMengKo site is confirmed as mid- or early Jurassic. So, the BaoManJie site should be mid- or early Jurassic.

The matrix REE concentration ratio of BaoManJie dinosaur footprint site showed up in the range of late Triassic to early Jurassic DaWa Layer. Does this strongly suggest that the formation of this site is of late Triassic to early Jurassic, not Cretaceous formation? Otherwise, how could it locate in this

线段是虚线，还有一条含有虚线细红色—▲—线条。蓝色的这两条线，就是取自于楚雄苍岭饱满街恐龙脚印的围岩和附近的水波纹围岩，它们可说是完全相同,而且坐落在“大洼层”围岩的范围内，红色那一条是双柏县河门口恐龙脚印的围岩,也都在这个区域内。河门口的地层，已确定为早中侏罗纪的，那么饱满街地层，看来也应该是早中侏罗纪的。

range? Is that possible that the Cretaceous matrix would have almost identical REE fingerprints as that of 70 or 80 million years earlier? We are waiting to explore further. Stratigraphy scholars marked this footprints and water wave mark layer as the Cretaceous formation is much far away from our conclusion of late Triassic to early Jurassic. This conflict is waiting for further study.

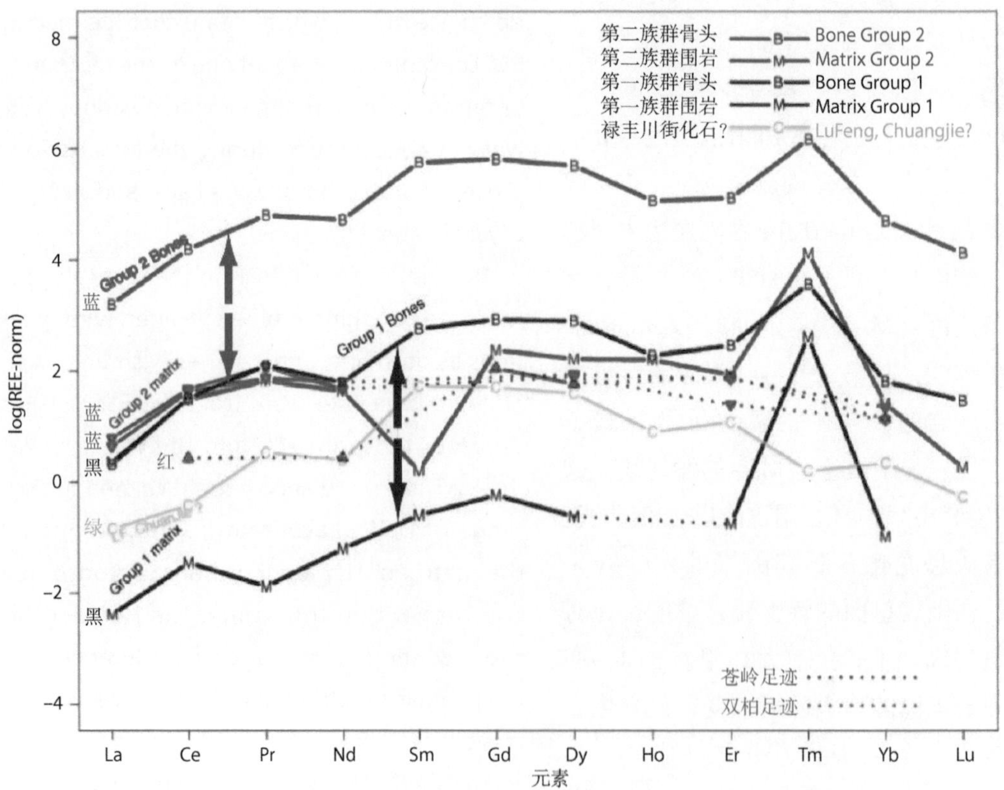

▲ 图 6.2.7 云南楚雄州内恐龙地层围岩稀土元素蜘蛛线图。垂直双箭头线段标示第二族群、第一族群之化石与围岩曲线差距，以全国土壤（A 层）稀土元素背景值基本统计量（CBA）为正常化基准

▲ Figure 6.2.7 Spidergrams of Matrix of Dinosaur Strata of ChuXiong Prefecture, YunNan; Vertical double arrow lines mark the distances between the bone and matrix. Normalized with REE Chinese Background Average (CBA)

楚雄市苍岭饱满街恐龙脚印岩层内的稀土元素浓度比例，竟然出现在晚三叠早侏罗纪“大洼层”的范围内，这是否强烈暗示，这里的地层是晚三叠到早侏罗的地层，而非白垩纪的地层呢？要不然，怎么会坐落在这个范围内？难道白垩纪的围岩会和早在七八千万年前的地层，有几乎完全相同的稀土元素指纹？这一点，有待探讨。地层学者把这一大块地区这些脚印和水波纹的地层，划分为白垩纪地层，与本研究所推论为晚三叠纪/早侏罗纪相去甚远。这个矛盾，有待研究。

4. 结论

本文透过采集样本、化学和统计分析恐龙化石与围岩内之稀土元素，提供了一个云南楚雄州内含恐龙地层信息的初窥窗口，研判推论暂称为“大洼层”、“姜驿层”、和“川街层”；然而本文所用到的样本，仅仅姜驿地区比较完整，其余楚雄州内各盆地内的恐龙点，如禄丰县、牟定县、武定县、双柏县等，所采取的样本数较少，未能充分完整描述恐龙地层的全貌，来日应该在各恐龙出土地点，采集更多样本分析，以便建立完整的云南恐龙地层资料，作为将来发现新恐龙点、推论其地层的参考。

恐龙化石内稀土元素统计分析与相关地质含意，本文仅做出相对恐龙地层研判推论的第一步。自初步从“铈异常(Cesium Abnormality)”与“氟铀浓度比率 (F/U Concentration Ratio)”来说，这几个恐龙地层当时的生态埋藏环境，类

4. Conclusion

This paper was done, including specimen collection, chemical, and statistical analyses of REE in dinosaur bones and matrix. It provides a window to see the dinosaur layers in ChuXiong Prefecture of YunNan province, temporarily named as DaWa Layer, JiangYi Layer, and ChuanJie Layer. However, most of the analyzed specimens were from JiangYi area. In comparison, specimens from the dinosaur sites in various basins of ChuXiong Prefecture, such as Lufeng County, MoDing County, WuDing County, and ShuanBo County were less. The results could not represent a complete comprehensive picture of the dinosaur strata. In the future, more specimens shall be collected at each dinosaur site to establish a complete YunNan Dinosaur Strata Database for future exploration of new dinosaur site, and references for other studies.

On the issue of statistical analyses and the related stratigraphic meaning of REE inside dinosaur bone, this paper deals the first step of stratigraphy relativity. If looked from the points of view of Cesium Abnormality and F/U Concentration Ratio, these dinosaur burial layers in those days were similar to reducing shallow water swamp area. Further studies such as the redox ecological environment, Ce abnormality analysis, diagenesis analysis, etc. (Metzger et al. 2004), have to be done in the future so that complete information of this critical dinosaur ecology and evolution can be available.

似于还原态的浅水沼泽地区。进一步推论，如埋藏时期环境的氧化/还原生态状态、铈异常分析、成岩作用分析等等(Metzger, et al., 2004)，有待来日进一步研究，以期对此重要早期恐龙生态环境乃至恐龙物种演化等相关信息，建立更完整的资料。

致谢

本项研究得到云南楚雄彝族自治州博物馆钟仕民馆长与馆内诸多协助，特表最高谢意；美国蒙大拿州立大学地球科学系维立其欧博士指点从地质化学角度探讨此一世界上非常重要早期恐龙地区，致使本文出现，一并感激。

Acknowledgement

We thank Director Zhong, ShiMing and his staffs of ChuXiong Prefecture Museum. Also, thanks to Dr. David Varicchio of the Department of Earth Science of Montana State University provided plenty of help on the geochemistry.

参考文献 References

（1）王云、魏复盛编著 (1990)；土壤环境元素化学，336 页，表 18.9；全国土壤（A 层）稀土元素背景值基本统计量。中国环境科学出版社。

（2）方晓思，庞其清、卢立伍、张子雄、潘世刚、王育敏、李锡康、程政武 (2000) 云南禄丰地区下、中、上侏罗统的划分。北京地质出版社。

（3）百度百科 (2009)：下禄丰组 (Xialufeng Fm) 的时代属早侏罗世；分布于云南境内；命名地点在云南禄丰县城东北沙湾、张家坳一带；为一套陆相红色碎屑沉积，以暗紫色、紫红色砂质泥岩和灰绿色砂岩为主，含爬行类 *Lufengosaurus huenei*、*Bienotherium yunnanense* 等及双壳类 *Unio ningxiaensis, Sib-iriconcha shensiensis*，叶肢介 *Gomphocythere cf recticulata*、*Darwinula bella* 等；在其下部常夹有绿色页岩，含 *Podoza-mites lanceolatus*、*Neocalamites* 等植物化石；厚 1500 米；与下伏一平浪组呈假整合接触。

（4）台湾环境检验所（2003）“电感耦合等离子体发射光谱法” NIEA M104.01C 标准方法；环境检验所，中坜。

（5）陈述云、黄晓钟 (1993) 楚雄苍岭恐龙足印初步研究。云南地质，第 12 卷，第 3 期，第 266–276 页。

（6）陈述云、黄晓钟 (1994) 从楚雄苍岭恐龙足印探讨有关问题。云南地质，第 13 卷，第 3 期，第 285–289 页。

（7）Chen, P.J., Li, J.J., Matsukawa, M., Zhang, H.C., Wang, Q.F., and Lockley, M.G. (2006) Geological ages of dinosaur-track-bearing formations in China: *Cretaceous Research*, 27, 22–32.

（8）Gromet, P., Dymek R.F., Haskin, L.A., and Korotev, R.K. (1984) The "North American shale composite": Its compilation, major and trace element characteristics: *Geochimica et Cosmochimica Acta*, Vol. 48, 2469–2482.

（9）Lockley, M., White, J., W.D., Matsukawa, M., Li, J.J., Feng, L., and Hong, L. (2002) The first sauropod trackways from China: *Cretaceous Research*, 23, 363–381.

（10）Le.cuyer, C., Bogey, C., Garcia, J.-P., Grandjean, P., Barrat, J.-A., Floquet, M., Bardet, N., and Pereda-Superbiola, X. (2003) Stable isotope composition and rare earth element content of vertebrate remains from the Late Cretaceous of northern Spain (Lan‹o): did the environmental record survive?: *Palaeogeography, Palaeoclimatology, Palaeoecology*, 193, 457–471.

（11）Luo, Z.X.; Wu, X.C. (1994) The small tetrapods of the Lower Lufeng Formation, Yunnan, China: In the shadow of the dinosaurs: early Mesozoic tetrapods, edit by Nicholas C. Fraser, Hans-Dieter Sues, Cambridge University Press.

（12）Martin, J.E., Patrick, D., Kihm, A.J., Foit Jr., F.F., and Grandstaff D. E. (2005) Lithostratigraphy, Tephrochronology, and Rare Earth Element Geochemistry of Fossils at the Classical Pleistocene Fossil Lake Area, South Central Oregon: *The Journal of Geology*, volume 113, 139–155.

（13）Metzger, C.A., Terry Jr., D.O., and Grandstaff, D.E. (2004) Effect of paleosol formation on rare earth element signatures in fossil bone: *Geology*, 32(6), 497–500.

（14）Patrick, D., Martin, J.E., Parris, D.C., and Grandstaff, D.E. (2004) Paleoenvironmental interpretations of rare earth element signatures in mosasaurs (reptilia) from the upper Cretaceous Pierre Shale, central South Dakota, USA: *Palaeogeography, Palaeoclimatology, Palaeoecology*, 212, 277–294.

(15) Staron, R.M., Granstaff, B.S., Gallagher, W.B., and Granstaff, D.E. (2001) REE Signatures in Vertebrate Fossils from Sewell, NJ: Implications for Location of the K-T Boundary: *PALAIOS*, 16, 255–265.

(16) Trueman, C.N. (1999) Rare Earth Element Geochemistry and Taphonomy of Terrestrial Vertebrate Assemblages: *PALAIOS*, 14, p. 555–568.

(17) Trueman, C.N., and Benton M. J. (1997) A geochemical method to trace the taphonomic history of reworked bones in sedimentary settings: Geology, 25(3), 263–266.

(18) Trueman, C.N., and Palmer, M.R. (1997) Diagenetic Origin of REE in Vertebrate Apatite: A Reconsideration of Samoilov and Benjamini, 1996: *Palaios*, 12(5), 495–497.

(19) Trueman, C.N., Benton, M.J., and Palmer, M.R. (2003) Geochemical taphonomy of shallow marine vertebrate assemblages: *Palaeogeography, Palaeoclimatology, Palaeoecology*, 197, 151–169.

6.3 化石内的碳酸盐

只是化石光谱中的方解石?

化石里面含有碳酸盐 (CO_3^{2-}) 很普遍，在最纯态它最强的红外波峰在 ~1440 cm^{-1}，如图 6.3.1 绿框内黑色箭头所示，它是较宽带，稍微拖着往高波数 1500~1750 cm^{-1} 尾巴，在一般碳酸钙红外线范围 (800~4000 cm^{-1})，下一个显著波峰在 ~900 cm^{-1}和一个在约1800 cm^{-1}很小的泛音 (overtone)波折。

但是在自然界出土的方解石，如图 6.3.2 所示，可看到很多非碳酸盐有机物质杂物的波峰。从埋藏学的角度来说这可理解，当碳酸阴离子 (CO_3^{2-}) 与其他阳离子（如 Ca^{2+}, Mg^{2+}, Fe^{2+}等）沉淀、形成诸碳酸化合物结晶（方解石、霰石、菱镁矿、菱铁矿等等），有些大分子分解的有机残渣被绑入了结晶里面，因此天然的碳酸盐矿物含有有机物，在红外线光谱出现各种有机波峰，Verma等报告说，磨成粉末未处理过的红鲍鱼珍珠母，加热六小时，这些有机波峰随着温度提升而逐渐消失，到了600℃，几乎都消失殆尽，如第5章图 5.2.4 所示。

对于化石内的碳酸盐来说，图 6.3.2 很明确地显示有机杂物，而非如图 6.3.1 纯碳酸钙。在图 6.3.2，碳酸根波峰如预期地出现在约1450 cm^{-1}，同时也看到很多甚至更强或无法忽略的波峰/带。碳酸钙在约900 cm^{-1} 的波峰，出现于纯式和自然界方解石样本光谱中，然而在自然界形成碳酸盐，波峰强度没有纯碳酸钙

6.3 Carbonates in Fossils

Just Calcite In the Spectra of Fossils?

Carbonates (CO_3^{2-}) are common inside fossils. In its pure form, it shows the strongest IR peaks at ~1440 cm^{-1} as shown in black arrow in the green box of Figure 6.3.1. It's kind of broadband with slightly more tailing toward higher wave-number in the range of 1500–1750 cm^{-1} range. In the regular IR range (800–4000 cm^{-1}) of calcium carbonate, the next noticeable peak is at ~900 cm^{-1} and tiny overtone kink at ~1800 cm^{-1}.

On the contrary, when naturally formed calcite is encountered, such as the Montana calcite in Figure 6.3.2, many non-carbonate spectral contributions of impurities which include organic matters are seen. This is understandable from the taphonomic point of view. When carbonate anions (CO_3^{2-}) participated with various cations (such as Ca^{2+}, Mg^{2+}, Fe^{2+}, etc.) forming various carbonate compound crystals (Calcite, Aragonite, Magnesite, Siderite, etc.), some decomposed organic debris from macromolecular were incorporated inside the crystals. Thus, naturally formed carbonate minerals containing these organics will show various organic peaks in the IR spectra. Verma reported when powdered and undisturbed nacre was heated for 6 hours, these organic peaks gradually diminishing along with higher temperature and almost totally gone at 600 °C, as shown in Figure 5.2.4 of Chapter 5.

For the carbonates inside fossils, it is a case of organic impurities in Figure 6.3.2, rather than the pure form of calcium carbonate as Figure 6.3.1. In Figure 6.3.2, carbonate peaks at ~1450cm^{-1} as expected, and can be seen along with many other even stronger or not-negligible peaks/bands.

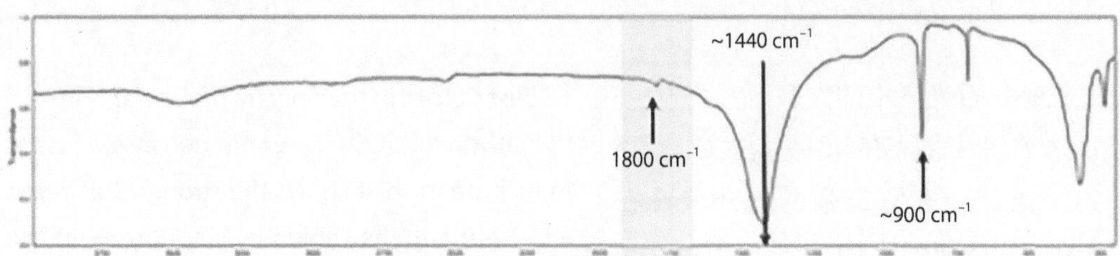

▲ 图 6.3.1　纯方解石（碳酸钙）红外线光谱

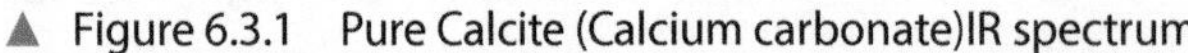

▲ Figure 6.3.1　Pure Calcite (Calcium carbonate)IR spectrum

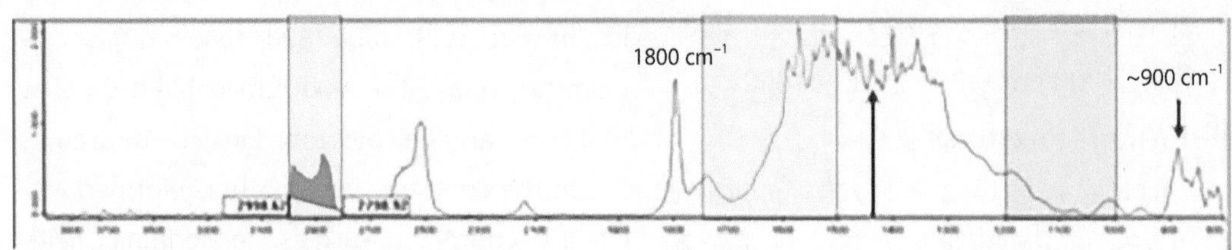

▲ 图 6.3.2　蒙大拿方解石红外线光谱

▲ Figure 6.3.2　Montana calcite IR Spectrum

那么强。在 1500~1750 cm^{-1} 范围内，纯碳酸钙的贡献很小，在这范围的波峰来自质子化蛋白质羧基 amide-I 和 amide-II 的对称与非对称伸缩。

图 6.3.2 内，在 1800 cm^{-1} 尖锐的波峰和在大约 2920 cm^{-1} 处的宽带，也被指派为有机杂物，亦即酸性蛋白质的裂解产物。在 1800 cm^{-1} 尖锐的波峰是酸性蛋白羧基的伸缩，而 2920 cm^{-1} 则是蛋白质 C—H 键的伸缩。1800 cm^{-1} 波峰也可能是酰卤 (acyl halides, R—COX)，酰卤可能由如蛋白质降解的氨基酸之氨基 (—NH_2) 被地下水或海水的卤离子取代而来。因此，当解读化石的红外线光谱，碳酸盐存在不可被忽略，碳酸盐是化石组成的成分。然而化石含有不只碳酸盐，还含有被保存的有机残留物。

The calcium carbonate peak at ~900 cm^{-1} shows up in both pure form and natural calcite. However, in the naturally formed carbonates, the intensity is not as strong as that of pure calcium carbonate. In the region from 1500 cm^{-1} to 1750 cm^{-1}, pure calcium carbonate has a tiny contribution. This region also has bands due to amide-I and amide-II asymmetric and symmetric stretching of protonated carboxylic group of protein.

The sharp peak at around 1800 cm^{-1} and broadband at around 2920 cm^{-1} of Figure 6.3.2 are also assigned to the organic impurities, i e., degradation products of acidic proteins. The sharp peak at 1800 cm^{-1} is the stretching of carbonyl of acidic protein while 2920 cm^{-1} as stretching of C-H bonds of proteins. The 1800 cm^{-1} peak could also come from acyl halides (R-COX), which could be formed by the replacing of amine functional groups (—NH_2) such as

实体化石除了以磷灰石群为主的硬骨头和牙齿之外，也先撇开软组织、外骨骼被保存下来的化石之外，还有一大类是以碳酸盐为基质背景的，最常见的就是珊瑚、贝壳类、海胆等水生动物。这些水生物的壳基本上是碳酸钙，然而碳酸盐矿物都有很好的解理。从理论上来说，应该很容易脆裂，稍微一碰就裂开了，不容易保存下来，何况只要埋藏环境稍微有点酸性，碳酸盐很容易被溶解掉，但是事实并非如此，现生的贝壳类海胆等，活得好好的，它们的化石也很多。那么原本容易解理的碳酸盐基质壳，是怎么胶结成那么硬的化石呢？在这些生物的壳内，到底有哪些有机物质当胶水，把容易解理的碳酸盐绑得紧紧的？这些“胶水”有没有被保存在化石里面？如果有的话，是怎么一种情况？是哪些有机物被保存下来了？碳酸盐在这些化石保存有机残留物扮演了什么角色？

6.4 痕量元素

（维基百科）痕量元素是某种化学元素，其浓度（或其他测量）很低，准确的定义则依科学领域而有不同。在分析化学中，“痕量元素”指的是“某一元素以原子规模测量在样本中的平均浓度小于百万分之 100”，或是这个元素在每克样本中含量少于 100 μg (μg，1 μg= 0.000001 g)。在生物化学上，“痕量元素”指的是，只需要微小量的某种食品元素，就能使组织适当地成长、发展、完成组

amino acids from the degradation of proteins with halide ions (Cl^-, Br^-, and I^-) from the ground- or sea-water. So, when interpreting fossil IR spectra, the existence of carbonates should not be neglected. Carbonates are parts of fossil constituents. However, fossil contains more than just carbonates. It contains preserved organic remains.

Devendra Verma, Kalpana Katti, Dinesh Katti, Photoacoustic FTIE spectroscopic study of undisturbed nacre from red abalone, Spectrochimica Acta Part A 64 (2006) 1051–1057

Besides the hard bones and teeth which are apatites in the body fossils, and neglect the soft tissue and exoskeleton, there are many fossils using carbonates as body parts, such as corals, seashells, and sea urchins marine animals. The shells of these marine creatures are the carbonates. However, carbonates have perfect cleavage. So, in theory, they can be cleaved off easily by a small bump, and not easy to be preserved. Moreover, if the burial environment has just a little bit of acidic, carbonates can be easily dissolved. However, the fact is not like this. These extant marine animals live very well, and so many fossils preserved. So, the question is why and how the easy cleavage carbonate-base shells turned into such hard fossils? Within the shells of these creatures, what kind of organic matter is used as 'glue' to bind the easily cleavable carbonates together? Did these 'glues' get preserved in fossils? If the answer is yes, then, how the situation is? What organic matter got preserved? What is the role of carbonates in these fossils for the organic remain preservation?

6.4 Trace Elements

(WiKi) A trace element is a chemical element

织的生理机能，如镁是个痕量元素。在地球化学上“痕量元素”某种化学元素是在岩石中的含量少于百万分之 1000 或 0.1%，这术语主要用于“火成岩岩石学”上。

就以人体生物化学的角度来说，矿物质是没有连接到碳原子的无机化学元素。如果身体需求量每天超过 100 毫克 (mg) 的 11 种，称为矿物质，若需求量少于此量者，则称为痕量元素。痕量元素每天需求量通常每天只要几毫克或几微克，缺乏或不平衡微量元素可能导致生理问题。

以下几种痕量元素为已知人体所需：铬、钴、铜、碘、铁、镁、钼、硒、锌，而以下几种痕量元素之量则未确定：氟、砷、硼、镍、矽、锡。从广义的角度来说，这些痕量元素，也该是较高等脊椎动物维持生理活动所需要的。但是，在个体活着的时候，这些痕量元素的浓度都很低，介于毫克和微克等级。然而，当古代生物石化之后，受到掩埋条件的影响，化石内这些元素的含量，可能会有大幅度的变化，如稀土元素，在脊椎动物活体内，几乎不存在，但是在骨头化石内，却有显著的浓缩吸附，永远卡在磷灰石群的晶格内。因此透过比对现生动物体内和化石内痕量元素，可提供一些埋藏学或死亡原因的信息。

以今天现有可用的痕量元素分析方法、设备与灵敏度等来说，可以说兵器库内可供挑选的武器多多，任君选择，比方说电感耦合等离子体质谱的话，在

whose concentration (or other measures of amount) is very low (a “trace amount”). The exact definition depends on the field of science. In analytical chemistry, a trace element is one whose average concentration of less than 100 parts per million (ppm) measured in the atomic count or less than 100 micrograms per gram. In biochemistry, a trace element is a dietary element that is needed in very minute quantities for the proper growth, development, and physiology of the organism. For example, magnesium is a trace metal. In geochemistry, a trace element is one whose concentration is less than 1,000 ppm or 0.1% of a rock’s composition. The term is used mainly in igneous petrology.

In the human body biochemically speaking, minerals are inorganic chemical elements not attached to a carbon atom. If the body requires more than 100 milligrams of eleven minerals each day, the substance is labeled a mineral. If the cellular body requires less than this, it is labeled a trace mineral. Trace minerals are generally needed in quantities of only a few milligrams (mg) or micrograms (μg) per day. Missing or imbalance of trace elements could result in physiological problems.

Our human body needs the following element: Cr, Co, Cu, I, Fe, Mg, Mo, Se, and Zn. And the quantity of the following elements are not sure yet: F, As, B, Ni, Is, and Sn. From a broader point of view to say, these trace elements are also needed by the higher level vertebrae to maintain their physiological function. However, when the being is still alive, the amount of these trace elements are very low, in the range of milligram or microgram. However, when the ancient creature died and went thru the fossil-

理想状况下可同时分析八十种元素（同位素），而且灵敏度可达10^{15}分之一 (part per quadrillion, ppq)，而且可以是（溶液）破坏性或（激光剥蚀）半破坏性的检测。所以取得数据并非主要的关键，关键在于如何解读所得到的数据以及在古生物化学方面的内涵。如下一节所提到，透过 XRF 分析检测始祖鸟骨头化石内的锌，证明它是肉食性动物。

6.5　元素分析

所有化学物质组成单位是元素，了解某块化石内含有哪些元素，可以提供很多重要的信息，这些信息可以用来推论诸多埋藏学 (Taphonomy) 过程。亦即从生物死亡被掩埋之后，到底发生了什么变化？除了原本生物形态体积的变化之外，更重要的是发生了哪些化学变化？哪些元素从地下水跑进生物躯体内，在酸碱值、温度压力等环境下，置换取代了哪些东西？此外，有哪些元素是原本生物体就有的、哪些被外来元素置换了？

举个实际的情况来考虑：脊椎动物的骨头和牙齿，其主要的矿物质成分是磷灰石群，生物性的磷灰石群 (biogenic apatites) 在该动物活着的时候处于介稳状态 (metastable)，动物死亡后，维持介稳状态的能量来源没有了，在动物死亡后，原本生物骨头内的生物性磷灰石，含有相对浓度较高的碳酸根、钠与其他离子的碳酸磷灰石 (carbonatoapatite)，

ization process, the contents of these elements influenced by the burial condition could have significant changes. For example, in the living creature, there is almost no REE inside the body. However, in the bone fossils, REE was absorbed and concentrated and remain inside the apatite crystal lattice forever. Thus, by comparing the amount of trace elements in the extant and extinct animals, some taphonomy or cause of death information can be derived.

As today, the method, instrument, and sensitivity for analyzing trace elements are plenty in the arsenal warehouse for picking. For example, in the ideal condition, the ICP-MS can analyze 80 elements (isotopes) simultaneously, and the sensitivity can be as high as part per quadrillion (ppq), and select destructive (solution) or semi-destructive (Laser Ablation) way. So, obtaining data is not the main factor. The critical issue is how to interpret the data and the meaning of these data in the PaleoChemistry point of view. As shown in the next section, thru XRF and found Zn in the bone of *Archaeopteryx* to prove it was meat-eater.

6.5　Elemental Analyses

Elements make all the chemical compounds. Understanding what elements inside a given fossil can provide very vital information. This information can be used to understand the taphonomy process. That is what happened after the organism died and buried. What took placed? Besides the morphological and volume changes, the organism went thru what chemical reactions. What elements went into the body of the organisms. Under what pH, temperature,

此化合物具有较大溶解度和反应性，因而比较不稳定，而在形成化石的重新结晶过程中，它们会转变成无机、热稳定性的氟磷灰石 (fluorapatites) 等矿物群，生物性磷灰石内的钙、钠、碳酸根和羟根等等，被岩层内溶解在水中的氟、稀土元素和微量元素等离子取代，吸附并锁在磷灰石的晶格内，此过程大约在动物死亡后几千年到一万年内完成。这些稀土元素一旦被固定于化石后，非常稳定，不会再从化石中溶解出来、流失掉或被其他元素置换，因而一直保留着在石化过后生物遗骸内，形成一个封闭系统。稀土元素被吸附的比率，取决于当时地层内化学组成与孔隙水情况，因而保留了该地层特有的“指纹”，亦即化石内的化学成分，记录了当时的成岩流体 (diagenetic fluids) 成分，可用来探索古环境的情况，再者相同地层之化石，不论生物品种，只要属于相同地层，都有相同或类似的“稀土元素指纹 (rare earth element fingerprint)”，亦即透过分析和比对这些化石里面的“稀土元素蜘蛛线 (spidergram)”，亦即化石内所含、经正常化后的稀土元素对数相对比例曲线，可用来反推某块化石属于那个地层，即使曾经被迁移过或再埋藏的化石也不受影响，故透过分析骨头化石内的稀土元素，可让某化石和原本地层对比。到目前的研究指出，透过此法判定相对地层，一旦石化过程完成后，不只不受不同生物种的限制，适用的范围也可相距数百公里，亦即只要是来自相同地层内

and pressure, what elements had been replaced by what? Furthermore, what elements were native and what was replaced?

Using a real situation for consideration: Apatites are the main minerals inside the vertebrae bones and teeth. While the animal was still alive, they were metastable biogenic apatites. However, after the death of the animal, the energy to maintain metastable state vanished, the native biogenic carbonatoapatites which contains a relatively higher concentration of carbonate, sodium and other ions have more solubility and reactivity, and hence unstable. During the fossilization, they would transform to inorganic, temperature stable such as fluorapatites and related minerals. The calcium, sodium, carbonate, and hydroxyl ions were replaced by the fluorite, rare earth elements and trace elements into the crystal lattices. This process took places about several thousand to ten thousand years after the animal was buried to complete. Once the rare earth elements were fixed inside the fossil, they were very stable and will not dissolve and washed out. They stayed inside the bone/teeth remains of the animal as a closed system. The absorption ratio of rare earth elements depends on the groundwater chemical composition and thus preserved the "fingerprint", ie., it recorded the chemical composition of the diagenetic fluids of that time and place, which can be used to explore the paleoenvironment. Furthermore, as long as the same stratum, fossils of that same stratum disregard to species will have the same or very similar Rare Earth Element Fingerprint. Which means by analyzing (after normalization) and comparing the REE spidergram of these fossils, it can identify a given piece of fossil belongs to

的生物化石，都会有相同或类似的稀土元素指纹。进行骨头/牙齿内的稀土元素分析，可透过电感耦合等离子体质谱或者激光剥蚀电感耦合等离子体质谱完成。

此外，在活着的脊椎动物骨头／牙齿内，几乎没有稀土元素存在于骨头内。因此，骨头化石内的稀土元素，是在死亡之后石化过程中取得的。又总稀土元素的含量和骨头化石内的磷含量，相关系数高达 0.93 以上，这表示说，骨头里面的稀土元素总含量越大，表示磷灰石群矿物量也越多。实际的例子是，禄丰龙成龙的镧系总稀土元素浓度，是胚胎的十一倍，而铀含量是五倍多。这一点，可从比较禄丰龙成龙与胚胎骨头内的总稀土元素/磷含量比例，明显看得出来。就以此来说，通过稀土元素做地层定年，镧系稀土用于相对定年，锕系的铀（和铅）同位素做绝对定年，必须考虑到所用的样本成长阶段的均一性，不能有部分样本是成年骨头，混杂着部分样本是胚胎或刚孵化出来的，因为从胚胎到成年骨头内磷含量，随着成长时间增加。

另外，一个有趣的元素分析可提供的重要信息，就是透过X射线荧光/同步辐射X射线荧光取得不同元素的（二维）分布图（图 6.5.1），从不同元素的分布情况，可以得到很多有用的信息。比方说肉食性的动物，在其骨头化石内会有锌，而植食性动物骨头内，就没有锌存在，因为植物性的蛋白质不含锌，只有动物性蛋白质才有，动物吃了含锌蛋白

which stratum, even if had transported or re-buried. Thus, it is possible to compare a given fossil to original stratum by REE analysis of fossil bones. Furthermore, studies showed relative dating by this method is not bound by different species, and up to several hundred kilometer ranges. As long as they were from the same stratum, they would have the same or very similar REE fingerprint. ICP-MS or LA-ICP-MS can conduct the actual analyses of REE in bones or teeth.

In the bones and teeth of living vertebrae, there is almost no REE inside. Thus, the REE inside the bones/teeth were acquired postmortem. The correlation between total amount of REE and Phosphor inside the bone is as high as 0.93, meaning the more total REE, the more apatites. The case can be seen clearly from the fact that the total La-REE contents of adult Lufengosaurus are about 11 times that of embryonic bones, while uranium concentration five times more. Base on this, while doing relative dating by La-REE and U(/Pb) for absolute dating, the specimens of the same growth stage, or at least closer shall be used instead of a mixture of adult and embryonic or neonate bones since P content increases along with the growth.

Another interesting elemental analysis can provide important two-dimensional elemental distribution information by X-ray Fluorescence (XRF) or synchrotron radiation XRF (sr-XRF) (Figure 6.5.1). From the distributions of various elements, much useful information can be obtained. For example, zinc existed inside the bones of the meat-eating animals, while the plant-eating animals do not have zinc in their bones. This is caused by the fact that plant proteins do not contain zinc. Only the animal pro-

▲ 图 6.5.1 始祖鸟骨头化石透过二维 X 射线荧光取得元素分布，绿色背景是锌，红色背景是碳酸钙。证明始祖鸟是吃荤的动物

▲ Figure 6.5.1 Two dimensional element distribution of *Archeopteryx* fossil by XRF, green is Zn and red is background is carbonates. This proves the *Archeopteryx* is a meat-eater

质，亦即肉食性动物，在其骨头内就会含有微量的锌，透过X射线荧光测试，很容易就可判断出来。

此外，元素分析，除了上述和以下要说的同位素分析之外，X射线荧光 / 同步辐射X射线荧光，还有一项很有用的信息：可以测出某元素不同的氧化价与其二维分布。举一个实际的例子来说，在快二亿年前禄丰龙恐龙胚胎有机残留物的研究过程中，我们论文已经被接受，才取得本项资料，证明这些胚胎骨头内，还保存着胶原蛋白，因此也无法放入那篇封面论文。请看图 6.5.2，这是使用斯坦福同步辐射 (SSRL/SLAC)X射线荧光所取得的有机硫分布图，背景蓝绿色是磷和钙（磷灰石），在此可以明显看见骨头和牙齿化石内，到处有红点的有机

teins contain zinc. The animal ate the zinc contain proteins, i.e., bones of carnivores will have a trace amount of zinc. By using XRF, this can be detected easily.

Besides the elemental distribution analyses mentioned in the previous paragraph and isotopes analyses in the following, XRF/sr-XRF can provide yet another very useful information: the oxidation state of a given element. A real example: While we were working on the organic remains of the ~200 million years old *Lufengosaurus* embryonic bones, our paper was accepted. I got this new information which was too late to include in our paper and for the cover. The new information is another evidence of collagen preservation inside the embryonic bones. Figure 6.5.2 shows the organic sulfur distribution from the srs-XRF done at SSRL/SLAC. The blue and green background represents phosphor and cal-

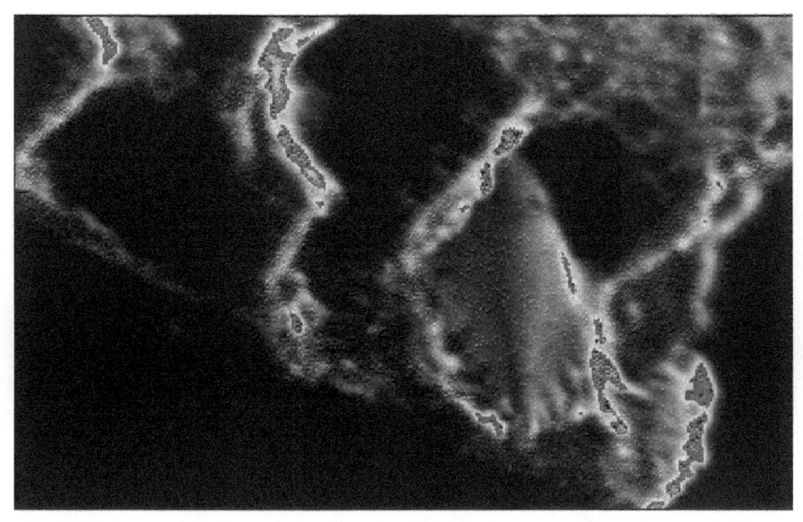

◀ 图 6.5.2 禄丰龙胚胎骨头内有机硫的分布（红点）

◀ Figure 6.5.2 Organic sulfur distribution (red dots) inside Lufengosaurus embryonic bone

硫。为什么这是另一个恐龙胚胎骨头内还保存着胶原蛋白的有力重要证据呢？硫在化石与围岩中是个常见的元素，无机硫矿物如黄铁矿（硫化铁）、石膏（硫酸钙）等，经常存在化石本身或/和附近的围岩内，可是有机硫只有可能从含硫的氨基酸而来，蛋白质（胶原蛋白）就是长链的氨基酸聚合物，其中有含硫的氨基酸：半胱氨酸 (cysteine)、甲硫氨酸 (methionine)、胱氨酸 (cystine) 等，这些有机硫和无机硫的X射线荧光不同，可区分出来。当看到禄丰龙胚胎骨头上处处红点的有机硫，我们就有了很好的证据，这些有机物是被保存下来了。

类似的方法，元素分析也用于考古学方面，如死海经卷，通过经卷墨水之氯 / 溴比例，判断经卷出处。研究者分析死海经卷的墨水，其中的发现之一，氯和溴比例指出该墨水是在坤然 (Qumran) 附近制造的，分析同时也发现，

cium (as apatites) while the everywhere red dots are organic sulfur. Why is this yet another significant evidence for collagen preservation? Sulfur is a commonly seen element inside fossil and matrix, such as inorganic sulfur minerals pyrite (iron disulfide) and gypsum (calcium sulfate). However, organic sulfur can only come from sulfur-contain amino acids. Proteins (collagen) are the long-chain polymer of amino acids, including cysteine, methionine, cystine, etc. The X-ray fluorescence of organic and inorganic sulfur is not the same and can be easily distinguished because there were different oxidation states or bonding. So when so many red dots of organic sulfur showed up, we have strong evidence to say that organic remains were preserved.

A similar method of elemental analysis also been used in archeology, such as Dead Sea Scroll thru the comparison of chlorine to bromine for where it came from. Researchers have analyzed the ink of Dead Sea Scrolls, finding, among other things, a ratio of chlorine to bromine that indicates the ink was created near

有一种特别的墨水，有时候称之为“红墨水”用于至少一份经卷上。

(http://www.livescience.com/58505-how-to-study-dead-sea-scrolls.html?utm_source=ls-newsletter&utm_medium=e-mail&utm_campaign=20170403-ls)。

6.6 同位素分析

在南美智利等沙漠地区，人类考古学者往往发现在某个地区，有少数木乃伊的服饰，显然和绝大部分的不一样，判断应该是来自其他地方，但是死在这里。因此，接下来的其中一个有趣问题是，这个已经变成木乃伊的人，来自何处？能知道来自何处，就可以探讨古代不同族群之间的往来。不过，如何鉴定呢？考古学家进行了跨领域交叉学科的研究，从木乃伊的牙齿里面的钙/锶同位素比例，可推论此人的饮食习惯，比方说，住在山上依赖狩猎吃动物者和住在平原吃玉米者，以及住在海边吃海鲜者，牙齿里面的钙/锶同位素比例，都会有不同。通过如此的研究，可得到一些该木乃伊来处的线索。

同样的原理，是否在恐龙或其他陆生脊椎动物化石的研究，也是如此呢？理论上来说，脊椎动物牙齿的组成与成分，大致相同，牙齿牙冠部分，包括珐琅质、珐琅/牙本质界面、牙本质等三层构成，主要的矿物成分是磷灰石群，钙与锶同样是碱土族二价元素，因此，食物中的钙与锶含量比例，会反映到牙齿

Qumran. Analysis has also suggested that a particular type of ink, sometimes called “red ink”, was used on at least one of the scrolls.

(http://www.livescience.com/58505-how-to-study-dead-sea-scrolls.html?utm_source=ls-newsletter&utm_medium=email&utm_campaign=20170403-ls).

6.6 Isotopes Analyses

Archaeologists study the mummies from the desert area of Chile, South America found that some of the mummies were not local and must be from somewhere else and died there due to the differences of their cloth. So an interesting question is where these mummies came from? Understanding where from can help to understand how the different groups interact. But how? Archaeologists used multiple disciplines to study them by analyzing Ca/Sr isotopes, which can be used to identify their diets, such as meat-eating mountain people, maze-eating plain, and coastal seafood people. The Ca/Sr isotopes ratios would not be the same. So such analysis can shine some vital information about where they came from.

With the same principle, can studies of dinosaurs or other land vertebrae be the same? In theory the composition of vertebrae teeth are very similar. The crown portion of a tooth have enamel, enamel-dentin-junction and dentine. The main mineral is apatites. Calcium and Strontium are group II Alkaline earth elements, thus, the ratio of Ca/Sr in their diet will reflect in their teeth. Furthermore, the Ca/Sr isotopes ratio in the apatites of the bones shall be able for inference of their diet.

的构造里面。也是同样的道理，骨头里面磷灰石群的钙/锶同位素比例，也应该可用于推论该动物的食性。

再者，曾经有研究指出，通过分析氧同位素比例，可以推论恐龙的体温。恐龙灭绝距今已有数千万年，要想获得它们的体温数据非常困难。但加州大学洛杉矶分校的Robert Eagle和他的研究团队找到了一种独特的方法：通过分析恐龙蛋壳化石中碳酸盐同位素的组成，就能确定雌性恐龙在排卵期间的体温。恐龙蛋的蛋壳产自恐龙的输卵管，它们位于恐龙身体中非常深的部位，因此蛋壳形成的温度能反映恐龙身体的核心温度。罗伯特·伊格尔的研究小组对一组现代鸟类和爬行类动物的蛋壳进行了同位素测定，以此来为恐龙体温的测定提供参照。研究的对象是距今大约7000万至8000万年前晚白垩世的长颈泰坦巨龙和体型较小的窃蛋龙，但结果让研究人员深感意外：两者在体温上差异较大，前者比后者更高。研究人员分析称，蜥脚类恐龙的体温高也许是由于它们体型大，例如，人们已经知道一些现代大型棱皮龟就因为体型大而具有较高的体温；不过，目前并没有发现恐龙有明显的温血动物或冷血动物特征；要准确回答这一问题，还尚需时日。

美国研究发现不同体型的恐龙体温差异大

2015-10-20 11:05:26, 2015-10-20 11:05:26, https://read01.com/kLMB4n.html

Furthermore, there were some studies showed by analyzing the oxygen isotopes ratio, dinosaur body temperature can be deducted out. Dinosaurs were extinct for many deca-thousand years. In order to know their body temperature is challenging. However, Robert Eagle and his team found a unique way to analyze the isotopes composition in the dinosaur eggshells to identify the body temperature of female dinosaurs during ovulation. The dinosaur eggshell comes from the oviduct deep inside the body. Thus, the formation temperature of the eggshell can represent the core temperature of dinosaurs. Robert Eagle's team using isotopes analyses to measure a pair of eggshells of extant birds and reptiles as a reference for dinosaur body temperature measurements. The dinosaurs studied were 70–80 million years old, late Cretaceous long neck *titanosaurid* and smaller body *oviraptorid*. The results were unexpected by the researchers. The body temperature difference between them is more prominent than expected. The front ones are higher than the later ones. According to the researchers, the higher sauropod body temperature could be caused by their vast body. We already know that *Dermochelys* has bigger body size and higher body temperature. It will take more times for researches to have a clear answer to the question of if dinosaurs were warm- or cold-blood.

American researchers found huge body temperature variation among different body size dinosaurs

Article Source

Isotopic ordering in eggshells reflects body temperatures and suggests differing thermophysi-

文章来源：华夏经纬网

看牙大发现　恐龙体温近似人类

https://tw.news.yahoo.com/看 牙 大发现-恐龙体温近似人类-105313054.html

（台湾“中央通讯社”华盛顿23日综合外电报道）科学家想出测量恐龙体温的方法，结果意外发现恐龙体温原来与人类体温可能相差无几。

当然，想知道恐龙体温，不能只是异想天开地把体温计插入恐龙舌下，何况这类庞然大物早在数百万年前就灭绝了；因此科学家退而求其次，研究可以反映体温的恐龙牙齿；他们发现，长颈腕龙体温约为38.2℃，体积较小的圆顶龙体温约为36.8℃，接近人类正常体温37℃。他们的研究今天刊登于“科学杂志”（*Science*）网站；但这项研究也将无法解决恐龙究竟是“恒温动物”，还是“冷血动物”的长期争辩。

当首度发现恐龙遗迹时，相关理论是它们行动笨拙且为冷血动物，但近年来科学界的共识偏向为恒温动物，行动更加活跃，如同电影“侏罗纪公园”（*Jurassic Park*）里的迅猛龙一般；加州理工学院首席研究员Robert A. Eagle简报表示：“分析结果真正让我们排除它们像鳄鱼一样是冷血动物的可能性。”但他也说：“这并不一定意味着这些大恐龙拥有像哺乳动物和鸟类一样的高代谢率；它们可能本来就是巨温性生物（Gigantotherms），因为体形如此庞大，因此能够维持体温。”

但非研究团队一员的澳大利亚阿

ology in two Cretaceous dinosaurs

Robert A. Eagle, Marcus Enriquez, Gerald Grellet-Tinner, Alberto Pe rez-Huerta, David Hu, Thomas Tutken, Shaena Montanari, Sean J. Loyd, Pedro Ramirez, Aradhna K. Tripati, Matthew J. Kohn, Thure E. Cerling, Luis M. Chiappe & John M. Eiler, NATURE COMMUNICATIONS · OCTOBER 2015, DOI: 10.1038/ncomms9296

Dinosaur body temperature is closer to human

("CNA" News) Scientists figured out the method to measure dinosaur body temperature and found that it is not much different from human body temperature.

Of course, to measure dinosaur body temperature cannot be done by just sticking a thermometer under its tongue, not to mention they already extinct many many million years ago. So the scientists find another way to measure dinosaur body temperature by looking dinosaur tooth. They found that giraffe body temperature is about 38.2°C, and the smaller body size *Camarasaurus* was about 36.8°C, close to the human body of 37°C. Their research was published in Science website today. However, this research can not resolve the long debate of if dinosaurs were warmblood or cold blood animal.

When dinosaur was first found, the prevailing thoughts were that they were slow moving and cold blood. However, in recent years, the general consensus of scientists are leaning toward warm blood and agile, as the *Velociraptor* in the movie Jurassic Park. Robert A. Eagle, chief researcher of Cal. Tech., explained in a briefing: "Our analyses remove the possibility of cold blood like the crocodile for sure." He also said that this does not mean that these giant dino-

得雷德大学（University of Adelaide）Roger Seymour说："关于恐龙代谢率的辩论，毫无疑问将会持续下去，因为永远无法直接测量，古生物学家将会时时寻找支持特定观点的证据，忽视反证。"（译者：台湾通讯社卢映孜）

以化石内生物矿物之同位素(^{13}C-^{18}O)顺序决定恐龙体温

Robert A. Eagle, Thomas Tütken, Taylor S. Martin, Aradhna K. Tripati, Henry C. Fricke, Melissa Connely, Richard L. Cifelli, John M. Eiler; Science 22 Jul 2011: Vol. 333, Issue 6041, pp. 443-445 DOI: 10.1126/science.1206196

在此我们使用和别的方法不同的"二元同位素测温法(clumped isotope thermometry)"来检测六个恐龙点大型侏罗纪恐龙的体温，这个技术基于分析碳酸盐矿物热力学偏爱碳(^{13}C)和氧(^{18}O)的重同位素结合(^{13}C-^{18}O)，或称为"二元(Clump)"。不像惯用的氧同位素测温法，二元同位素测温法应用，不必依赖知道或假设矿物形成水内氧同位素组成。以此法所测量的参数Δ_{47}数值，为从牙齿生物磷灰石（通式$Ca_5(PO_4, CO_3)_3(OH, CO_3, F, Cl)$）中碳酸盐成分从22~37℃所释放出来的二氧化碳，有一个理论模型预估能有更大的温度范围；这个方法可以重建现生和化石哺乳动物以及外温动物(ectotherms)所期望的身体温度，精确(accuracy)到约1℃，准确(precision)到1~2℃；与仪器测量爬行类体温指出，二元同位素测量牙齿，可能表示外温动

saurs had the same high metabolic as mammals or birds. They could be gigantotherms. Because of the vast body size, thus they could maintain body temperature.

Roger Seymour of the University of Adelaide, who is not involved in this research said: "The metabolism of dinosaurs will continue because we can not measure directly. Paleontologists will find evidence of supporting a particular point of view, and neglect the oppose."

Dinosaur Body Temperatures Determined from Isotopic (^{13}C-^{18}O) Ordering in Fossil Biominerals

Here we apply a different approach to this problem using clumped isotope thermometry to determine the body temperatures of giant Jurassic sauropods by analyzing material from six sites. This technique is founded on the thermodynamic preference of rare heavy isotopes of carbon (^{13}C) and oxygen (^{18}O) to bond with each other (^{13}C-^{18}O), or "clump," in carbonate containing minerals. Unlike the well-established oxygen isotope thermometer, application of clumped isotope thermometry is not dependent on knowing or assuming the oxygen isotope composition of the water from which a mineral grew. The parameter measured in this approach is the Δ_{47} value of CO_2 liberated from the carbonate component of tooth bioapatite [generalized as $Ca_5(PO_4, CO_3)_3(OH, CO_3, F, Cl)$]. Bioapatite Δ_{47} values follow a temperature dependence indistinguishable from inorganic calcite ($CaCO_3$) over the range 22–37°C, and a theoretical model predicts this should be the case even over a greater range of temperatures. This approach is capable of reconstructing the expected body temperatures of modern and fossil

物的平均身体温度，而非最高体温，其最高体温在某些情况下，类似于哺乳动物。

mammals and ectotherms with an accuracy of ~1°C and a precision of 1–2°C. Comparison with instrumental measurements of reptile body temperatures indicates that clumped isotope measurements of teeth probably reflect average body temperatures in ectotherms, rather than peak body temperatures (which could be similar to those seen in mammals in some cases).

第7章　有机残留物与无机物交互作用

Chapter 7　Interactions between Organic Remains and Inorganics

最后这章叙述有机残留物与无机物的交互作用，并以实例打开眼界展望将来。

The last chapter describes the interaction of organic remains and inorganics and uses some examples to enlarge our vision for the future.

7.1　在胶原白保存中的赤铁矿

7.1　Hematite in Collagen Preservation

无机物可能在保存有机残留物的过程中扮演关键的角色，就以原本在古生物体内的大分子有机物来说，如夹在磷灰石群背景内的胶原蛋白，在漫长的石化过程中，无法排除长链的胶原蛋白没有裂解成为较短的多肽(peptides)，甚至多肽降解为单体氨基酸的可能性，也无法排除这些较小分子的重新聚合(repolymerization)，再度组合成较长链的大分子，而在这些可能当中无机元素或无机化合物可能扮演催化剂的作用。

蛋白质 ⇄ 多肽 ⇄ 氨基酸

举一个实际的例子，Marry Schweitzer团队利用将恐龙骨头化石用乙二胺四乙酸(ethylenediaminetetra-acetic acid, EDTA)溶解去钙，进行有机残

Inorganics may play a vital role in the preservation of organic remains. Speaking of macro organic molecules inside the ancient organism, such as the collagen in the apatites matrix, it's not possible to exclude that the long chain proteins did not break down to shorter peptides, or even further down to amino acids, and also can not rule out these smaller molecules to re-polymerize to long chain big molecules. In such, the inorganics (or particular element) plays the role as the catalyst.

Protein ⇄ Peptide ⇄ Amino Acid

As a real case: Mary Schweitzer team used EDTA decalcification method for studies of organic remains collagen. In which they propose that (Goethite, FeO(OH)) plays the crucial role in the preservation of collagen. However, in our

留物胶原蛋白的论述，其中提出针铁矿 (goethite, FeO(OH)) 对于骨头化石内胶原蛋白的保存，扮演了重要的角色。虽然这个论点，被我等的研究论文指出错误，我们用在原位 (*in situ*) 没有任何化学反应方式所检测到的是赤铁矿 (hematite, Fe_2O_3)，她的针铁矿论述不是完全正确，因为针铁矿很可能是在她们团队溶解去钙的过程中，赤铁矿被水解成为针铁矿 (FeO(OH))，这种情况在大自然界，可见于针铁矿假象赤铁矿样本（晶体形状保存着赤铁矿的形态，但是内部成分已经转换成针铁矿）: $Fe_2O_3 + H_2O \rightleftarrows 2FeO(OH)$。

虽然Schweitzer推论的矿物种不对，可是她所提出来的铁离子在胶原蛋白保存所扮演的催化剂角色，应该没有错误。既然铁离子在第一类胶原蛋白的保存上，扮演某种关键角色，那么其他的离子，会不会在其他有机残留物的保存上，也扮演类似的角色？举个实际的例子：带羽毛恐龙与鸟类的羽毛颜色，曾有论文说是铜离子；有机物黑色素小体 (melanosomes) 的保存和铜离子似乎有密切关系；在 1999 年，从沙漠鸟面龙 *Shuvuuia deserti* 的标本检验出含有会衰变的 β 角蛋白 (β-keratin)，明显地缺乏 α 角蛋白 (α-keratin)，β 角蛋白是构成鸟类羽毛的主要蛋白质；2010 年在中华龙鸟 (*Sinosauropteryx*) 的丝状痕迹发现黑色素小体的存在；数种恐龙与早期鸟类的羽毛痕迹研究，发现这些物种的羽毛痕迹存在黑色素小体，研究人员在

paper, we pointed out that it may not be the case. We conducted our works *in situ* and found (Hematite, Fe_2O_3). We thought her goethite statement is not exactly right because the hematite could be hydrolyzed to goethite during their decalcification. This can be found in nature as goethite pseudomorph hematite (the shape of hematite crystal was preserved, but the content already changed to goethite): $Fe_2O_3 + H_2O \rightleftarrows 2FeO(OH)$.

Even though Schweitzer's conclusion is not right of the mineral species, the iron ions as catalyst role shall still be valid. So it looks like that iron ions play a specific role in the preservation of collagen type I, then, would other ions play a similar role in the preservation of organic remains? For example, the hair/feather of dinosaurs and bird are known to have color pigments preserved. There are papers stated that copper ions are closely related to the preservation of melanosomes organic remains. In 1999, degradable β-keratin, not α-keratin was found in *Shuvuuia deserti*. β-keratin is the main protein that constitute bird feathers. In 2010, in the fibrous trace of *Sinosauropteryx*, melanosomes were found. In the trace of hairs and feathers of several dinosaurs and early birds, feather trace were melanosomes also found. Some researcher discovered there are two melanosomes: Eumelanosome and Phaeomelanosome. Then, in 2011, it was found that trace metals could be the brown and black color marker in the fossilized feathers and other tissues. It is a consistent method to identify eumelanosome in fossils.

化石中发现了两种黑色素小体：真黑色素 (eumelanosome) 和褐黑色素 (phaeomelanosome)；2011 年的另一项研究表明，痕量金属可作为化石中的羽毛及其他组织内的棕色和黑色的着色标志，这是识别化石中真黑色素小体的一种前后结果一致的强而有力方法。

7.2　碳酸盐胶合

在整个中生代的地球大气与温度（图 7.2.1），若与当下来比，可以这么说，整个中生代地球的平均气温高出很多，有人估计整个中生代的平均气温比现在高出十几摄氏度，甚至几十摄氏度，一年当中只有两个季节："热季" 和 "更热季"，降水量比现在更丰沛，植被非常茂盛，要不然就无法养活那么多食量惊人的恐龙。再者地球南北极都没有覆冰盖，整个地球处于 "暖期"，相对地，当今南北极都有很厚的覆冰盖，我们活在地球 "冰期"。此外整个大气层的平均二氧化碳浓度，也为当今的四五倍，这

7.2　Carbonate Cementing

During the whole Mesozoic Era, the earth temperature and atmosphere (Figure 7.2.1) in comparison to the current time, the average temperature then was much higher than current, some estimated more than 10 °C. There were only two seasons in a year: hot and even hotter. There were more rains and more vegetation flourish. It had to be so to sustain the enormous diet demand for dinosaurs. At that time, no ice on north and south poles. The whole earth was in a warm period. In contrast, it is thick ice cover on both poles now. We are living in the ice age. Furthermore, the average CO_2 concentration then was about 4 or 5 times more than current. This sketches out the background environment of dinosaurs.

Thus, when we study PaleoChemistry, we can NOT ignore the reality of earth past and use the contemporary (ice age) condition to judge the old (warm age), particularly the influence of high CO_2 concentration during the fossilization process. When carbon dioxide dissolves in water, it will turn into carbonic acid. Therefore,

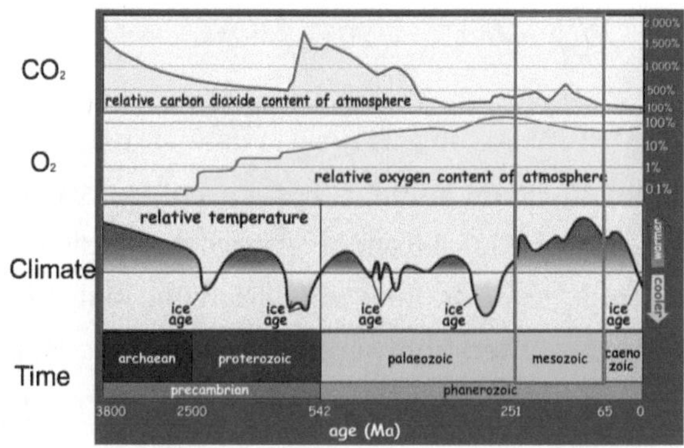

Compiled Geological Climate Chart by Terry Bottrill of UTM

◀ 图 7.2.1　地球气候与大气层历史

◀ Figure 7.2.1　Earth History and Climate

是整个恐龙所存活中生代的背景概况。

因此，当我们研究古生物化学的过程中，不可忽略地球过去的实际情况，而“以今（冰期）论古（暖期）”，特别是大气层中的二氧化碳高浓度，在石化过程中所产生的影响。当二氧化碳溶解到水中成为碳酸，因此，当大气层中的二氧化碳浓度高，可以预期水中的碳酸浓度也会比现在较高。较高浓度的碳酸，除了本身进入化石内之外，也会因为其弱酸性，而溶解更多的元素，一起带入化石之内，化石内普遍含有碳酸盐，是很常见的情况。

这种碳酸盐胶合现象，可以从探讨肉食恐龙小锯齿的红外线光谱看得很清楚：第5章图 5.2.5 显示二氧化碳气体的物理吸附 (2345 cm^{-1}) 和化学吸附 (1640 cm^{-1})。图中上层几张照片所示大气中二氧化碳的物理和化学吸附深入范围，都只达到珐琅牙本质界面 (Dentin-Enamal-Junction, DEJ) 而已，而没有进入更深到牙本质里面去。图下面的红外线影像显示，二氧化碳的物理吸附作用，加上两种碳酸磷灰石，形成了一个保护层（白色曲线），亦即碳酸盐胶合作用，好像一堵围墙，保护着牙本质里面的有机残留物，可见于最右下角的烷基波峰范围 (2800~3000 cm^{-1})。

7.3 自然界的骗局

化石或无机构造？科学家深入早期生物

when the high carbon dioxide concentration in the atmosphere, it is expected that higher carbonic acid in the water. Higher carbonic acid (weak acid) will dissolve more elements into fossils besides itself will become a part of the fossil. Carbonate minerals are prevalent inside fossils.

The carbonate cementing phenomena can be seen from the IR spectra of our study on the serrated teeth of meat-eating dinosaur teeth. Figure 5.2.5 of Chapter 5 shows the physio-absorption of CO_2 at 2345 cm^{-1}, and chemio-absorption at 1640 cm^{-1}. The top images show that physio- and chemio-absorption of carbon dioxide penetration reached the Enamel-Dentin-Junction (DEJ) and not into the dentin area. The images of infrared below showed the physio-absorption of carbon dioxide combine with two types of carbonatoapatites to form a protective layer as indicated by the white line. This is the carbonate cementing as a wall to protect the organic remains (see alkyl functional group peaks at 2800–3000 cm^{-1}) inside the dentine.

7.3 Nature Deception

Fossil or inorganic structure? Scientists dig into early life forms

Date: March 15, 2017 Source: Florida State University

Fossil-like objects grew in natural spring water abundant in the early stages of the planet, an international team of researchers has discovered. However, they were inorganic materials that resulted from simple chemical reactions. An international team of researchers discovered that inorganic chemicals could self-organize into complex structures that mimic primitive life

https://www.sciencedaily.com/releases/2017/03/170315144400.htm

一个国际研究团队宣称，在星球早期，在自然泉水中，会长出像化石的东西，但是它们只是从简单化学反应产生的无机物质。一个国际研究团队发现，无机化合物能自我组织成为模仿地球原始生命的复杂结构。

佛罗里达州立大学化学教授Oliver Stein-bock和在格纳达西班牙研究议会 Consejo Superior de Investigaciones Científicas (Spanish Research Council) 的 Juan Manuel Garcia-Ruiz 教授在《科学进展》(*Science Advances*) 发表一篇论文，显示在星球早期的自然泉水中，成长出像化石的东西，但是它们只是简单化学反应产生的无机物质。这使得鉴定地球最早期显微化石更加棘手，也需要重新审视在其他星球寻找生命的方法方式。

Garcia-Ruiz 说："无机大分子，从形态和化学成分来说，可能和古代生命无法分辨"，科学家过去在实验室有看到这种迹象，但现在 Oliver Steinbock 和 Garcia-Ruiz 的研究明确指出，在自然界也有此现象。为了这个工作，科学家采集分析了北加州内泉 (Ney) 的很独特苏打水，当今这种泉水全世界只有少数几个，但是在地球存在早期，分布很广泛。只要加入一种很常见的化合物——钙或钡盐，这种水就产生小的构造，如管子、螺旋、虫样东西，很像早期的生物，这种水也产生复杂的像海贝壳珍珠母的矿物结构，这些无机结构与真正化石的类

on Earth.

Florida State University Professor of Chemistry Oliver Steinbock and Professor Juan Manuel Garcia-Ruiz of the Consejo Superior de Investigaciones Científicas (Spanish Research Council) in Granada, Spain published an article in Science Advances that shows fossil-like objects grew in natural spring water abundant in the early stages of the planet. However, they were inorganic materials that resulted from simple chemical reactions. This complicates the identification of Earth's earliest microfossils and redefines the search for life on other planets and moons.

"Inorganic microstructures can potentially be indistinguishable from ancient traces of life both in morphology and chemical composition," Garcia-Ruiz said. Scientists had seen hints of this in past lab work, but now through Steinbock and Garcia-Ruiz's research, it is clear that this also happened in nature. To do this work, the team of scientists collected and analyzed an extreme form of soda water from the Ney Springs in Northern California. Today this type of water is found in only a few spots worldwide, but it was widespread during the early stages of Earth's existence. By the addition of just one other ubiquitous chemical -- calcium or barium salt —, this water produces tiny structures, such as tubes, helices, and worm-like objects that are reminiscent of the shapes of primitive organisms. The water also generates complex mineral structures that are similar to nacre -- the shiny substance of seashells. The similarities between actual fossils and these inorganic structures go beyond appearance and extend to their chemical nature. This will make it even more complicated for scientists examining early evidence of

◀ 图 7.3.1　分子模型（球棍）图示，新的研究指出无机化合物能自我组织成复杂结构，看来像地球原始生物

◀ Figure 7.3.1　Illustration of molecule model (stock image). New research suggests that inorganic chemicals can self-organize into complex structures that mimic primitive life on Earth. Credit: © artemegorov / Fotolia

似性，远超过外观并延伸到它们的化学本质，这个使得科学家检验地球早期生命更为复杂。

Steinbock说：“我们的发现，披露出简单生物形态和复杂无机结构不平凡的类似性，使得鉴定地球和其他行星最早显微化石更为困难。此发现很迷人，如果我到火星，我怎么鉴定化石？我要怎样说服自己，它曾经活着？将来科学家得更警惕，任何看起来像生物的，不一定是生物。”

故事来源 *Story Source: Materials provided by Florida State University. Original written by Kathleen Haughney.*

life on Earth.

“Our findings reveal an unusual convergence of simple biological shapes and complex inorganic structures and make the job of identifying earliest microfossils on Earth and life on other planets even harder,” Steinbock said. “It is fascinating. How could I identify a fossil if I went to Mars? How could I convince myself that it was once alive? In the future, scientists will need to be even more alert that everything that looks like life is not necessarily life.”

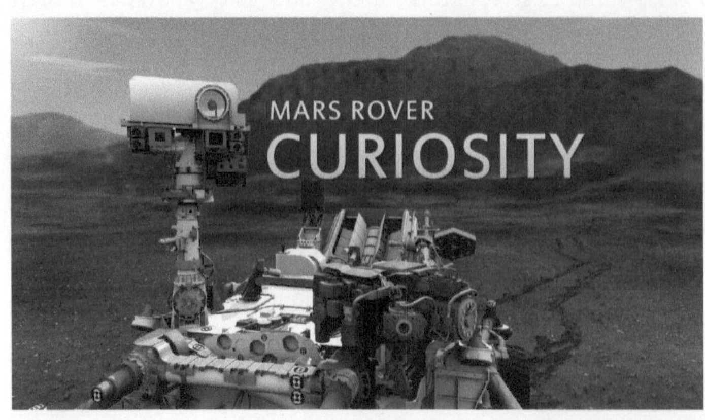

◀ 图 7.4.1　火星好奇号漫游车

◀ Figure 7.4.1　Curiosity Mars Rover

7.4 我们能向火星好奇号学习吗?

这一节绝对不是长他人威风，而是对于科学家成就的礼赞，能够把整个分析化学实验室诸多仪器分析设备，做成这么精致这么小，令人赞叹。本节资料取自美国国家航空暨太空总署(NASA)。

在火星上的化学分析实验室(SAM)

找寻可能的过去生命；石化脂质、脂肪酸、氨基酸、胺甚至是核碱基

美国国家航空暨太空总署发射到火星的“好奇号(Curiosity)”(图7.4.2)，不只是因为它是由一位华裔妹妹命名，而是从古生物化学的角度来说，真是顶尖顶好的成就。它主要的使命在于寻找火星上是否曾经有生命存在的证据，通过地球遥控，操作好奇号里面的火星样本分析设备，进行必要而结果能确信的分析，令人惊讶的是，这个只有微波炉大小的火星样本分析设备(图7.4.3)，竟然能够在如此紧密的空间内，装下这么多能受得了火星严酷温度环境，还能做出这么精密的科学分析，取得可用可信的数据。

好奇号是一辆美国国家航空暨太空总署火星科学实验室辖下的火星探测车，主要任务是探索火星的盖尔撞击坑大风坑(Gale Crater)，为美国国家航空暨太空总署火星科学实验室计划的一部分。好奇号在2011年11月26日北美东部标准时间10:02于卡纳维尔角空军基地进入火星科学实验室太空飞行器，

7.4 Can We Learn from Curiosity on Mars?

This section is not for praising other people's prestige, but a very high glorification for the scientist achievements. They packed the instruments of a whole analytical lab into such a small compact size. It is amazing. Information from NASA.

Chemical analysis lab on Mars, SAM (Sample Analysis at Mars)

Searching for evidence of past life; fossilized lipids, fatty acids, amino acids, amines and possibly even nucleobases

The NASA's Curiosity (Figure 7.4.2) is not only because named by a Chinese girl living in the States, but as a marvelous achievement from the point of view of PaleoChemistry. Its primary mission is to search the evidence to see if life existed on Mars. Thru remote control fro earth to operate and conduct various analytical instruments inside. These analyses must be reliable. More astonishing is that so many analytical instruments were pack tightly in this microwave oven size (Figure 7.4.3). Moreover, it has to be able to sustain the harsh Mars environment without sacrifice the data integrities of scientific analyses.

Curiosity is a car-sized robotic rover exploring Gale Crater on Mars as part of NASA's Mars Science Laboratory mission (MSL). Curiosity was launched from Cape Canaveral on November 26, 2011, at 15:02 UTC aboard the MSL spacecraft and landed on Aeolis Palus in Gale Crater on Mars on August 6, 2012, 05:17 UTC. The Bradbury Landing site was less than 2.4 km (1.5 mi) from the center of the rover's touchdown target after a 560 million km (350 million mi) journey. The rover's goals include the investigation of the Mar-

▲ 图 7.4.2　好奇号自拍 (NASA)

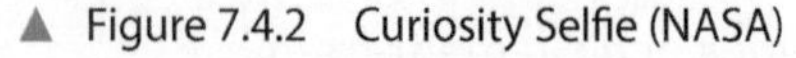

▲ Figure 7.4.2　Curiosity Selfie (NASA)

▲ 图 7.4.3　好奇号的火星样本分析设备

▲ Figure 7.4.3　Sample analysis facility of Curiosity (NASA)

并成功在 2012 年 8 月 6 日协调世界时 05:17 于大风坑的伊奥利亚沼着陆 (Aeolis Palus in Gale Crater)。好奇号经过 56300 万公里的旅程，着陆时离预定着陆点布莱德柏利降落地只相差 2.4 公里。好奇号的任务包括：探测火星气候及地质，探测盖尔撞击坑内的环境是否曾经能够支援生命，探测火星上的水及研究日后人类探索的可行性。

好奇号的设计将是计划中的火星 2020 探测车任务设计基础，2012 年 12 月，好奇号原本执行 2 年的探测任务被无限期延长。2014 年 6 月 24 日，好奇号在发现火星上曾经有适合微生物生存的环境之后，执行满一个火星年的探测任务。好奇号的一个主要目标，是在 2012 年 8 月登陆红色星球后，在其着陆

tian climate and geology; assessment of whether the selected field site inside Gale Crater has ever offered environmental conditions favorable for microbial life, including investigation of the role of water; and planetary habitability studies in preparation for future human exploration.

Curiosity's design will serve as the basis for the planned Mars 2020 rover. In December 2012, Curiosity's two-year mission was extended indefinitely. NASA's Mars Curiosity rover will complete a Martian year—687 Earth days—on June 24, having accomplished the mission's primary goal of determining whether Mars once offered environmental conditions favorable for microbial life. One of Curiosity's first major findings after landing on the Red Planet in August 2012 was an ancient riverbed at its landing site. Nearby, at an area known as Yellowknife Bay, the mission met its significant goal of determining whether

点找到老河床，在一个叫做耶洛奈夫湾(Yellowknife Bay)的地区附近，它达成了它主要目标，确定火星大风坑是否适合简单的生命存在，这是个历史性的“是”答案，来自两块泥岩板，它用钻机取样，接着对这些样品进行分析，显示该地点曾经是个温水的湖床，这是生命必需不可或缺的元素成分和地球上某些微生物要用的化学能源。如果火星曾经有生物，这里是个很好家园。

生物性

(1) 探定有机碳化合物的本质和库存种类；

(2) 调查生命的化学建构元件（碳、氢、氮、氧、磷和硫）；

(3) 鉴定可能代表生物存在的特征（生物痕迹和生物分子）；

地质和地质化学

(4) 调查火星表面、接近表面地质材料的化学、同位素、和矿物组成；

(5) 解读形成岩石土壤的过程。

一般的样品分析策略从高分辨率相机开始，寻找有兴趣的特征。如果找到特定表面令人感兴趣，好奇号可以用红外激光器蒸发一小部分，检查所得到的光谱特征以查询岩石的元素组成。如果该特征很有趣，它将使用长臂把样本送到显微镜和X射线光谱仪，进一步仔细观察。如果样品需要进一步分析，好奇号可以钻入岩石，将粉末样品送到内部的火星样本分析设备或化学矿物分析(CheMin)实验室。

火星样本分析设备实际上有三种仪

the Martian Gale Crater ever was habitable for simple life forms. The answer, a historic "yes," came from two mudstone slabs that the rover sampled with its drill. Analysis of these samples revealed the site was once a lakebed with mild water, the essential elemental ingredients for life, and a type of chemical energy source used by some microbes on Earth. If Mars had living organisms, this would have been a good home for them.

Biological

(1) Determine the nature and inventory of organic carbon compounds;

(2) Investigate the chemical building blocks of life (carbon, hydrogen, nitrogen, oxygen, phosphorus, and sulfur);

(3) Identify features that may represent the effects of biological processes (biosignatures and biomolecules);

Geological and geochemical

(4) Investigate the chemical, isotopic, and mineralogical composition of the Martian surface and near-surface geological materials;

(5) Interpret the processes that have formed and modified rocks and soils.

The general sample analysis strategy begins with high-resolution cameras to look for features of interest. If a particular surface is of interest, Curiosity can vaporize a small portion of it with an infrared laser and examine the resulting spectra signature to query the rock's elemental composition. If that signature is intriguing, the rover will use its long arm to swing over a microscope and an X-ray spectrometer to take a closer look. If the specimen warrants further analysis, Curiosity can drill into the boulder and deliver a powdered sample to either the SAM or the CheMin analytical laboratories inside the

器组合，包括四极质谱仪 (quadrupole mass spectrometer)、气相色谱仪 (gas chromatograph) 和可调激光光谱仪 (tunable laser spectrometer)。四极质谱仪和气相色谱仪可以联合成气相色谱质谱仪，气相色谱进行分离，四极质谱仪则做有机物的绝对鉴定；可调激光光谱仪取得二氧化碳中精确的碳和氧同位素比例，也测量甲烷的微量程度和它的碳同位素，它也找寻并测量其他与生命有关的轻元素，如氢、氧和氮。

质谱仪将元素和化合物按质量分离以进行鉴定和测量；气相色谱仪把土壤和岩石样品加热蒸发，然后将所得气体分离成各种组分进行分析；激光光谱仪测量大气中诸如甲烷，水蒸气和二氧化碳的碳、氢和氧的各种同位素的丰度，这些测量精确到千分之十几。

因为这些化合物对生命是必不可少的，所以它们的相对丰度是评估火星能否支持过去或现在生命的重要信息。在火星科学实验室的火星分析实验设备 (sample analysis at Mars, SAM) 中，用来调查当前和过去火星生命所需的分子和化学元素，火星分析实验设备寻找有机化合物碳化学、除了碳之外的轻元素和行星变化痕迹的同位素之化学状态。

好奇号 ChemCam 的化学元素第一道激光光谱（“Coronation” Rock，2012 年 8 月 19 日）。ChemCam 能够记录多达 6144 种不同波长的紫外线、可见光和红外光；启动等离子体球 (plasma balls) 检测在 240 nm 和 800 nm 之间的可见光、

rover.

Actually a suite of three instruments, include a Quadrupole mass spectrometer, gas chromatograph, and tunable laser spectrometer. The QMS and the GC can operate together in a GC-MS mode for separation (GC) and definitive identification (QMS) of organic compounds. The TLS obtains precise isotope ratios for C and O in carbon dioxide and measures trace levels of methane, and its carbon isotope. Sample Analysis at Mars also looks for and measures the abundances of other light elements, such as hydrogen, oxygen, and nitrogen, associated with life.

The mass spectrometer separates elements and compounds by mass for identification and measurement. The gas chromatograph heats soil and rock samples until they vaporize, and then separates the resulting gases into various components for analysis. The laser spectrometer measures the abundance of various isotopes of carbon, hydrogen, and oxygen in atmospheric gases such as methane, water vapor, and carbon dioxide. These measurements are accurate to within ten parts per thousand.

Because these compounds are essential to life as we know it, their relative abundances are an essential piece of information for evaluating whether Mars could have supported life in the past or present. The Sample Analysis at Mars (SAM) Suite Investigation in the MSL Analytical Laboratory is designed to address the present and past habitability of Mars by exploring molecular and elemental chemistry relevant to life. SAM addresses carbon chemistry through a search for organic compounds, the chemical state of light elements other than carbon, and isotopic tracers of planetary change.

近紫外和近红外范围进行，好奇号的ChemCam在火星上第一次初步激光测试，发生在2012年8月19日，是在着陆点N165（“Coronation”岩石）附近的布莱德柏利岩石上进行的。

2017-08-24 (https://www.seeker.com/space/planets/nasas-curiosity-rover-strengthens-case-that-mars-was-once-habitable?utm_source=facebook&utm_medium=social&utm_campaign=seeker)

美国太空总署的好奇号，在过去五年收集的资料，让科学家建立一个大风坑和好奇号所在的夏普峰 (Mount Sharp) 最底层的详尽历史描述，本次任务的岩石研究显示，这个地方曾经是个泥泞的湖床，充满了水。最新的研究更明确地指出，此处曾经是个可居住的环境，好奇号采集岩石的诸多矿物，也显示，随着火星在几百万年前开始失去大气层、行星水分丧失于太空的古环境变化细节。

美国国家航空暨太空总署詹森太空中心的研究员，也是该科研主要领导者Elizabeth Rampe的新闻稿说：“我们到大风坑调查夏普峰底层，有些从水中沉淀的矿物和不同的环境，这些地层大约在35亿年前形成和地球开始有生命同一时期；我们认为火星早期类似于地球早期，所以它的环境应该是可居住的。”研究者特别研究在夏普峰底部的四个样本，使用好奇号的钻子采样，并用其上的化学矿物实验室仪器来分析，特别关注一层叫做湖沉积的泥岩，这是湖泊沉

Curiosity ChemCam's first laser spectroscopy of chemical elements ("Coronation" Rock, August 19, 2012). ChemCam can record up to 6,144 different wavelengths of ultraviolet, visible and infrared light. Launch detection of plasma balls between 240nm and 800nm of visible light, near-ultraviolet, and near-infrared ranges. On Mars, Curiosity's ChemCam conducted the first preliminary laser test on August 19, 2012, at Bradbury rocks near the Landing N165 ("Coronation" Rock).

2017-08-24 (https://www.seeker.com/space/planets/nasas-curiosity-rover-strengthens-case-that-mars-was-once-habitable?utm_source=facebook&utm_medium=social&utm_campaign=seeker)

Data gathered by NASA's Curiosity rover over the past five years have allowed scientists to construct a detailed portrait of the history of Gale Crater and the lowermost layers of Mount Sharp where the rover has been traversing. Rocks studied during the mission have shown that this site was once a muddy lakebed, filled with water. The latest research suggests with even more certainty that this was once likely a habitable environment. The diversity of minerals in the rock samples collected by Curiosity are also revealing details about the ancient environmental changes that occurred as Mars started to shed its atmosphere millions of years ago and much of the water on the planet's surface was lost to space.

"We went to Gale Crater to investigate these lower layers of Mount Sharp that have these minerals that precipitated from water and suggest different environments," said Elizabeth Rampe, a NASA exploration mission scientist at Johnson Space Center and lead author of a

积形成的。

7.5 三叶虫在头部下蛋？

古老三叶虫可能从头部排出精卵。

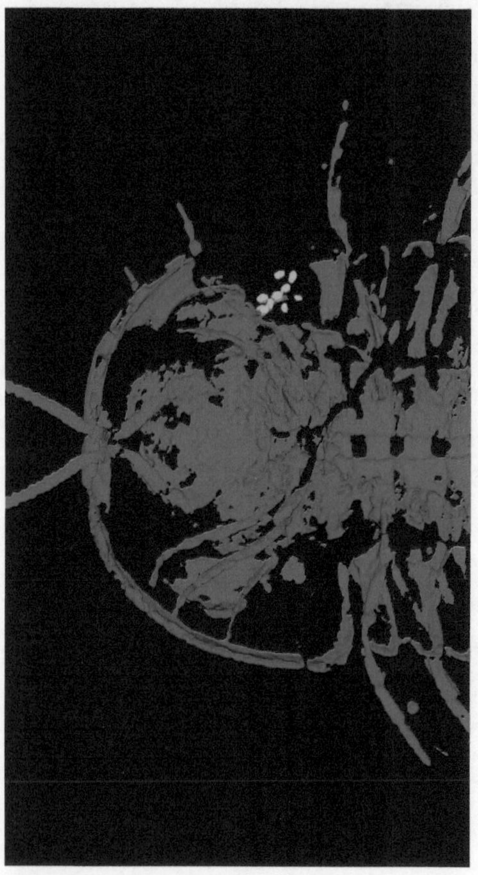

▲ 图 7.5.1 三叶虫在头部的卵

▲ Figure 7.5.1 The eggs of trilobite in the head area

三叶虫，三区块性像螃蟹的生物，在早期古生代是主要生物，其量众多，成为大部分收藏家的入门收藏品。可是古生物学家一直到现在，都很少找到这些灭绝节肢动物如何繁殖，最近学者们研究三分节虫 *Triarthrus eatoni* 化石样本

new study, in a press statement. "These layers were deposited about 3.5 billion years ago, coinciding with a time on Earth when life was beginning to take hold. We think early Mars may have been similar to early Earth, and so these environments might have been habitable." The researchers looked specifically at four samples that were collected from the lower layers of Mount Sharp using the rover's drill and studied with the onboard chemistry lab, the Chemistry and Mineralogy (CheMin) instrument. They looked specifically at the mineralogy of a layered mudstone called lacustrine, which is formed by lake sedimentation.

7.5 Eggs of Trilobite in the Head?

Ancient trilobites may have released sperm and eggs from their heads.

By Carolyn Gramling Feb. 23, 2017 , 4:30 PM

Trilobites—three-sectioned, crablike critters that dominated the early Paleozoic—are so abundant that they have become the gateway fossil for most collectors. However, paleontologists have found little evidence of how the extinct arthropods reproduced—until now. Researchers studying a fossil specimen of the trilobite *Triarthrus eatoni* spotted something odd just next to the animal's head: a collection of small (about 200 micrometers across), round objects (in light green). Those, they determined, are actually eggs—the first time anyone had observed fossil trilobite eggs right next to the critters themselves. The structures were exceptionally well preserved, the eggs and exoskeletons of the trilobites replaced with an iron sulfide ore called pyrite. They came from the Lorraine

时，在此动物的头部附近，发现了一堆大约直径 200 微米奇怪圆形（图片中淡绿色）东西，他们研判这些是真正的卵，头一次有人在邻近它的头部发现三叶虫卵化石。这些结构保存得很好，三叶虫的外骨骼和卵被黄铁矿取代。这些三叶虫从整个美国东北地区奥陶纪（4.85 亿年前至4.44 亿年前）洛林 Lorraine 组岩石出土，这个地方多年来一直都是采集三叶虫者的麦加圣地，因为化石都是黄铁矿化。卵的位置是暗示性的，研究人员在3月份《地质》(*Geology*）发表论文。这个假说提出，三叶虫从头部某处的性器官排出精卵，就如今日的鲎；一个发现此的可能原因是，三分节虫 *T. eatoni* 在三叶虫世界相对奇特，在三叶虫世界育雏行为相对奇怪：这个物种喜欢严峻低氧，而且可能比其他物种更会盯着它们的卵。但是作者提出，终于把三叶虫透过性交繁殖的假说推翻掉，这个挑战性但是没引起注意的假说，是因为经常发现一个叠在另一个个体上的样本而来的，而实际上，三叶虫更像下卵的动物，就如鲎群聚，一个叠在另一个身上，公鲎为了竞争将母鲎刚排出来的卵受精。

Group, a rock formation that spans much of the northeastern United States and dates to the Ordovician period (about 485 million to 444 million years ago); it has long been a mecca for trilobite hunters because of the pyritization. The placement of the eggs is suggestive, the researchers report in the March issue of Geology: They hypothesize that trilobites released their eggs and sperm through a genital pore somewhere in the head—much like modern horseshoe crabs do today. One possible reason for the rarity of the find may be that the brooding behavior of was relatively unusual in the trilobite world: The species tended to prefer a harsh low-oxygen environment, and may have kept a closer eye on their eggs than other trilobite species. But, the authors note, one idea this finding does lay to rest is that trilobites might reproduce via copulation—a titillating but little-regarded hypothesis based on the fact that trilobites are sometimes found clustered on top of one another. Instead, trilobites were most likely spawners—and, in fact, that clustering behavior may be another parallel to horseshoe crabs, which can climb on top of one another in competition to fertilize released eggs.

Posted in: ArchaeologyPlants & Animals, DOI: 10.1126/science.aal0820

7.6 硬壳来自海水成分变化、新发现古菌分子化石

硬壳来自海水成分变化

来源：爱丁堡大学

在俄罗斯西伯利亚有多玛 (Yudoma) 河的田野工作。

7.6 Hard Shell Came from Change of Seawater Composition and New Archaeal Molecular Fossils

Skeletons developed as chemistry of oceans changed

这个研究指出，随着海水的成分变化，生物硬壳出现于5.5亿年前。科学家说，在远古海洋钙和镁浓度改变，氧气浓度改变，促使海洋生物从原本的软体动物演化成有硬壳生物。团队表示：一直到现在，对于硬壳最早是如何形成的，所知不多。硬壳的成分是碳酸钙。过去认为，软体动物经过大灭绝，而让硬壳动物繁盛。然而爱丁堡大学的研究者发现，最早有坚硬身体部位的生物，和很近亲的软体动物同时存在。

研究团队研究从西伯利亚石灰岩出土的一些化石，这些岩石是好几亿年前高浓度碳酸钙海水造成的。他们得出结论：有硬体的生物，首先出现于高浓度碳酸钙的环境，让生物体发展原始的硬体。大约一千万年后，地球生物多样性快速增加，称之为寒武纪爆发，硬壳生物开始繁盛；团队说，受到猎食者的威胁，促使生物在碳酸盐浓度不是那

◀ 图7.6.1　土星之土卫六

◀ Figure 7.6.1　Titan of Saturn

Date: April 4, 2017

Source: University of Edinburgh

This is fieldwork at the Yudoma River in Siberia, Russia.

Skeletons and shells first came into being 550 million years ago as the chemical make-up of seawater changed, a study suggests. Ancient marine life may have developed from soft-bodied animals into creatures with hard body parts as oxygen levels rose and calcium and magnesium levels in ancient oceans changed, researchers say. Until now, little was known about how skeletons and shells first evolved, which are made of calcium carbonate , the team says. Previous theories suggested that soft-bodied organisms had undergone a mass extinction, which allowed organisms with skeletons and shells to flourish. However, researchers at the University of Edinburgh have found that the earliest lifeforms with hard body parts co-existed with closely related soft-bodied species.

The team examined a range of fossils unearthed from limestone rocks in Siberia, which formed millions of years ago from seawater with high levels of calcium carbonate. They concluded that hard-bodied lifeforms were first to present only in such environments where high levels of calcium carbonate allowed organisms to develop primitive hard parts. Around 10m years later, the diversity of life of Earth increased rapidly -- a period known as the Cambrian explosion -- and hard-bodied life began to thrive. An increased

么丰富的环境，长出硬体。团队说：硬身体部件的发展，称之为生物矿化(biomineralization)，是软体动物演化大跃进。

本项研究发表于 *Proceedings of the Royal Society B* 期刊，是与 Lomonosov Moscow State University 合作的。爱丁堡大学地质科学学院Rachel Wood教授说："动物如何产生硬壳，是生物演化的一个重大事件，我们现才开始了解这个演化的过程。"

资料来源Story Source:

Materials provided by University of Edinburgh.

期刊资料 Journal Reference:

Rachel Wood, Andrey Yu Ivantsov, Andrey Yu Zhuravlev. First macrobiota biomineralization was environmentally triggered. Proceedings of the Royal Society B: Biological Sciences, 2017; 284 (1851): 20170059 DOI: 10.1098/rspb.2017.0059

美国国家航空暨太空总署检测到土卫六可能支持生命的分子（泰坦星：即土卫六，是环绕土星运行的一颗卫星）。

https://www.galaxymonitor.com/nasa-detects-molecule-that-could-support-life-on-titan/

美国国家航空暨太空总署科学家明确检测到土星月另土卫六大气层内的丙烯腈化合物。

丙烯腈（$CH_2{=}CH{-}C{\equiv}N$）是有刺激性气味，极易燃的有机液体。

美国国家航空暨太空总署科学家

threat from predators led lifeforms to develop new, more complex hard parts in environments that were less carbonate-rich, the team says. The development of hard body parts, through a process called biomineralization. marked a significant evolutionary advance from the previous world of soft-bodied life, the team says.

The study is published in the journal Proceedings of the Royal Society B. The research was carried out in collaboration with Lomonosov Moscow State University. Professor Rachel Wood, of the University of Edinburgh's School of GeoSciences, who led the study, said: "How animals produced shells and skeletons is one of the major events in the evolution of life. We are only now starting to understand the processes underlying this revolution."

NASA Detects Molecule That Could Support Life On Titan

https://www.galaxymonitor.com/nasa-detects-molecule-that-could-support-life-on-titan/

NASA scientists have definitively detected the chemical acrylonitrile in the atmosphere of Saturn's moon Titan.

Acrylonitrile ($CH_2{=}CH{-}C{\equiv}N$) is an Irritating odor easily combustible organic liquid.

NASA scientists have definitively detected the chemical acrylonitrile in the atmosphere of Saturn's moon Titan, a place that has long intrigued scientists investigating the chemical precursors of life. On Earth, acrylonitrile, also known as vinyl cyanide, is useful in the manufacture of plastics. Under the harsh conditions of Saturn's largest moon, this chemical is thought to be capable of forming stable, flexible structures similar to cell membranes. Other researchers have previously suggested that acrylonitrile

明确检测到土星月另土卫六大气层内的丙烯腈化合物；土卫六长久以来，就让科学家很想研究生命的化学前驱物。在地球，丙烯腈，也称为 2-丙烯腈 (2-propenenitrile) 是制造塑胶的原料。在土星最大月亮严酷的条件下，这个化合物可形成类似于细胞膜的稳定弹性结构，有些科学家以前提过，丙烯腈是土卫六大气层的成分，但是从未报告过明确在于一大堆有机、高碳、分子混合大气中，检测到此化合物。

现在，美国国家航空暨太空总署研究员，透过在智利阿塔卡玛巨型毫米/次毫米矩阵 (Atacama Large Millimeter/submillimeter Array, ALMA) 检测到丙烯腈的化学指纹，他们发现在土卫六的平流层，亦即导致棕橘色大气层灰濛部分，有大量的这个化合物。在美国国家航空暨太空总署位于马里兰州葛林贝特 (Greenbelt) 的Goddard太空航行中心 (Goddard Space Flight Center)，2017-07-28 科学进步期刊论文作者主要科学家 Maureen Palmer 说："我们有明确的证据显示土卫六的大气层有丙烯腈，我们认为这种原始物质大量地到达其表面。"

地球上的动植物细胞，在土卫六无法存在，那边的地表温度是−290 ℉ (−179 ℃)，湖泊边缘是液态甲烷。在 2015 年，大学的科学家研究在土卫六云内是否有任何有机分子，在如此恶劣的条件下，形成类似于地球上生物双层酯质薄而可弯曲的构造。双层酯质是细胞膜的主要部件，它分隔了细胞内部和外

is an ingredient of Titan's atmosphere, but they did not report an unambiguous detection of the chemical in the smorgasbord of organic, or carbon-rich, molecules found there.

Now, NASA researchers have identified the chemical fingerprint of acrylonitrile in Titan data collected by the Atacama Large Millimeter/submillimeter Array (ALMA) in Chile. The team found large quantities of the chemical on Titan, most likely in the stratosphere — the hazy part of the atmosphere that gives this moon its brownish-orange color. "We found convincing evidence that acrylonitrile is present in Titan's atmosphere, and we think a significant supply of this raw material reaches the surface," said Maureen Palmer, a researcher with the Goddard Center for Astrobiology at NASA's Goddard Space Flight Center in Greenbelt, Maryland, and lead author of a July 28, 2017, paper in Science Advances.

The cells of Earth's plants and animals would not hold up well on Titan, where surface temperatures average minus 290 degrees Fahrenheit (minus 179 degrees Celsius), and lakes brim with liquid methane. In 2015, university scientists tackled the question of whether any organic molecules likely to be on Titan could, under such inhospitable conditions, form structures similar to the lipid bilayers of living cells on Earth. Thin and flexible, the lipid bilayer is the main component of the cell membrane, which separates the inside of a cell from the outside world. This team identified acrylonitrile as the best candidate. Those researchers proposed that acrylonitrile molecules could come together as a sheet of material similar to a cell membrane. The sheet could form a hollow, microscopic sphere that they dubbed an "azotosome". This sphere could serve as a tiny

在世界。科学家找到了最好的候选者。这些研究者建议丙烯腈分子可聚合，成为类似于细胞膜的层状材料，这层材料可形成内部细微空间，他们称之为固氮质（azotosome），这种小球可作为小的储藏和运输容器，就如双层酯质所能的那样。

美国国家航空暨太空总署天文生物学研究机构所设立的Goddard太空生物中心（Goddard Center for Astrobiology）主任 Michael Mumma说："能够形成稳定的膜，来分开内部与外部环境非常重要，因为它提供保存化学物质长久到可以产生互动。如果丙烯腈可形成类似于膜的构造，对于土星月亮土卫六来说，是形成其生命的重要步骤。"郭大团队测定出在土卫六大气层中有很多丙烯腈，浓度达十亿分之2.8，这个化学物，可能在其平流层（高度至少为125英里，200公里）很丰富，接下来丙烯腈进入较低冷的大气层，凝结下雨到表面。科学家计算在土卫六第二大湖Ligeia Mare 可能堆积的丙烯腈，这个湖的大小，就如地球上美加五大湖中的休伦湖和密西根湖合并起来那么大，在土卫六的生命中，科学家估计在每毫升（四分之一茶匙）液体内，会有足够的丙烯腈。这相对于在地球上，海洋边缘每毫升海水大约含有一百万个细菌。

研究团队关键性合并了 11 个原本阿塔卡玛巨型毫米 / 次毫米矩阵用来校正望远镜列阵光量的高解析数据集，检测到土卫六的丙烯腈。通过合并数据，

storage and transport container, much like the spheres that lipid bilayers can form.

"The ability to form a stable membrane to separate the internal environment from the external one is important because it provides a means to contain chemicals long enough to allow them to interact," said Michael Mumma, director of the Goddard Center for Astrobiology, which is funded by the NASA Astrobiology Institute." If membrane-like structures could be formed by vinyl cyanide, it would be an important step on the pathway to life on Saturn's moon Titan." The Goddard team determined that acrylonitrile is plentiful in Titan's atmosphere, present at concentrations up to 2.8 parts per billion. The chemical is probably most abundant in the stratosphere, at altitudes of at least 125 miles (200 kilometers). Eventually, acrylonitrile makes its way to the cold lower atmosphere, where it condenses and rains out onto the surface. The researchers calculated how much material could be deposited in Ligeia Mare, Titan's second-largest lake, which occupies roughly the same surface area as Earth's Lake Huron and Lake Michigan together. Over the lifetime of Titan, the team estimated, Ligeia Mare could have accumulated enough acrylonitrile to form about 10 million azotosomes in every milliliter, or quarter-teaspoon, of liquid. That is compared to roughly a million bacteria per millilter of coastal ocean water on Earth.

The key to detecting Titan's acrylonitrile was to combine eleven high-resolution data sets from ALMA. The team retrieved them from an archive of observations initially intended to calibrate the amount of light being received by the telescope array. In the combined data

Palmer和她同僚检测出三条符合丙烯腈的指纹光谱线。这个发现，基于其他研究者通过美国国家航空暨太空总署太空船卡西尼（Cassini）号质谱仪，在十年前推论丙烯腈存在。该论文资深作者Goddard科学家Martin Cordiner说："检测到这个难搞天文生物学重要化学物质，令科学家很兴奋，他们急切想判断生命是否能在如土卫六冷酷的环境中出现。这个发现，对于我们理解太阳系形成的化学复杂性非常重要。"

set, Palmer and her colleagues identified three spectral lines that match the acrylonitrile fingerprint. This finding comes a decade after other researchers inferred the presence of acrylonitrile from observations made by the mass spectrometer on NASA's Cassini spacecraft. "The detection of this elusive, astrobiologically relevant chemical is exciting for scientists who are eager to determine if life could develop on icy worlds such as Titan," said Goddard scientist Martin Cordiner, senior author on the paper. "This finding adds an important piece to our understanding of the chemical complexity of the solar system."

7.7 展望未来

正如本书开宗明义所说的，古生物化学是融合当下古生物学与化学的一个交叉学科新领域，其背后的科研思维是"看进骨头内"，改变传统古生物学"只看骨头外面"。古生物化学从化石里面成分看到外部，把化石相关信息看透透，试着用化学之"化"来解读、充实古生物变成化石之"化"。这是一个崭新的领域，也是俗语说的，我们正起航向未有航海图的水域。

从另一个角度来说，从过去到如今，并非完全没有人在这个领域里面某个小角落摸索过，如二十世纪九十年代就有人发表过在七千万年前暴龙腿骨内发现胶原蛋白，2013年笔者团队在二亿年前禄丰龙胚胎骨头化石找到有机残留物（第一类胶原蛋白）等等。但是这些研究都属于零星，这边一点那边一点，深度与广度都稍嫌粗略，未曾有系统性

7.7 Looking Forward

As stated at the beginning of this book, PaleoChemistry is a new interdisciplinary field of science by combining current paleontology and chemistry. The guiding thought is "Look into the Bone," which changes the "Look outside the Bone" of conventional paleontology. PaleoChemistry looks into the chemical composition inside and outside of the fossil to gain a much complete knowledge, and trying to use the "change" of "Study of Change (the term of 'Chemistry' in Chinese)" to interpret and enhance the "change" of ancient life turned into "Changed Rock (the term of 'Fossil' in Chinese)". This is a brand new territory. In other words, we are sailing into uncharted water.

Looking from another angle, from the past to now, it is not that someone did something in this domain. For example, in the 90 of last Century papers were published on finding collagen in *T. rex* femur, and in 2013 our team found organic remains (collagen type I) inside the

的完整探讨。毕竟，学化学者的化石知识不足，学古生物的又不懂化学，所谓本位主义，隔行如隔山啊！本书将过去这些附属于古生物学的零星成果综合起来，提出一个全盘性新概念，从更基础更广泛开放的心胸，来探讨古生物更深层的奥秘，并期望能在地球生命演化领域，做出重大的贡献。

从实际的角度来说，过去受到科研仪器设备的限制或有人想从古生物化学的角度来探讨，比方说，多种完全破坏性的“湿化学 (Wet Chemistry)”分析方法，根本无法做到本质就非常不均匀化石的“在原位 (*in situ*) 同点 (*in loco*)”分析。然而这几年来，随着各种先进的科研仪器，特别是同步辐射和中子扫描等强烈照射源，还有各种分析化学科研仪器的精进，古生物化学所需用到的分析方法与仪器灵敏度，都有了极大幅的增进。因此如何善用这些科研武器库内的先进威力设备，成为现代科研人员的考验。

我们相信，在不久的将来，有更多科研者，在“看进骨头内”思维的引导下，充分利用这些先进设备，探讨更多化石与其邻近围岩成分，揭开古代生物更多的奥秘，更深入地了解这颗蓝色星球生命演化的含义，让我们更加爱护它。

举一个实际的小例子来说，当我们可以揭开几亿年化石里面，还保存着有机残留物的保存机制，不管是在磷灰石群背景或碳酸盐背景的，只要能得到确

~200 million years old *Lufengosaurus* embryonic bones. However, these are sporadic, a piece here, and a piece there. The depth and width were not optimized. No systematic exploration of this subject matter was taken. After all, chemists may not have enough paleontological knowledge and vice versa. Self-departmentalism separates each other apart as different mounts. This book compiles these paleontological bits and presents a comprehensive new concept to start from the most fundamental level with an open mind to explore the deeper meaning of the mystery of ancient life, and hope to have more contributions to the life evolution of this Earth.

Speaking from the practical point of view, in the past, researchers were limited by the research instruments and facilities even if somebody wanted to explore from the PaleoChemistry angle. For example, the total destructive wet chemistry analyses cannot perform *in situ in loco* analyses on the very heterogeneous fossils. However, within these recent years, more state of art instruments and facilities are available, such as strong synchrotron and neutron radiation source with many advanced analytical chemical instruments. The sensitivity of instrumentation and methodology for PaleoChemistry study improved so much. Therefore, how to use these powerful weapons in our weaponry becomes a severe challenge for contemporary researchers.

It is our belief that in the coming future, more researchers under the guidance of "Look into the Bone" and fully utilizes these state of art instruments and facilities to explore the compositions of the fossils and immediate surrounding matrix to reveal more mysteries of these ancient

认机制，这就有可能对人类做出很可观的贡献。在此机制的正面应用之一，比方说，把它用到食物的保存，那会产生多大的影响啊？如今地球人口超过七十亿，富有国家每年每天倒掉多少食物？在此同时每天，非洲有多少小朋友饿死？如果能将食物的保存稍加延长给他们吃，那会救活多少人命？曾经有位朋友说，他以前在美国某家比萨 (Pizza) 公司工作，公司生产含青椒的比萨，从出厂到客户肚子里，不能超过 12 天，否则就得销毁，所以好些远处的运送，必须靠空运，成本很高，但是，如果能将这些比萨的保存期从 12 天延长到 18 天，则全美国境内都可用 18 轮卡车运送，公司光是省下昂贵的空运，每个月就可省下几千万美元的运输费用。从反面的应用来说，既然知道保存机制，就可研发出破解之道，就如有一把锁，就可作出开锁的钥匙，把这个机制反过来用，没有人容忍垃圾掩埋场盖在家里后院，但是人们的垃圾即便有各种处理的方法，还是无法大幅度地减少掩埋体积，垃圾掩埋场与居民的抗争，是个很头痛的问题！因此，如果能把垃圾的一部分消化掉，减少垃圾体积，就可以缓和经常要盖新垃圾掩埋场的窘状。

另外一个例子，在本书第5章5.1节有提过，骨头化石里面的第一类胶原蛋白，正是中药龙骨的主要有效成分，若能就此进一步探讨，应该可以开发出整系列的中药材，前途无法预料。中医用中药材，已经几千年了，不要忘记我们

life and to understand the meaning of life evolution of this blue planet so that we can love and care about it more.

As a small example, when we unveiled the fact that organic matter can be preserved inside the hundreds of million years old fossils, it does not matter if they were in apatites or carbonates background, and if we can be sure the mechanism of this preservation, it can produce a very significant contribution to humanity as a whole. For example, if we use this mechanism on the positive side for food preservation. How big will the impact that is? There are more than 7 billion human beings on this Earth now. In rich countries, how many tons of food was thrown into garbage dumps every year or every day? At the same time, how many African children starve to death? If the preservation time can be extended just a little bit longer and sent to them, how many children can be saved? Once a friend told me that he used to work for a pizza company. All the pizza containing green pepper have to be in the customer's stomach within 12 days from the factory or have to be destroyed. So, for the distant places, they have to ship them by expensive air cargo. However, if the preservation can be extended from 12 to 18 days, the pizza can be transported by 18 wheeler trucks to anywhere in the USA. That will save several ten-millions every month just for the transportation cost. When this mechanism is used on the opposite side, since we have the mechanism of how to preserve, we can also find a way to digest the matter, like if there is a lock, a key can be made to unlock it. So, to use this mechanism can be used on the other side for garbage reduction. Nobody wants the garbage dump in

的老祖宗就是靠此存活下来，才会有今日的你我，但是对于中药的研究，尚未走上完整科学化，很多中药的成分还不清楚，其药理药效还只能靠中医师的经验。然而屠呦呦拿到诺贝尔奖的青蒿素，救了多少人啊？又如冬虫夏草过去只能到高山耸岭拼老命去采集，每年摔死多少采集者？但是近年来有人研发出人工培养的菌和虫，把这个贵重中药价格普及化，惠及了很多普通百姓呀。

古生物化学的潜力，在此朦胧阶段，我们还只能看到隧道远处的一个小光点。相信，当我们同心协力走下去，这一小点光会从一小点，变成一大面，终而把我们都笼罩在其中；所以，在此呼吁，也绝对诚心欢迎大家的投入，本书开了路，大家来发扬光大。

再提一点，虽然笔者对于考古学是外行的。但是，古生物化学所用的方法和设备，也都可以用同样的“看进骨头内”思维来进行，提升跨领域交叉学科的科研水准啊！如把“看进骨头内”这个词改两个字成“看进骨董内”，不是也行吗？对于考古学来说，利用现代化的科研方法和设备来探讨鉴定骨董，使用客观的科学数据取代人为经验的误差，在骨董的真伪鉴定，绝对是必要的，也可还原真相。毕竟现代的高仿已经到了一个地步，光是靠人眼与经验来鉴定，实在不可靠啊！张大千不就是高仿中的高手吗？

his backyard. There are various ways to reduce the volume of the garbages now. However, these cannot meet the demand of piling up garbage volume. This is a tough and confrontation issue for so many societies nowadays. So, if parts of the garbage volume can be digested and reduce the total volume, it can avoid the embarrassment of needing new garbage dumps.

Another example mentioned in Section 1 of Chapter 5, “Making the dino bone soup”, the collagen type I should be the effective ingredient of the “Dragon Bone” of Chinese medicine. If this can be further studied, a whole series of new Chinese medicines could be developed. As of now, it is not predictable. The Chinese medicines were used for several thousands of years. Do not forget our ancestors survived by that. That is also why we are here today. However, the pharmacognosy study of Chinese medicines is not on a comprehensive scientific road yet. Too many Chinese medicine compositions are still unclear and have to rely on the experience of Chinese medical doctors. However, how many lives were saved by the artemisinin by the Nobel laureate Ms. Youyou Tu? Also, for the *Cordyceps sinensis*, in the past, how many people die for collecting them in the high mountains? However, in recent years, it can be cultivated in the lab, which reduces the price and benefits so many people.

The potential of PaleoChemistry at this hazy stage, we just see a tiny light dot at the far end of the tunnel. When we walk together, this tiny light dot will become a big plane, and eventually soak us in the full brightness. Thus, we are very sincerely asking and inviting everybody to get on board on our boat. This book started and waiting to see more people to carry forward.

One further point. To the archeology, the author is an outsider. However, the same methodology and facility used in PaleoChemistry can be used under the same guideline of "Look into the Bone" to elevate the multi-discipline to a new height. For example, it is perfectly OK to change the phrase "Look into the Bone" to "Look into the Antique." For the archeology field to use the modern scientific research methodology and facility to authenticate a given antique is necessary to restore the truth. After all, there are so many "high imitators" out there. By using human eyes and experience for authentication is not very reliable. Mr. Chang, Da-Chien was the best of the best high imitator!

7.8 一个综合实例

在本书的最后，笔者提出一个最近发表实际综合多领域的古生物化学科研课题，作为撰写论文的范例。

7.8 A Real Combined Example

At the very end, the author presents a recently published real multi-discipline PaleoChemistry research topic as an example for writing inter-discipline paper.

Structure and evolutionary implications of the earliest (Sinemurian, Early Jurassic) dinosaur eggs and eggshells

Authors: Koen Stein[1,2*°], Edina Prondvai[3,4], Timothy Huang[5,6], Jean-Marc Baele[7], P. Martin Sander[8,9], Robert Reisz[5,6,10*]

Affiliations:

1 Earth System Science - AMGC, Vrije Universiteit Brussel, Pleinlaan 2, 1050 Brussels, Belgium.

2 Royal Belgian Institute of Natural Sciences, Directorate 'Earth and History of Life', Rue Vautier 29, 1000 Brussels, Belgium.

3 Evolutionary Morphology of Vertebrates, Ghent University, K.L. Ledeganckstraat 35, 9000 Gent, Belgium.

4 MTA-ELTE Lendület Dinosaur Research Group, Eötvös Loránd University, Pázmány P. s. 1/C, 1117 Budapest, Hungary.

5 International Center of Future Science, and Dinosaur Evolution Research Center of Jilin University, Changchun, Jilin Province, China.

6 National Chung Hsing University, Taichung 402, Taiwan.

7 Department of Geology and Applied Geology, Faculty of Engineering, University of Mons, Place du Parc 20, 7000 Mons, Belgium.

8 Steinmann Institute of Geology, Mineralogy, and Paleontology, Division of Paleontology, University of Bonn, Nussallee 8, 53115 Bonn, Germany.

9 Natural History Museum of Los Angeles County, Dinosaur Institute, 900 Exposition Boulevard, Los Angeles, CA 90007, USA.

10 Department of Biology, University of Toronto Mississauga, Mississauga, Ontario L5L 1C6, Canada.

* Correspondence to: kstein@vub.be; robert.reisz@utoronto.ca ° Lead Contact.

Abstract

One of the fossil record's most puzzling features is the absence of preserved eggs or eggshell for the first third of the known 315 million year history of amniote evolution. Our meagre understanding of the origin and evolution of calcareous eggshell and amniotic eggs in general, is largely based on Middle Jurassic to Late Cretaceous fossils. For dinosaurs, the most parsimonious inference yields a thick, hard shelled egg, so richly represented in the Late Cretaceous fossil record. Here, we show that a thin calcareous layer (≤ 100 μm) with interlocking units of radiating crystals (mammillae) and a thick shell membrane already characterize the oldest known amniote eggs, belonging to three coeval, but widely distributed Early Jurassic basal sauropodomorph dinosaurs. This thin shell layer strongly contrasts with the considerably thicker calcareous shells of Late Jurassic dinosaurs. Phylogenetic analyses and their Sinemurian age indicate that the thin eggshell of basal sauropodomorphs represents a major evolutionary innovation at the base of Dinosauria and that the much thicker eggshell of sauropods, theropods, and ornithischian dinosaurs evolved independently. Advanced mineralization of amniote eggshell (≥150 μm in thickness) in general occurred not earlier than Middle Jurassic and may correspond with a global trend of increase in atmospheric oxygen.

Introduction

The origin of the amniote egg is a topic of great significance because it represents one of the major evolutionary innovations in vertebrate evolution, allowing the group to complete their invasion of the terrestrial landscape and sever their reproductive cycle from the aquatic medium (1). However, paleontological studies of this pivotal event have been greatly hampered by the poor early record of fossil eggs (2, 3). Recent attempts to fill the gaps in fossil eggshell phylogeny still leave at least 125 million years of amniote evolution between the appearance of amniotes in the fossil record and the first appearance of preserved terrestrial eggs or eggshells (4–12). The oldest known eggs or eggshells have been reported (7–12) from three Sinemurian (195–192 Ma) sauropodomorph dinosaurs, *Massospondylus* from the Elliot Formation of South Africa, *Lufengosaurus* from the Lufeng Formation of Yunnan, China, and *Mussaurus* from the Laguna Colorada Formation of Argentina (Figure 1). Within the context of their respective localities, some of these materials have been examined to a limited extent. In a study on prenatal remains of *Lufengosaurus*, some of the authors of the current study previously provided a very brief description of its eggshell, and noted its extreme thinness (12). Other authors working on *Massospondylus* initially discarded them as crocodile eggshells (13), and later as having a diagenetically altered microstructure (9). *Mussaurus* eggshell has to our knowledge never been described in a formal publication. These remains are the earliest confirmed amniote eggshells recorded in the fossil record. Due to their rarity, fragmentary nature, and great geographic distance from each other, they were never studied from the perspective of the evolution of amniote eggshell. Here we aim to understand their microstructural features and try to elucidate when and how the earliest mineralized eggshells could have evolved. To accomplish our goal, we utilized petrographic sections, analytical chemistry tools and computational statistical methods (description in Materials and Methods section). This study contributes to our understanding of the evolution of rigid shelled eggs; a key trait in the evolutionary success of archosaurs.

Results

Eggshell structure

The calcareous layer of *Lufengosaurus* eggshells (C2019 2A233) ranges from 60–90 μm in thickness. They consist of crocodile eggshell-like wedge- and crown- shaped shell units that are relatively wide compared to the calcareous layer thickness (Figure 1 a-f). Polarized light microscopy suggest that the outer surface of the eggshell is unaltered (Figure 1f, Supplementary Information). The very thin crystalline layer (~10 μm) topping the eggshell units is phosphatic in nature (Fig S1, S2), and looks scalloped with shallow pits and low ridges, not necessarily matching eggshell unit borders. These surface irregularities or tubercles are of such small dimensions that the surface of the calcareous layer looks smooth (Fig S2a), and it remains unclear if they match the ornamentations seen in younger dinosaur eggshells. The bulk of the units, corresponding to the mammillary cones, is formed by a calcite radial ultrastructure (sensu 15) with interlocking crystalline units (Figure 1c-f). The patchy cathodoluminescence texture suggests some radial crystal wedges experienced recrystallization, but most of the original microstructure is conserved (Figure S1b. No tabular structures or horizontal accretion lines can be observed. The growth centre of the units is embedded in a phosphorus-rich (Figure S1, S2), thick fibrillar layer (60–75 μm) representing the eggshell membrane (Figure 1e). Pore spaces are rare and difficult to discern (Figure 1e). Due to the fragmentary nature of the materials, and because pores were not always unambiguously identifiable, it was not possible to make an estimation of pore density. However, pore distribution does not appear to be consistent with the presence of tubercles or depressions on the outer surface. A tangential section through the membrane shows clusters of crystals with flower-like arrangements (Figure S3). The lack of a thick palisade layer and the overall thinness of the calcareous shell clearly distinguish *Lufengosaurus* eggshells from avian and other younger dinosaurian eggshells.

The South African *Massospondylus* (BP/1/5254, BP/1/5347) calcareous eggshell layer is slightly thicker (80–100 μm) than that of *Lufengosaurus*. The eggshell units are very difficult to discern (Figure 1i,j). In the past, these units have been interpreted as wedge-shaped (8). Our cathodoluminescence analysis (Figure 1k; Figure S1) shows very high luminosity of calcite in the eggshell units, which supports the idea that these structures are the result of diagenetic alteration of the original microstructure (9) (Figure 1k; Fig S1). Nonetheless, eggshell is present in *Massospondylus* eggs from several different localities in South Africa, and of similar thickness as in *Lufengosaurus*, and some features remain recognizable. The outer surface of the eggshell, as in *Lufengosaurus*, is rugged with low tubercles and shallow depressions. Occasional pores are distributed unevenly throughout the shell surface (Figure 2c, d). Below the calcareous layer, a dark, isotropic layer (50 – 90 μm thick, cross polarized light) merges with, or entirely obscures the mammillary cones (Figure S3). We identify this layer as a remnant of the eggshell membrane, given its position relative to the calcareous layer and its chemical similarity with the *Lufengosaurus* shell membrane (rich in phosphate and calcite, Figure S1-S3). A shell membrane is also preserved in some of the complete eggs with the embryos (Fig 2a).

The Argentinian *Mussaurus* eggshell (PVL 5965) is severely affected by diagenesis. Only few sparse and widely scattered calcite crystals, similar in size and shape to the radiating crystals in the mammillary cones of *Lufengosaurus* eggshell units, remain of the calcareous layer (Figure 1g, h). The eggshell membrane is preserved as a thick (150 – 180 μm) phosphatic layer with little internal structure.

Calcareous layer to membrane thickness ratios may vary due to incomplete preservation of the membrane. They range from ~1:1 in *Lufengosaurus* and ~1.5:1 *Massospondylus*, but remain uncertain in *Mussaurus* due to the loss of an intact, coherent calcareous layer.

The identity of all three taxa is unquestionable based on the presence of embryonic remains (10–12, 16, contra 13). We thus reconstruct basal sauropodomorph eggshell as having a thin calcareous layer, composed of low, wide mammillary cones (approximate width to height rations of 1:1) attached to a membrane of at least similar thickness (Figure 3).

Taphonomic and evolutionary implications

The phylogenetically informed regression analysis of mineralized eggshell thickness versus egg mass in a wide taxonomic range of extant and extinct egg laying amniotes (Figure 4a) revealed a significant positive relationship between egg size (mass) and shell thickness, with considerable phylogenetic signal ($\lambda=0.86$; $p<0.001$; see SI). The regression function is largely determined by the taxa with rigid-shelled eggs (non-avian dinosaurs, birds and crocodiles). Negative outliers, in which the size of the eggs and their shell thickness are well below the regression line, are the extant and fossil groups with known or inferred flexible shelled-eggs, such as marine turtles, squamates, and pterosaurs (Figure S4). Interestingly, *Lufengosaurus* and *Massospondylus* plot with these negative outliers emphasising the pronounced thinness of their calcareous eggshell relative to their egg mass (Figure 4). However, due to the interlocking nature of the crystal units, these basal sauropodomorphs most likely had rigid eggshell. This interpretation is supported by the preservational characteristics of all egg fragments recovered from the various sites. All retain their curvature, and even though the eggs of *Massospondylus* are somewhat crushed, they show the typical cracking and fragmenting associated with rigid structures (Figure 1, 17, 18). This observation contrasts with the preservational characteristics of soft- shelled fossil material, now abundantly preserved for the pterosaur *Hamipterus* (19).

A time-calibrated cladogram of archosauromorphs suggests the ancestral state for dinosaurs is the thin-shelled condition (Figure 4b, Supplementary Information). Maximum likelihoods

(ML) of ancestral character states imply low ratios of eggshell thickness to egg mass are plesiomorphic in Dinosauria. These calculations are based on the logical assumption that the archosauromorph root node represents a poorly mineralized eggshell (cf. 2, 3), the value of which is derived from the lowest observed value among the extant amniote taxa (*Pelusios sinuatus*) (see Supplementary Information). Reconstructed relative eggshell thicknesses for the base of the dinosaur tree are very close to those of the early sauropodomorphs described here. The ancestral state reconstruction suggests independent eggshell thickening events in all major archosauromorph clades, but also within different dinosaur clades. Evolutionary reversals are also demonstrated (cf. *Pelusios sinuatus* and *Carretta carretta*). Finally, it is important to note that these thickening events generally occurred after the Sinemurian (~195 Ma).

Discussion

Our detailed examination of the eggs of these basal sauropodomorph dinosaurs shows that all have an extremely thin mineralized eggshell layer. Different diagenetic settings of their respective localities affected the original microstructure to different degrees, with *Lufengosaurus* having the best, and *Mussaurus* the least preserved details (Figure 1, Supplementary Information). The structural characteristics of these Early Jurassic dinosaur eggshells are unlike those in any other known dinosaur. The extreme thinness could have resulted from decalcification during egg incubation, as seen in some *Massospondylus* eggs containing advanced stage embryos. However, this is unlikely because there is no sign of resorption craters at the base of the crystal units, and the recent collection of a complete *Massospondylus* nest with undeveloped embryos makes this unlikely (8, Figure 2a). In addition, the similar thinness, eggshell unit characteristics, and outer surface ornamentation suggest that the calcareous shell layers are similar to their original thickness in both *Lufengosaurus* and *Massospondylus*. A low ratio of calcareous layer to eggshell membrane thickness is usually associated with flexible-shelled eggs of extant amniotes (20), however, in line with our other observations, we conclude that these early dinosaurs had thin, albeit rigid-shelled eggs (Figure 3), a highly unusual, unexpected condition. The semi-arid depositional conditions (7, 11, 12) and relative thinness of the eggshell suggest the eggs needed to be protected from dehydration (21, 18, 22). Hence, as in many other dinosaurs (23, 24) and most modern-day non-avian reptiles, the eggs were most likely buried in the nest, although this hypothesis needs further support by more complete data on pore density and relative eggshell porosity. Previous studies have pointed to a combination of nesting site fidelity, colonial nesting, and parental care in these early sauropodomorphs (11, 12). It is thus possible that through their behavioural ecology, these sauropodomorphs created taphonomic conditions that allowed the preservation of such delicate structures.

Eggshells with a comparatively thick membrane but thin calcareous layer are in sharp contrast with heavily mineralized dinosaurian eggshells commonly found during the Cretaceous

(13, 15). Ancestral state reconstruction of this feature may be affected by lack of information from earlier reptilian clades and crucial basal taxa, such as early ornithischians. Nonetheless, the lack of pre-Middle Jurassic rigid fossil eggshells (3, 5, 25), the different mammillary ultrastructure in crocodilian and dinosaurian eggshells, and the aragonitic nature of turtle eggshells, provide strong support for the hypothesis of independent eggshell thickening events in these reptiles (see Supplementary Information for further results and discussion of the ancestral state analysis). This scenario also favours the independent origin of extended eggshell growth in the dinosaurian clades Ornithischia, Sauropodomorpha and Theropoda. All known dinosaurian eggshells, including those described here, possess mammillae with radiating calcite crystals (14), therefore, mammillated eggshell with calcite radial ultrastructure can be considered a dinosaurian synapomorphy.

It seems straightforward to assume a flexible non-mammillated → flexible mammillated → rigid mammillated succession of eggshell structural evolution. However, it does not have to be so strictly sequential or directional, as there may be an intricate interplay between biological and environmental factors shaping eggshell structure and composition (20). Diversity in eggshell micro- and ultrastructure in different reptilian clades points to numerous convergences, secondary losses and reversals. Turtles demonstrate this evolutionary complexity by revealing conditional aragonite/calcite composition of the calcareous layer (20, 26) and multiple eggshell-softening events (27) (Figure 4b), even with complete loss of mammillae in the pleurodiran *Pelusios sinuatus* (28).

Our ancestral state reconstruction shows an independent thickening of the calcareous layer in several archosauromorph clades during the Jurassic. Interestingly, this does not seem to be directly related to increase in body size, and hence egg size. However, the occurrence of the earliest strongly mineralized archosauromorph and turtle eggshells in the Middle and Late Jurassic (5, 29) coincides with the recovery to modern day atmospheric oxygen values (Figure 4b; 30). The GEOCARBSULF model suggests atmospheric oxygen levels dropped during the Permian and Triassic from an all-time high (32–33%) in the Late Carboniferous to an all-time low (15%) in the Early Jurassic (30). Such models calculate Phanerozoic atmospheric oxygen levels by representation of nutrient cycling and estimation of productivity, or by isotope mass balance (30–32). Estimated pO_2 may vary depending on the used model, nonetheless, a negative excursion in the Hettangian (201–199 Ma) clearly precedes a general trend of atmospheric oxygen increase in the Sinemurian, 199–191 million years ago (Figure 4B, 30, 32, 33).

In modern reptiles, oxygen restriction is known to play an important inhibiting role on eggshell growth and other aspects of embryonic development (34, 35). Furthermore, Plaeosaurus, a Norian to Rhaetian basal sauropodomorph from Central Europe and Greenland (Figure 1) is phylogenetically close to the materials presented here and known from abundant remains (e.g. 36), but hitherto no eggs have been found. Eggs predating the Early Jurassic would likely be very difficult to find. Nonetheless, there is no evidence of any fossil eggs preserved during the 120 million years of amniote evolution that would predate the findings described here, anywhere around the globe and in any type of depositional system. We suggest that egg physiology and low atmospheric oxygen levels may have inhibited eggshell thickening before the end of the Early Jurassic, when atmospheric oxygen levels started to rise again. However, it should be stated that this remains a hypothesis and further testing it is beyond the scope of the current study.

Material and methods

Thin sectioning

The eggshell from the Early Jurassic DaWa locality in the Lower Lufeng Formation of Yunnan, China, is documented from a 3–4 cm long calcareous nodule containing numerous eggshell fragments (but no bones) (Figure 1). The material is housed in the Chuxiong Prefectural Museum under catalogue no. C2019 2A233. Uncut shell fragments can be identified from their high Ca and P content with μXRF (Figure S2). Radial and tangential petrographic sections were made from the sample (Figure 1a-e; Figure S3). The eggshells were found in a 10–20 cm thick monotaxic bonebed. The layer solely contains dislocated basal sauropodomorph embryonic elements ascribed to *Lufengosaurus* (12). In contrast to the DaWa locality, specimens from the Rooidraai locality in South Africa are complete eggs with well-preserved embryos inside (Figure 2; 10, 11, 16). Thin sectioned *Massospondylus* eggshell (Figure 1i,j; Figure S3) was not directly obtained from a nest with embryos, but sampled from an adjacent nest in the same horizon containing embryonic remains ascribed to *Massospondylus* (10). The *Massospondylus* material is housed at the Bernard Price Institute of Palaeontology of the University of Witwatersrand under catalogue no. BP/1/5254 and BP/1/5347. Despite the taphonomical difference between Lufeng and Rooidraai, the two localities are similar in geology, temporal range, environment, and faunal assemblages (10, 11, 12). The *Mussaurus* eggshell (Figure 1g-h) fragment was sampled from a nest containing eggs with embryos (specimens stored at the Instituto "Miguel Lillo", Tucuman, catalogue no. PVL 5965). The specimen was collected from the Early Jurassic of the Laguna Colorada

Formation of Patagonia, Argentina by researchers of the Museo Paleontológico Egidio Feruglio in Trelew, near the original *Mussaurus* embryo discovery site (7).

Light microscopy and SEM

Fossil eggshell specimens were thin sectioned in the Steinmann Institut (University of Bonn) and studied under single plane polarizers (ppl) and cross- polarized light (xpl) under a Leica DMLP and a Zeiss Axioskop compound microscope. Photos of sections were taken with a Leica 425 firecam and Zeiss Axiocam. Scanning electron microscopy images were taken with a JEOL JSM 6300 (Tokyo, Japan).

Cathodoluminescence

Cathodoluminescence imaging (Figure 1, Figure S1) was performed using a Cambridge Image Technology (CITL) Mark 5 cathodoluminescence system (Hatfield, UK) at University of Mons, Belgium. Beam conditions were 15 kV acceleration voltage and 500 μA beam current. The cold cathode electron gun produced an unfocussed elliptical beam of ca. 60 mm^2, which results in a current density of 8 $\mu A/mm^2$. Helium was used instead of air in order to improve beam stability. The cold cathode electron gun produced an unfocused beam of a few mm in diameter. Spectral cathodoluminescence imaging was achieved by inserting narrow bandpass optical filters within the lightpath. Filtering at 880 nm allowed observing the emission of Nd^{3+} which substitutes Ca^{2+} in apatite. In this mode, the strong yellow-red cathodoluminescence of calcite is suppressed and the infrared cathodoluminescence of apatite is enhanced. Filtering at 640 nm isolates the emission of Sm^{3+} but is also influenced by the strong cathodoluminescence of calcite, which is activated by Mn^{2+} at ca. 605 nm. The cathodoluminescence images were captured with a high-sensitivity, Peltier-cooled digital color camera. For spectral imaging, the camera was used in 2x2 binning mode in order to capture monochromatic images. Cathodoluminescence spectra were recorded using a CITL OSA2 optical spectrometer with a Peltier-cooled CCD detector and a spectral resolution of 4 nm. The spectra are corrected for background and ambient light (dark measurement) but not for system response.

μXRF and Raman spectroscopy

Identifying eggshell specimens with a thickness of only 100 to 200 μm proved sometimes equivocal under the microscope. Moreover, the identity of the membrane and calcareous shell was not always clear. Therefore, we employed spectroscopic methods to characterize chemical composition of the eggshell components. First we used μX-ray fluorescence (μXRF, M4 Tornado, Bruker Nano Technologies, Berlin, Germany) to identify major element distribution in fluorescence maps of fossil eggshell fragments (Figure S2). Element distribution maps show a relative counts signal after deconvolution. Only elements of interest (Ca, P, Fe, Si) are highlighted. Line scans (Figure S2c, e) show relative counts signal and were extracted from map data to demonstrate gradients of element composition along a chosen transect in the samples. μXRF results were cross referenced with Raman spectroscopy (Figure S2f, g). We used a fully integrated confocal Raman microscope (LabRAM HR Evolution, HORIBA Scientific, Kyoto, Japan) equipped with a high stability confocal microscope with XYZ motorized stage and a multichannel air cooled CCD detector (spectral resolution <1cm^{-1}, lateral resolution 0.5μm, axial resolution 2μm). Two lasers are mounted on the instrument: a HeNe laser (633nm) and a Solid state laser (532nm). Initially, the green laser was used to reduce signal noise, but due to overheating, and even burning of the sample, the red laser had to be installed. Both lasers were used in combination with a 50x objective. Intensity for spot measurements ranged from 2.5–25 mW.

Phylogenetic regression of eggshell thickness vs egg mass

To examine the relative rigidity of the early sauropodomorphs eggshells compared to the size of the eggs, we compiled a comprehensive dataset of calcareous eggshell thickness (mineralized calcite or aragonite layer thickness) and egg mass in a variety of fossil and extant egg-laying amniotes (two snakes, one lizard, six turtles, three crocodiles, two pterosaurs, 37 non-avian dinosaurs, 46 birds; see Table S1, Figure 4a). Besides the early sauropodomorph eggshells measured in this study, eggshell thickness data were collected from the literature (20–29, 38–43). Egg mass data were estimated from published size data (20, 21, 40–44) with the formulae $M = 5.60 \times 10^{-4} \times L \times B2$ (M, egg mass; L, maximum egg length; B, maximum egg breadth) for non-avian sauropsids (20) and $M = 5.48 \times 10^{-4} \times L \times B2$ for birds (45). Mass and dimensions of a *Lufengosaurus* egg are extremely difficult to estimate, but our values were based on a size comparison of embryonic remains with those of *Massospondylus* (10–12). Elements belonging to *Massospondylus* embryos are generally 1.5 times smaller in length than those of *Lufengosaurus*, translating in a three times larger egg volume and mass for *Lufengosaurus*.

The egg mass and shell thickness dataset was then used in regression analyses to investigate how eggshell rigidity is reflected in the relationship between eggshell thickness and egg mass across these taxa. All calculations were performed in R version 3.2.3 (2015 The R Foundation for Statistical Computing).

To account for the trait correlations resulting from phylogenetic interrelationships on the regression outcome, a phylogenetic tree containing all taxa (nexus file S1) in the dataset was constructed in Mesquite v3.04 (46), where topologies were based on the literature (47–53). Unknown divergence times and branch lengths were based on data of the age of the oldest fossil occurrence of eggshell taxa. For measuring phylogenetic signal in thickness of calcareous eggshell layer and egg mass, we used two different methods: Blomberg's K, and Pagel's λ (phylosig from package 'phytools'; 54) both of which gave a significant phylogenetic signal for both traits ($p \leq 0.001$). Relationship between eggshell thickness and egg mass was linearized by ln-transformation of both variables. Phylogenetic Generalized Least Squares (PGLS) regression (gls with Brownian motion evolution from package 'nlme' (55) as well as gls with Pagel's λ scaling parameter (corPagel) from package 'ape' (56) was performed on the ln-transformed dataset. Based on AIC values and log-likelihoods, the λ-model fitted our data better and therefore was chosen for the interpretation of our results. 95% confidence band was visualized for the regression line (ggplot in package 'ggplot2'; 57). Taxa with relatively thin shelled eggs were identified in the regression using three methods for outlier recognition: QQ-plot of residuals' normality, density plot of residuals, and boxplot.stats function (Figure S4).

Ancestral state reconstruction of eggshell features

After identifying the most likely physical properties of fossil eggshells by means of phylogenetic regression, we focused on known fossil archosauromorph eggshells, and computed the ancestral states of the ratio of calcareous layer thickness to egg mass in a variety of taxa. Only limited fossil specimens were available for this analysis, and were balanced with extant species of all known modern archosauromorph clades. Taxa and values are listed in Table S2.

Tree topology (nexus file S2) was compiled from literature data. The position of pterosaurs is based on 53 and 58 (but see 59, 60 for contrasting views). Choristoderes are placed as basal archosauromorphs (61, 62), and turtles are considered sister taxon to archosaurs, based on current molecular evidence (48, 63–65, but see 66 for a contrasting view). Divergence dates were collected from literature data (47–53, 58) and the Paleobiology Database.

Reconstruction of ancestral states was computed in R using the functions 'fastAnc' and 'contMap' of the phytools package (54), with the root node set to represent the ancestral poorly mineralized eggshell, a hypothetical value based on 2 and 3, and the lowest observed value among the extant taxa in our analysis (*Pelusios sinuatus*). The value for the root was set at 1.05 because it maximizes observable differences between reconstructed and observed states.

Acknowledgments

KS thanks the Fonds Wetenschappelijk Onderzoek Vlaanderen for funding and Ph. Claeys for helping improve an earlier version of the MS. EP was funded by the Bijzonders Onderzoeksfonds – Universiteit Gent (grant nr. 01P12815). RR was funded by Jilin University, and University of Toronto. We thank Dr. Diego Pol for providing the *Mussaurus* eggshell materials.

Author contributions

All authors contributed to the research. KS and EP wrote the manuscript. KS, EP, JMB, PMS, RR designed and performed experiments, contributed to writing. RR initiated and guided project, and together with TH excavated and provided fossil materials.

Declaration of interest

The authors declare no competing interests

Data availability statement

The datasets generated during and/or analysed during the current study are available from the corresponding authors on reasonable request.

References

1. Reisz, R. R. The origin and early evolutionary history of amniotes. Trends Ecol. Evol. 12, 218–764 (1997).
2. Hirsch, K. F. The oldest vertebrate egg? J. Paleontol. 53, 1068–1084 (1979) .

3. Sander, P. M. Reproduction in early amniotes. Science 337, 806–808 (2012).
4. Araújo, R. et al. Filling the gaps of dinosaur eggshell phylogeny: Late Jurassic Theropod clutch with embryos from Portugal. Sci. Rep.-UK 3, 1924; 10.1038/srep01924 (2013).
5. Garcia, G., Marivaux, L., Pélissié, T. & Vianey-Liaud, M. Earliest Laurasian sauropod eggshells. Acta Palaeontol. Pol. 51, 99–104 (2006).
6. Fernandez, V. et al. Evidence of Egg Diversity in Squamate Evolution from Cretaceous Anguimorph Embryos. PLoS One 10, e0128610 (2015).
7. Bonaparte, J. F. & Vince, M. El hallazgo del primer nido de dinosaurios triasicos, (Saurischia, Prosauropoda), Triasico Superior de Patagonia, Argentina [The discovery of the first nest of Triassic dinosaurs (Saurischia, Prosauropoda,) from the Upper Triassic of Patagonia, Argentina]. Ameghiniana 16, 173–182 (1979).
8. Grine, F. E. & Kitching, J. W. Scanning electron microscopy of early dinosaur eggshell structure: A comparison with other rigid sauropsid eggs. Scanning Microscopy 1, 615 (1987).
9. Zelenitsky, D. K. & Modesto, S. P. Re-evaluation of the eggshell structure of eggs containing dinosaur embryos from the Lower Jurassic of South Africa. S. Afr. J. Sci. 98, 407–408 (2002) .
10. Reisz, R. R., Scott, D., Sues, H.-D., Evans, D. C. & Raath, M. A. Embryos of an early Jurassic prosauropod dinosaur and their evolutionary significance. Science 309, 761–764 (2005).
11. Reisz, R. R., Evans, D. C., Roberts, E. M., Sues, H. D. & Yates, A. M. Oldest known dinosaurian nesting site and reproductive biology of the Early Jurassic sauropodomorph *Massospondylus*. P. Natl. Acad. Sci. USA 109, 2428–2433 (2012).
12. Reisz, R. R. et al. Embryology of Early Jurassic dinosaur from China with evidence of preserved organic remains. Nature 496, 210–214 (2013).
13. Carpenter, K. Eggs, nests, and baby dinosaurs: a look at dinosaur reproduction (Indiana Univ. Press, 1999).
14. Scotese, C. R. Atlas of Earth History, Volume 1, Paleogeography, PALEOMAP Project, Arlington, Texas (2001).
15. Mikhailov K. E. Fossil and recent eggshell in amniotic vertebrates: fine structure, comparative morphology and classification. Spec. Pap. Palaeontol. 56, 1–77 (1997).
16. Reisz, R. R., Evans, D. C., Sues, H.-D.& Scott, D. Embryonic skeletal anatomy of the sauropodomorph dinosaur *Massospondylus* from the Lower Jurassic of South Africa. J. Vertebr. Paleontol. 30, 1653–1665 (2010).
17. Hayward, J. L., Dickson, K. M., Gamble, S. R., Owen, A. W. & Owen, K. C. Eggshell taphonomy: environmental effects on fragment orientation, Hist. Biol. 23, 5–13 (2011).
18. Marsola, J. C. de A., Batezelli, A., Montefeltro, F. C., Grellet-Tinner, G. & Langer, M. C. Palaeoenvironmental characterization of a crocodilian nesting site from the Late Cretaceous of Brazil and the evolution of crocodyliform nesting strategies. Palaeogeogr., Palaeocl. 457, 221–232 (2016).
19. Wang, X., Kellner, A. W. A., Jiang, S., Cheng, X. & Wang, Q. Egg accumulation with 3D embryos provides insight into the life history of a pterosaur. Science 358, 1197–1201 (2017).
20. Packard, M. J. & Demarco, V. G. In Egg Incubation (eds. Deeming, D. C., Ferguson, M. W. J.) 53–70

(Cambridge Univ. Press, 1991).

21. Ar, A., Paganelli, C. V., Reeves, R. B., Greene, D. G. & Rahn, H. The avian egg: water vapor conductance, shell thickness and functional pore area. Condor 76, 153–158 (1974).

22. Deeming, D. C. & Ferguson, M. W. J. Methods for the determination of the physical characteristics of eggs of Alligator mississippiensis: a comparison with other crocodilian and avian eggs. Herp. J. 1, 458–462 (1990).

23. Sander, P. M., Peitz, C., Jackson, F. & Chiappe, L. Upper Cretaceous titanosaur nesting sites and their implications for sauropod dinosaur reproductive biology. Palaeontogr. Abt. A 284, 69–107 (2008).

24. Tanaka, K., Zelenitsky, D. K. & Therrien, F. Eggshell porosity provides insight on evolution of nesting in dinosaurs. PLoS ONE 10, e0142829–23; 10.1371/journal.pone.0142829 (2015).

25. Hirsch, K. F. Parataxonomic classification of fossil chelonian and gecko eggs. J. Vertebr. Paleontol. 16, 752–762 (1996).

26. Baird, T. & Solomon, S. E. Calcite and aragonite in the egg shell of Chelonia mydas L. J. Exp. Mar. Biol. Ecol. 36, 295–303 (1979).

27. Zelenitsky, D., Therrien, F., Joyce, W. & Brinkman, D. B. First fossil gravid turtle provides insight into the evolution of reproductive traits in turtles. Biol. Letters 4, 715–718 (2008).

28. Kusuda, S. et al. Diversity in the matrix structure of eggshells in the Testudines (Reptilia) Zool. Sci. 30, 366–374 (2013).

29. Lawver, D. R. & Jackson, F. D. A review of the fossil record of turtle reproduction: eggs, embryos, nests and copulating pairs. Bull. Peabody Mus. Nat. Hist., 55, 215–236; 10.3374/014.055.0210 (2014).

30. Berner, R. A. Phanerozoic atmospheric oxygen: New results using the GEOCARBSULF model. Am. J. Sci. 309, 603–606 (2009).

31. Mills, B., Belcher, C. M., Lenton, T. M. & Newton, R. J. A modeling case for high atmospheric oxygen concentrations during the Mesozoic and Cenozoic. Geology, 44, 1023–1026; 10.1130/G38231.1 (2016).

32. Schachat, S. et al. Phanerozoic pO_2 and the early evolution of terrestrial animals. P. Roy. Soc. B, 285, 20172631; 10.1098/rspb.2017.2631 (2018).

33. Royer, D. L., Donnadieu, Y. Park, J. Kowalczyk, J. & Goddéris, Y. Error analysis of CO_2 and O_2 estimates from the long-term geochemical model geocarbsulf. Am. J. Sci. 314, 1259–1283 (2014).

34. Hempleman, S. C. Adamson, T. P. & Bebout, D. E. Oxygen and avian eggshell formation at high altitude. Resp. Physiol., 92, 1–12; 10.1016/0034-5687(93)90115-Q (1993).

35. Owerkowicz, T. Elsey, R.M. & Hicks J.W. Atmospheric oxygen level affects growth trajectory, cardiopulmonary allometry and metabolic rate in the American alligator (Alligator mississippiensis). J. Exp. Biol., 212, 1237–1247; 10.1242/jeb.023945 (2009).

36. Galton, P. M. & Upchurch, P. In The Dinosauria, 2nd edition (eds. Weishampel, D. B., Dodson, P., Osmolska, H.) 232–258 (University of California Press, 2004).

37. Baele, J. M., Dreesen, R. & Dusar, M. Assessing apatite cathodoluminescence as a tool for sourcing oolitic ironstones. Anthropol. Præhist. 126, 57–67 (2016).

38. Deeming, D. C. Ultrastructural and functional morphology of eggshells supports the idea that dinosaur eggs were incubated buried in a substrate. Palaeontology 49, 171–185 (2006).

39. Ferguson, M. W. J. The structure and composition of the eggshell and embryonic membranes of Alligator mississippiensis. Trans. Zool. Soc. London 36, 99–152 (1982).
40. Grellet-Tinner, G., Wroe, S., Thompson, M. B. & Ji, Q. A note on pterosaur nesting behavior. Hist. Biol. 19, 273–277; 10.1080/08912960701189800 (2007).
41. Hirsch K. F. Contemporary and fossil chelonian eggshells. Copeia 382, 382–397; 10.2307/1444381 (1983).
42. Osborne, L. & Thompson, M. B. Chemical Composition and Structure of the Eggshell of Three Oviparous Lizards. Copeia, 2005, 683–692; 10.1643/CH-04–280R1 (2005).
43. Wang, X. et al. Sexually Dimorphic tridimensionally preserved pterosaurs and their eggs from China. Curr. Biol. 24, 1–8; 10.1016/j.cub.2014.04.054 (2014).
44. Unwin, D. M. & Deeming, D. C. Pterosaur eggshell structure and its implications for pterosaur reproductive biology. Zitteliana B28, 199–207 (2008).
45. Hoyt, D. F. Practical methods of estimating volume and fresh weight of bird eggs. Auk, 96, 73–77 (1979).
46. Maddison, W.P. & Maddison, D.R. Mesquite: a modular system for evolutionary analysis. Version 3.10 http://mesquiteproject.org (2016).
47. Evans, S. E. & Jones, M. E. H. In New Aspects of Mesozoic Biodiversity (Vol. 132) 27–44 (Springer, 2010).
48. Chiari, Y., Cahais, V., Galtier, N. & Delsuc, F. Phylogenomic analyses support the position of turtles as the sister group of birds and crocodiles (Archosauria). BMC Biol. 10, 65; 10.1186/1741-7007-10-65 (2012).
49. Fong, J. J., Brown, J. M., Fujita, M. K. & Boussau, B. A phylogenomic approach to vertebrate phylogeny supports a turtle-archosaur affinity and a possible paraphyletic Lissamphibia. PLoS ONE 7, e48990; 10.1371/journal.pone.0048990.t002 (2012).
50. Field, D. J. et al. Toward consilience in reptile phylogeny: miRNAs support an archosaur, not lepidosaur, affinity for turtles. Evol. Dev. 16, 189–196; 10.1111/ede.12081 (2014).
51. Jarvis, E. D. et al. Whole-genome analyses resolve early branches in the tree of life of modern birds. Science 346, 1320–1331; 10.1126/science.1253451 (2014).
52. Lloyd, G. T. et al. Dinosaurs and the Cretaceous Terrestrial Revolution. P. Roy. Soc. B 275, 2483–2490 (2008).
53. Nesbitt S. J. The early evolution of archosaurs: relationships and the origin of major clades. Bull. Am. Mus. Nat. Hist. 352, 1–292 (2011).
54. Revell, L. J. Phytools: an R package for phylogenetic comparative biology (and other things). Methods Ecol. Evol. 3, 217–223; 10.1111/j.2041–210X.2011.00169.x (2012).
55. Pinheiro, J. et al. Package 'nlme': Linear and Nonlinear Mixed Effects Models. CRAN repository: https://cran.r-project.org/web/packages/nlme/nlme.pdf (2016).
56. Paradis, E. et al. Package 'ape': Analyses of Phylogenetics and Evolution. CRAN repository: http://ape-package.ird.fr/ (2015).
57. Wickham, H. & Chang, W. Package 'ggplot2': An Implementation of the Grammar of Graphics. CRAN repository: http://ggplot2.org, https://github.com/hadley/ggplot2 (2016).
58. Hone, D. W. E. & Benton, M. J. An evaluation of the phylogenetic relationships of the pterosaurs among archosauromorph reptiles. J. Syst. Palaeontol. 5, 465–469; 10.1017/S1477201907002064 (2007).

59. Bennett, S. C. The phylogenetic position of the Pterosauria within the Archosauromorpha. Zool. J. Linn. Soc. 118, 261–309 (1996).

60. Bennett, S. C. The phylogenetic position of the Pterosauria within the Archosauromorpha re-examined. Hist. Biol. 25, 545–563; 10.1080/08912963.2012.725727 (2013).

61. Evans, S. E. The skull of Cteniogenys, a choristodere (Reptilia: Archosauromorpha) from the Middle Jurassic of Oxfordshire. Zool. J. Linn. Soc. 99, 205–237 (1990).

62. Jalil, N. E. A new prolacertiform diapsid from the Triassic of North Africa and the interrelationships of the Prolacertiformes. J. Vertebr. Paleontol. 17, 506–525 (1997).

63. Crawford, N. G. et al. More than 1000 ultraconserved elements provide evidence that turtles are the sister group of archosaurs. Biol. Letters 8, 783–786; 10.1098/rsbl.2012.0331 (2012).

64. Lee, M. S. Turtle origins: insights from phylogenetic retrofitting and molecular scaffolds. J. Evol. Biol. 26, 2729–2738; 10.1111/jeb.12268 (2013).

65. Lu, B., Yang, W., Dai, Q. & Fu, J. Using genes as characters and a parsimony analysis to explore the phylogenetic position of turtles. PLoS ONE 8, e79348; 10.1371/journal.pone.0079348 (2013).

66. Schoch, R.R. & Sues, H.-D. A Middle Triassic stem-turtle and the evolution of the turtle body plan. Nature 523, 584–587; 10.1038/nature14472 (2015).

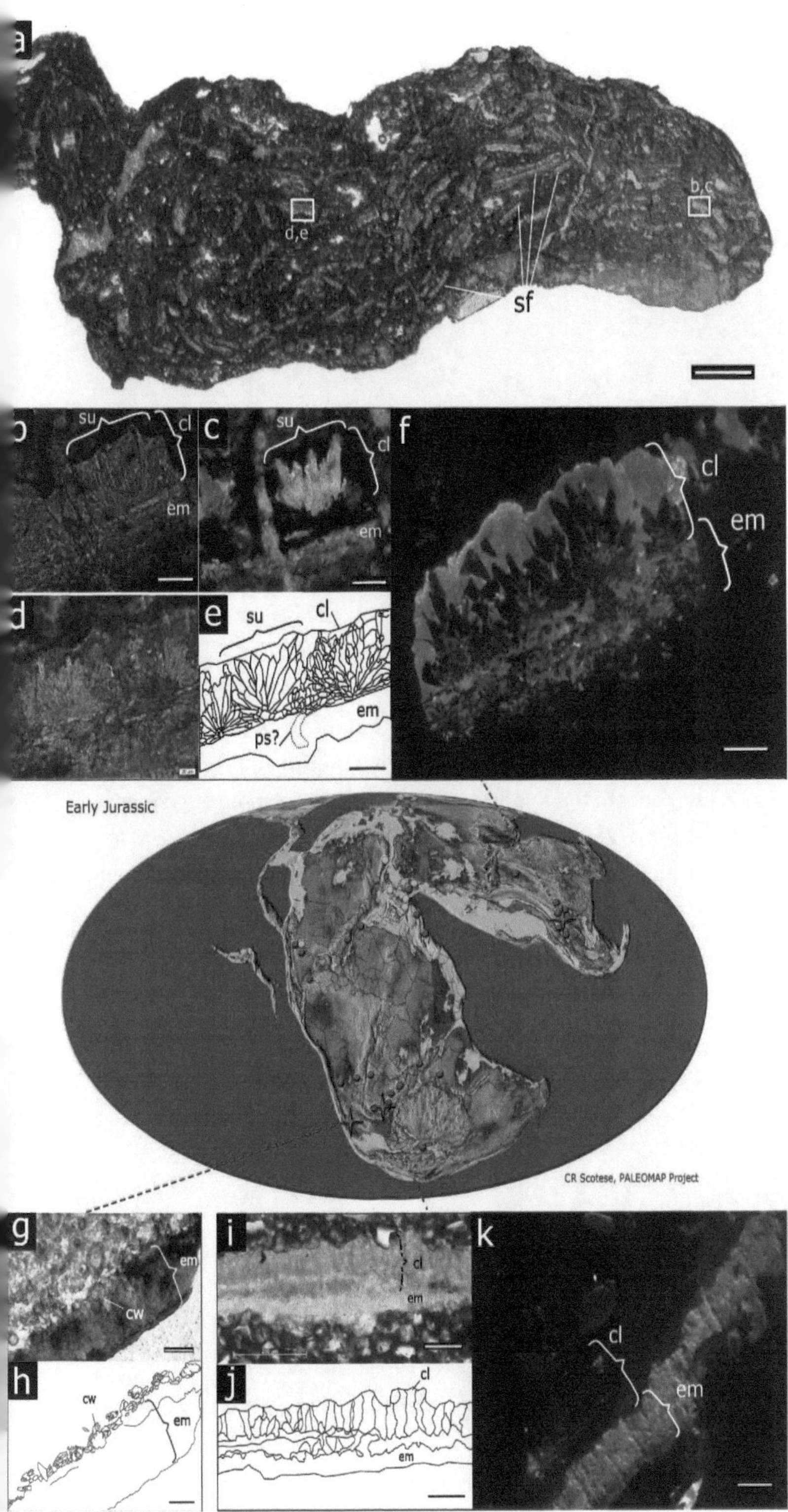

◀ Figure 1. Basal sauropodomorph eggshell microstructure and their respective Sinemurian localities (crosses) among the Rhaetian (green) to Sinemurian (red) global record of sauropodomorph fossil sites (circles). a-f, *Lufengosaurus* (Chuxiong Prefectural Museum, catalogue no. C2019 2A233), g, h, *Mussaurus* (Instituto 'Miguel Lillo", Tucuman, catalogue no. PVL 5965), i-k, *Massospondylus* (Bernard Price Institute of Palaeontology, University of Witwatersrand, catalogue no. BP/1/5254). a, section through nugget containing numerous *Lufengosaurus* eggshell fragments (plane polarized light, ppl). b, close-up (ppl) of a *Lufengosaurus* eggshell fragment, showing calcite crystals of the mammillary layer radiating from an organic core embedded in the eggshell membrane. c, as in b under cross polarized light (xpl), highlighting the calcite crystals of a mammillary cone. d, different xpl view with lambda waveplate, e. line drawing of d. f, cathodoluminescence view with 880 nm filter. g, *Mussaurus* eggshell, showing thick eggshell membrane, and distorted calcareous layer. h, line drawing of g. i, *Massospondylus* eggshell fragment (ppl), showing wedges in the calcareous layer, and a homogenous eggshell membrane. j, line drawing of i. k, cathodoluminescence view with 880 nm filter. Scale bars: in a: 1 mm, b-f, k: 50μm, g- j: 100 μm. Abbreviations: cl, calcareous layer; cw, crystal wedges of calcareous layer; em, eggshell membrane; ps, pore space; su, shell unit. See also Figure S1 to S3. (Map from 14 with permission).

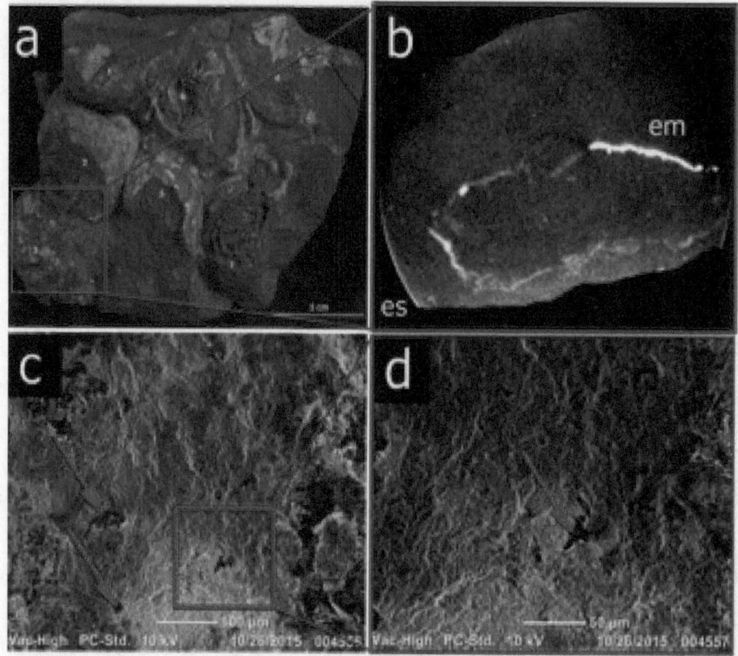

▲ Figure 2. Eggshell membrane and porosity in *Massospondylus* eggs (BP/1/5347). a, nest of *Massospondylus* eggs with preserved embryos. Note the presence of numerous cracks in the eggs, likely caused by postmortem crushing of the thin but hard eggshell. Eggshell membrane is exposed in egg number 4, just beneath the skull, and in egg number 7, just beneath the right scapula. b, CT scan of a complete egg in a, showing the eggshell (es) and the detached preserved eggshell membrane (em). c, outer surface SEM image of a *Massospondylus* eggshell fragment showing rare small and irregularly shaped pores occurring in random patterns (red arrows). d, enlarged view of boxed area in c. See also Figure S1 to S3.

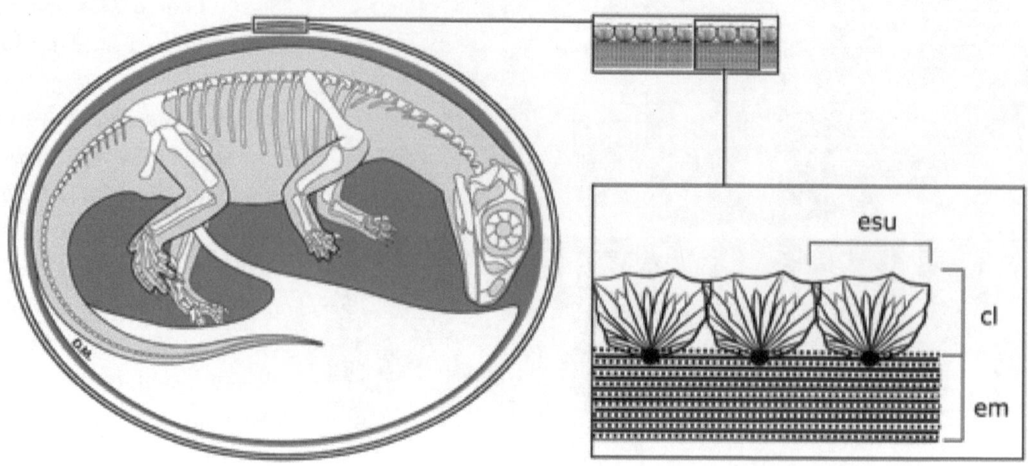

▲ Figure 3. Reconstruction of a basal sauropodomorph egg showing detail of the eggshell. Eggshell units (esu) form the calcareous layer (cl) and are embedded with organic cores in the eggshell membrane (em). See also Figure S1 to S3. Embryo reconstruction by R. David Mazierski with permission.

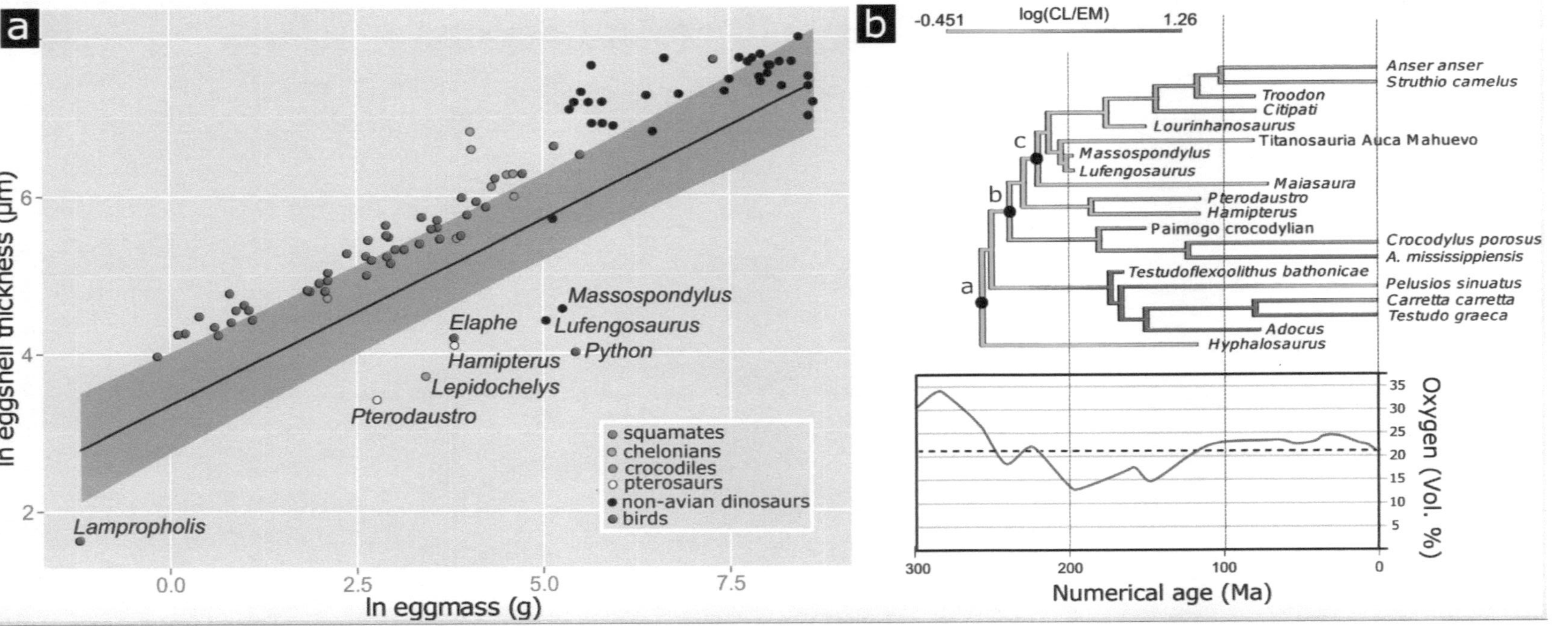

Figure 4. Relationship between eggshell thickness and egg mass in different egg- laying archosauromorphs and time calibrated maximum likelihood (ML) analysis of the ancestral states of relative eggshell thickness evolution. a, PGLS regression line and 95% confidence band on the ln-transformed dataset. *Massospondylus* and *Lufengosaurus* represent negative outliers (see SI) emphasizing the extreme thinness of the calcareous layer compared to other dinosaurs. b, ML ancestral state reconstruction of log-transformed calcareous layer thickness (CL) to egg mass (EM) ratios. Note that the root was set to represent the hypothesized ancestral flexible shelled condition. Nodes represent a, Archosauromorpha, b, Archosauria c, Ornithodira, d, Dinosauria, e, birds. Note the independent acquisitions of thick eggshell in choristoderes (represented by *Hyphalosaurus*), chelonians, crocodiles, pterosaurs and several dinosaur clades, as well as reversals in chelonians. From the Sinemurian (199 Ma) onwards, eggshells (e.g. *Testudoflexoolithus* and *Lourinhanosaurus*) show a significant calcareous layer thickness increase corresponding with atmospheric oxygen increase. See also Figure S4.